Geographies of Disruption

Tan Yigitcanlar · Tommi Inkinen

Geographies of Disruption

Place Making for Innovation in the Age of Knowledge Economy

 Springer

Tan Yigitcanlar
Queensland University of Technology
Brisbane, QLD, Australia

Tommi Inkinen
Centre for Maritime Studies,
Brahea Centre
University of Turku
Turku, Finland

ISBN 978-3-030-03206-7   ISBN 978-3-030-03207-4 (eBook)
https://doi.org/10.1007/978-3-030-03207-4

Library of Congress Control Number: 2018959262

© Springer Nature Switzerland AG 2019
This work is subject to copyright. All rights are reserved by the Publisher, whether the whole or part of the material is concerned, specifically the rights of translation, reprinting, reuse of illustrations, recitation, broadcasting, reproduction on microfilms or in any other physical way, and transmission or information storage and retrieval, electronic adaptation, computer software, or by similar or dissimilar methodology now known or hereafter developed.
The use of general descriptive names, registered names, trademarks, service marks, etc. in this publication does not imply, even in the absence of a specific statement, that such names are exempt from the relevant protective laws and regulations and therefore free for general use.
The publisher, the authors and the editors are safe to assume that the advice and information in this book are believed to be true and accurate at the date of publication. Neither the publisher nor the authors or the editors give a warranty, express or implied, with respect to the material contained herein or for any errors or omissions that may have been made. The publisher remains neutral with regard to jurisdictional claims in published maps and institutional affiliations.

This Springer imprint is published by the registered company Springer Nature Switzerland AG
The registered company address is: Gewerbestrasse 11, 6330 Cham, Switzerland

This book is dedicated to our beloved, brilliant and beautiful family members:
Susan, Ela, Selin
&
Pirjo, Hugo

Foreword

Any analysis of the geography of innovation across the globe tends to find an inherent spatial unevenness in the distribution of inputs, outputs and outcomes, especially with regard to innovation that can be depicted as disruptive and radically market changing. Such unevenness is particularly pronounced when examining urban innovation geographies. While some cities act as powerful engines and motors of innovation, many others lag and are marked by an embedded and ingrained culture, infrastructure and industrial structure that works against the generation of the forms of innovation communities that propel knowledge-based economic development.

In this highly comprehensive volume, Tan Yigitcanlar and Tommi Inkinen examine the geography on innovation from both a theoretical and practical perspective. They clearly show that the unevenness of development across cities stems from their ability to organise the flow of knowledge that underpins successful, and especially disruptive and radical, innovation. More generally, the idea of knowledge-based urban development underpinning this book has emerged as a means of systematically examining the role of knowledge and networks as key components of urban economic evolution (Yigitcanlar 2009). In essence, this approach takes the view that knowledge is a key construct for understanding the modes of urban development, with an explicit focus on identifying the knowledge assets of urban environments. Consequently, the development path of an urban economy is likely to be dependent upon the quality of the urban innovation ecosystem it possesses and the knowledge-based assets within it (Johnston and Huggins 2016).

Yigitcanlar and Inkinen's narrative and analysis indicates that increasing globalisation is leading to urban and city level becoming more important than nations in promoting and understanding innovation and economic growth. The innovative milieu of urban settings means that they are often singled out by scholars as being key territorial units within which endogenous forms of development flourish. In an evolutionary context, knowledge-based urban development involves multiple threads of relationships among its actors and resources at both a firm and spatial level, which interact in a complex manner. Fundamentally, a key driver of urban growth consists of the capability of organisations in a city to access and subsequently utilise appropriate economically beneficial knowledge.

Furthermore, the entrepreneurship required to catalyse innovation itself has a pronounced geographical dimension, with differences in start-up rates found across cities, as well as differences in the success of start-ups and entrepreneurial attitudes, all indicating the role of the urban environment in fostering entrepreneurship. Cities can become 'incubators of new ideas' and provide opportunities for entrepreneurship to take place, as well as for discovering the valuable new knowledge required for innovation and technological disruption. However, with global competition and information technology creating a proliferation of new potential solutions and pathways, innovation has become unpredictable and recombinant, with multiple competing solutions with different strengths in different contexts (Sabel and Saxenian 2008).

In general, the notion of a 'knowledge city' and the 'knowledge-based urban development' school of thought encompasses a view of 'knowledge' that is set within a framework seeking to integrate the economic, social and environmental assets and factors underpinning development. In essence, the knowledge-based urban development concept is rooted in the notion that the sources of high rates of economic performance and subsequent growth stem from the role that the production, distribution and use of knowledge play within and across economies, with networks playing a fundamental role. More recently, the roles of particular places within urban environments, which are variously described as urban innovation spaces, districts or precincts, are viewed as key catalysts of knowledge-based urban development. It is the harnessing of these urban innovation hubs that form an important feature of the reframing of contemporary urban economic development. Indeed, it appears that the most effective urban innovation districts have a high degree of network assets, along with more traditional physical and economic assets, producing a dynamic environment that strengthens proximity and knowledge spillovers (Katz and Wagner 2014).

There are a range of cities across the globe that have established successful innovations hubs and urban knowledge districts, a number of which are analysed by Yigitcanlar and Inkinen in this volume. Many more cities are also aspiring to become centres of knowledge-based urban development. It is undoubtedly impossible to replicate or clone successful cities and urban regions, or any other knowledge district, precinct or hub for that matter. However, there are many lessons that can be learned about how to improve the innovation capacity of urban locations, in particular, the role networks have played in making some cities with strong innovation hubs the centres of global growth. At the same time, it is important to recognise that such hubs should not be viewed as utopian because, like firms, the challenges involved in their creation and sustainability are contingent on a set of factors related to prevailing local and global forces. For instance, the past failure of many science parks as a policy response for generating high-technology activity can be strongly related to the fact that although they facilitated the co-location of companies, such policies ignored the processes through which this co-location could be activated into meaningful interaction and collaboration, particularly between industry and academia. It has long been argued that most science park developments are no more than high-tech fantasies, contributing little to increased links between industry and the academic world (Massey et al. 1992).

A key lesson for many innovation 'weak' cities is that the 'recipes' associated with apparent success stories such as Silicon Valley based on investing in ingredients such as technology parks, university research, engineers, venture capital and incubators alone is unlikely to be enough to produce the desired impact upon innovation. Rather than seeking to replicate the components and ingredients of successful cities, a more fruitful approach is to also connect with them through both firm-level and inter-city collaborations. In particular, there is a need to nurture global and local open search networks, and to define and invest in distinct capacities through processes of experimentation. In this sense, and as Yigitcanlar and Inkinen make clear, it is the spatial management of knowledge and innovation within and across cities that is a critical factor for long-term economic development. This requires a patient and long-term commitment by policymakers and other stakeholders, such as universities and key firms, with an interest in promoting an innovation-enhancing business environment.

Such spatial management of innovation should include policies that support firms to access the most appropriate and suitable knowledge for their innovation needs. The openness of firms and their engagement in a more systemic innovation process have been demonstrated to lead to higher levels of innovation (Laursen and Salter 2006). Therefore, reference to an era of open innovation is increasingly apparent within innovation studies along with the emergence of the idea that firms must combine internal and external knowledge for innovation (Chesbrough 2003a). As a result, innovation is now characterised as involving inter-organisational networks, with innovative outputs being the product of collaborative linkages with a range of external actors, including customers, suppliers and universities.

Given the growing importance of connectivity, city policymakers must become increasingly aware of the need for entrepreneurial firms to establish a portfolio of both sustained and more dynamic ephemeral knowledge sources. Without this balance, entrepreneurs and their firms run the risk of becoming locked into using outdated knowledge that undermines their innovative capabilities. Alongside this, there is an ongoing requirement for urban policy to ensure sufficient absorptive capacity and human capital within the base of entrepreneurial- and innovation-driven firms within a particular city. Therefore, urban innovation policies must be closely meshed with strategies to upskill the workforce and overall level human capital to allow firms to effectively identify, absorb and transform into innovation the wealth of knowledge potentially available to them.

In general, theories relating to the economic development, transformation and renewal of cities are largely based on a multiplicity of explanations relating to the location, agglomeration and organisation of firms, industries and capital. A host of factors relating to the availability of capital in the form of investment and resources, the skills of the workforce, the availability and capability of entrepreneurs and other agents of innovation, as well as the cooperation and collaboration achieved through ecosystems, are all offered as explanatory factors These approaches echo the notion of cities as 'Schumpeterian hubs' for recombining human capital in order to generate innovation (Wolfe 2017). As suggested above, notions relating to ecosystems and agency suggest that merely investing in capital may not be enough to secure innovation and high rates of economic growth, especially the type of transformative renewal that is required in industrially mature and economically lagging cities regions.

An emerging position within urban and regional innovation and development theories concerns the role of human behaviour and institutions in shaping and moderating this behaviour (Huggins and Thompson 2017). Principally, the focus is on entrepreneurial human behaviour and the means by which such behaviour drives innovation and subsequently economic development. It is argued that the type and nature of human agency existing within cities and regions at particular points in their development is a significant factor in explaining the capacity of these places to achieve economic transformation and renewal. While the configuration and capability of economic ecosystems—which may be conceptualised through notions such as clusters, innovation systems, districts and hubs—determine urban and regional development outcomes, at the micro-level it is the role of certain key human agents within cities and regions who actually shape the nature and evolution of these ecosystems. In other words, human agency is one of the key rooted drivers associated with more traditional explanatory causes of economic development and transformation and should be considered seriously when addressing the routes available to economically mature cities in their bid to foster renewal and transformation.

Indeed, it is argued that human agency is based on a rationality that is spatially bounded (Huggins and Thompson 2017; Pike et al. 2016). Fundamentally, cities themselves produce a spatially bounded rationality that determines the forms and types of human agency apparent in a given city, and subsequently the nature of knowledge, innovation and development. In particular, innovation, entrepreneurship and creativity are social processes that involve groups of people who build off one another historically, and are the products of cities, which act as the key organising unit for these activities and bring together firms, talent and the urban institutions necessary for creating a vibrant innovation environment (Florida et al. 2017). Similarly, the symbiotic relationship between key agents and their location is found in research examining the role of a limited number of 'star' scientists in promoting the innovation performance of certain cities (Zucker et al. 1998; Moretti 2012).

Based on this line of thinking, Yigitcanlar and Inkinen correctly point to the role of institutions, culture and human behaviour—especially entrepreneurial behaviour—as a factor in predicting the extent to which innovation aspirations can become realised. As they argue, innovation is a prerequisite component of the urban development process. In economically successful cities, innovation is harnessed, distributed and capitalised upon through ecosystems of connected agents who create the networks, knowledge and institutions that positively and openly evolve to sustain innovation and economic development. In mature and lagging cities, the requirement for innovation becomes paramount if they are to renew and transform their economies and shift away from economic systems that are likely to be steeped in processes of negative, rather than, positive lock-in.

Given these challenges, the economic renewal, transformation and formation of effective and efficient urban ecosystems is likely to be partly determined by the behavioural life of cities in terms of their underlying and dominant cultural and psychological traits (Huggins and Thompson 2017). These traits determine the forms of human agency to be found in particular cities, and this agency itself is a determinant of the economic ecosystems within them. Human agency necessarily comes in many forms and varieties, but a key form of agency necessary for innovative-led renewal and transformation concerns entrepreneurial agents. Supported by political and worker agency, entrepreneurs represent the key catalysts of change at the urban level. Throughout history, transformation has tended to be led by a core group of entrepreneurial agents that have taken a lead in positively evolving the economies in which they are physically situated, often through new generations of agent producing innovations that push further forward the technological frontier set by their predecessors (Hall 1998; Mokyr 2017).

In mature urban economies, it is all too often the case that these types of agent have migrated to other cities with stronger ecosystems and greater opportunities, or that such agents have not been nurtured in the first instance due to underlying psycho-cultural traits that are not conducive to innovation. This leaves these cities in an economic situation whereby they lack a critical mass in those industries and sectors through which value and innovation can be best achieved. Of course, there will continue to be activity in many sectors, often with a higher than average proportion of activity relating to foundational economic areas such as public services and non-tradable industries.

Some have suggested that one route to renewal for mature cities is to focus on creating innovation and value in these foundational sectors. To an extent, this is a plausible approach, with the potential capacity to trigger the renewal process. However, it is likely to be limited in terms of achieving long-term and sustained renewal unless there are key entrepreneurial agents operating within these industries that possess a mindset, willingness and power to disrupt existing economic systems and models. In other words, it is the capability to facilitate institutional and cultural change—which is likely to be generational—that is the centrepiece of urban economic transformation and innovation. Therefore, the persistence and reproduction of a cultural and institutional environment that actively works against an evolutionary trajectory embedded in creating innovative and knowledge-based economic systems remains the greatest challenge for many cities. The analysis presented in this volume represents an important step to addressing these issues in a systematic and long-term manner.

Cardiff, UK
Prof. Robert Huggins
School of Geography and Planning
Cardiff University
hugginsr@cardiff.ac.uk

References

Chesbrough, H. (2003a). *Open innovation: The new imperative for creating and profiting from technology*. Boston, MA: Harvard Business Review Press.
Florida, R. Adler, P. & Mellander, C. (2017). The city an innovation machine. *Regional Studies,* 51(1), 86–96.
Hall, P. (1998). *Cities in civilization.* New York: Pantheon Books.
Huggins, R. & Thompson, P. (2017). The behavioural foundations of urban and regional development: culture, psychology and agency, *Journal of Economic Geography.* https://doi.org/10.1093/jeg/lbx040.
Johnston, A., & Huggins, R. (2016). The spatio-relational nature of urban innovation systems: Universities, knowledge intensive business service firms, and collaborative networks. *Journal of Urban Technology,* 23(1), 29–52.
Katz, B., & Wagner, J. (2014). *The rise of innovation districts: A new geography of innovation in America.* Washington: Brookings Institute.
Laursen, K. & Salter, A., (2006). Open for innovation: The role of openness in explaining innovation performance among UK manufacturing firms. *Strategic Management Journal,* 27 (1), 131–150.
Massey, D. B., Quintas, P., & Wield, D. (1992). *High-tech fantasies: Science parks in society, science and space.* London: Routledge.
Mokyr, J. (2017). *A culture of growth: The origins of the modern economy.* Princeton: Princeton University Press.
Moretti, E. (2012). *The new geography of jobs.* New York: Houghton Mifflin and Harcourt.
Pike, A., MacKinnon, D., & Cumbers, A. (2016). Doing evolution in economic geography. *Economic Geography,* 92(2), 123–144.
Saxenian, A., & Sabel, C. (2008). Roepke lecture in economic geography: Venture capital in the "periphery": The new argonauts, global search, and local institution building. *Economic Geography,* 84(4), 379–394.
Wolfe, D. A. (2017). Innovation and creativity in city-regions. In R. Shearmur, C. Carrincazeaux, & D. Doloreux (Eds.) *Handbook on the geographies of innovation* (pp. 174–186). Cheltenham: Edward Elgar.
Yigitcanlar, T. (2009). Planning for knowledge-based urban development: Global perspectives. *Journal of Knowledge Management,* 13(5), 228–242.
Zucker, L. G., Darby, M. R., & Armstrong, J. (1998). Geographically localized knowledge: Spillovers or markets? *Economic Inquiry,* 36(1), 65–86.

Preface

The first two decades of the twenty-first century have witnessed a rebirth of the city as an engine of innovation. This renaissance has been an organic response to technological and societal pressures, opportunities and norms. This is a sharp reversal from the latter half of the twentieth century, which saw the decay and erosion of the city as a place of economic value creation. In fact, today the urbanisation of human experience is a dominant phenomenon and a reality of the new millennium.

On the one hand, the twenty-first century is identified as the *Century of Cities* (e.g. Landry 2000) as the world urbanisation rate is currently over 50% and it is expected to reach 75% by the mid-century. This figure is already well over 80% in some countries such as Australia and the UK. Today, cities represent a world of opportunities, being responsible for over 80% of the gross national product, which suggests a powerful link between urbanisation and economic prosperity (Mori and Christodoulou 2012). On the other, the twenty-first century is acknowledged as the *Century of Knowledge* (e.g. Drucker 1994) as today over 50% of gross domestic product in developed countries moved consistently from material base to knowledge base. Naturally, the convergence of these two emerging conditions of human civilisation—cities and knowledge—at the dawn of the new millennium marked the twenty-first century as the *Century of Knowledge Cities* (see Carrillo 2006).

The changing and highly challenging conditions of the twenty-first century have been significantly impacting our economy, society and built and natural environments. Today, the generation of knowledge, mostly in the form of science, technology, engineering, mathematics and arts, is seen as a panacea for the adaptation to changes and management of challenges. Making space and place that concentrate on knowledge generation, exchange and marketing has become a priority for many nations and their prominent cities. The concepts of 'knowledge city' and 'knowledge-based urban development' (KBUD) are seen as powerful instruments to facilitate an urban turnaround and bring prosperity and sustainability to our cities. Knowledge-based urban development is also seen as a powerful approach to form the *Geographies of Disruption*—such as knowledge cities and knowledge precincts.

Furthermore, innovation, creativity and knowledge generation are dominantly urban phenomena that require a certain scale and intensity of knowledge infrastructure as well as vibrant urban life with a full mix of diversity and tolerance. Marketable knowledge generation is also dependent on a large pool of talented labour power and consumption, which is critical to form a functional urban region suitable for 'knowledge precinct' development. In such landscape, cities concentrate on extensive global networks as intense mediums of exchange for knowledge precincts to flourish. Additionally, knowledge workers, primary creative sources of knowledge precincts, prefer inspiring cities with a thriving cultural life, an international orientation and high levels of social and cultural diversity. A big city with an evidence of world city formation accommodates high-quality urban services (i.e. high-quality residential areas, cultural districts, recreational facilities, connectivity to global air transport networks and so on) and a diversified economic base including extensive supplier and distribution networks and specialised services (Yigitcanlar et al. 2008b).

Knowledge precincts can be regarded as the spatial nexus of knowledge-based urban development that chiefly refers clustering of R&D activities, high-tech manufacturing of knowledge-intensive industrial and business sectors linked by mixed-use environment including housing, business, education and leisure within an urban-like setting. The working definition of such areas differs from country to country and context to context (i.e. high-tech/knowledge/innovation cluster, knowledge/innovation hub, knowledge or innovation precinct/district, digital/knowledge village and so on), more or less indicating a clustering of high-tech enterprises with a commercial mix of urban life and culture, predominantly within central urban locations (Yigitcanlar et al. 2008b). Examples of successful knowledge precincts include 22@Barcelona (Barcelona), One-North (Singapore) and Strijp-S (Eindhoven). These new generation urban knowledge precincts also contribute significantly to the knowledge city transformation efforts of their cities.

Against this brief backdrop, this book examines the theoretical and practical expansions of geographies of disruption or in other words geographies of innovation—i.e. knowledge cities and knowledge precincts. The book aims to underline the growing importance of knowledge and innovation for the competitiveness of cities and their regions in the largely uncharted territory of research on innovation geographies of knowledge cities and knowledge precincts and their knowledge-based development, and then, to provide an in-depth and comprehensive understanding in both theoretical and practical aspects on the knowledge-based urban development and its implications and prospects on our cities and their innovation hubs of knowledge precincts. With these aims in mind, the book attempts to address the following questions:

- Why is pursuing a goal to form geographies of disruption important for cities of the global knowledge economy era?
- What are the key conditions and policy drivers to transform cities into geographies of disruption?

- What is the role of place making in the creation process of geographies of disruption?
- How can lessons learned from global best practices be evaluated for and considered by other cities?
- How can the performance of knowledge cities and their knowledge precincts be compared internationally?
- How can a holistic long-term vision be developed for cities aspiring to create geographies of disruption?

In order to achieve these aims and address the research questions, the book is structured under four main parts and elaborated briefly as below:

Part I Theory of Innovation Geography
This part of the book aims to develop a thorough understanding of the conceptual foundation of the field of 'innovation geography'. The part lays forth the main terminology and theoretical concepts related to (a) Innovation, (b) Technology and (c) Spatiality applied in geography and regional studies. These concepts include innovation systems, location and space, and regional development (focusing on cities). The chapter has its academic foundations on economic geography and takes into account social, spatial and institutional development domains. Part I of the book provides the academic background for the following Part II that focuses on the practices and efforts done in cities that aim to develop their position in a global economy. The following main questions are answered in the chapter: (a) How to conceptualise innovation geography? (b) What are the key characteristics depicting innovation geography? (c) How have these concepts and characteristics applied in economic geography? (e) How do innovative locations and spaces enhance their performance in innovation?

Part II Practice of Innovation Geography
This part of the book aims to develop a thorough understanding of the practical elements used and implemented in cities in order to promote innovative development. The foundation lies in theoretical background presented in Part I, and this part intends to locate innovative cities and open up practices and examples of how to understand 'innovative geography' in a real-world setting. Examples are drawn from various locations and cities globally. Thus, Part II addresses the following research questions: (a) How are innovative geographies realised in different contexts and countries? (b) What are the elements depicting these innovative locations? (c) What commonalities these locations have in relation to innovation and technology? (d) What lessons may be learned from the experiences gained in these locations?

Part III Place Making for Knowledge Cities
This part of the book aims to develop a thorough understanding of the underlining conceptual and practical issues on the transformation of cities into knowledge cities, and planning, designing and fostering existing and emerging knowledge precincts. The part intends to place knowledge cities under the microscope scrutinising their potential and progress from the lens of knowledge-based urban development—

focusing on economic, societal, spatial and institutional development domains—targeting the cities and their regions. Part III of the book focuses on the knowledge city development attempts cities to address the following research questions: (a) Why is pursuing a goal to form geographies of disruption important for cities of the global knowledge economy era? (b) What are the key conditions and policy drivers to transform cities into geographies of disruption? (c) How can lessons learned from global best practices be evaluated for and considered by other cities? (e) How can the performance of knowledge cities be compared internationally? (f) How can a holistic long-term vision be developed for cities aspiring to create geographies of disruption?

Part IV Place Making for Knowledge Precincts
This part of the book aims to develop a thorough understanding of the underlining conceptual and practical issues on the planning, designing and fostering existing and emerging knowledge precincts. The part intends to place knowledge precincts under the microscope scrutinising their potentials and progress from the lens of knowledge-based urban development and place making. This part of the book attempts to address the following research questions: (a) What is the role of knowledge precincts in knowledge city formation? (b) What are the key conditions and policy drivers to plan and develop thriving knowledge precincts? (c) How can lessons learned from global knowledge precinct best practice be evaluated for and considered by other cities? (d) How can the performance of knowledge precincts be compared internationally? (e) What are the key issues concerning planning and design of knowledge precincts in a citywide holistic and integrated perspective?

The findings of a comprehensive study reported in this book indicate a number of the key strategies and actions for cities that are aiming to become 'geographies of disruption'. The followings are among the critical strategies and actions for transforming cities successfully into prosperous knowledge cities:

- Establishing strong political and societal will through effective governance, leadership, stewardship and community engagement;
- Setting up a dedicated agency to orchestrate and promote knowledge-based urban development and knowledge city transformation through an enhanced intergovernmental and interdepartmental collaboration;
- Developing a strategic vision and dynamic and long-term development planning with a specific knowledge-based urban development and knowledge city focus;
- Stocktaking to understand the city's existing achievements and potentials, and benchmarking to find out what other cities have been doing and capable of;
- Providing adequate financial support, incentives and strategic investments to foster entrepreneurial culture, and empower citizens as knowledge creators and innovators;
- Forming urban innovation engines through knowledge and innovation space development (e.g. knowledge precincts) as an integrated part of the urban fabric to trigger, speed up, host and facilitate innovation;
- Investing in and encouraging triple helix model (i.e. public-private-academic) partnerships in knowledge-intensive projects and investments;

- Supporting and closely working with universities and R&D institutions that are the primary sources of research excellence, innovation, growth and prosperity;
- Utilising effective strategic city branding, marketing and alliance building activities to increase the international and domestic reputation to attract talent and investment;
- Creating and preserving the unique identity and image of the city based on endogenous assets to increase the appeal for attracting and retaining knowledge workers;
- Enhancing the international and multicultural character of the city by encouraging and increasing the societal openness and tolerance to diversity;
- Investing in the further development of human and social capitals to foster the talent base and intellectual wealth of the city, and create a cohesive knowledge society;
- Providing sustainable, high-quality, diverse and affordable housing and urban infrastructure, services and amenities to all citizens.

The findings reported in the book also indicate a number of key strategies and actions for urban precincts that are aiming to become 'geographies of disruption'. The followings are among the critical strategies and actions for transforming innovation hubs of cities successfully into prosperous knowledge precincts:

- Engaging public sector—i.e. three-tier government—as the committed principal initiator and enabler in the emergence and development processes of knowledge precincts is crucial;
- Founding a dedicated agency for overseeing the knowledge precincts' lifecycle —and establishing coordination and collaboration with other public departments and private and academic sectors—is beneficial;
- Developing a long-term pathway and a roadmap—through policymaking, planning and investment—for sustainable emergence and development of knowledge precincts is critical;
- Establishing a tradition of building upon local assets and strengths, which will also determine areas to specialise in (so-called smart specialisation), is essential;
- Creating a vibrant and diverse entrepreneurial environment—supported by venture capital funding, and public-sector incentives—is necessary;
- Constructing a vibrant urban environment—supported by world-class hard and soft infrastructures and amenities—is paramount;
- Designing public places and participation mechanisms that encourage and increase knowledge communities' interaction with local communities and having say about the growth and future directions of their knowledge precincts is essential;
- Investing further in networking and pull factor creation to attract (and retain) investment and talent in knowledge precincts—that also includes branding and marketing—is needed;

- Keeping a close eye on the progress of competitor cities and their knowledge precincts—while networking and collaborating with them—through constant benchmarking exercises is useful;
- Undertaking further comprehensive investigations of existing, emerging and prospective knowledge precincts is needed.

The authors of the book believe that, in the light of these generic recommendations, cities have the opportunity to take a leadership role in developing an effective and efficient planning and development policy for the city and the region and their existing, planned and future knowledge precincts. This would significantly strengthen the efforts to become a 'geography of disruption'—or prosperous knowledge city and place cities (and their metropolitan regions) with their knowledge precincts in the global map for attracting and retaining investment and talent.

Brisbane, Australia Tan Yigitcanlar
Turku, Finland Tommi Inkinen

References

Carrillo, F. J. (Ed.) (2006). *Knowledge cities: Approaches, experiences and perspectives.* New York: Routledge.
Drucker, P. (1994). *The post-capitalist society.* New York: Harper-Collins.
Landry, C. (2000). *The creative city.* London: Earthscan.
Mori, K., & Christodoulou, A. (2012). Review of sustainability indices and indicators: Towards a new City Sustainability Index (CSI). *Environmental Impact Assessment Review,* 32(1), 94–106.
Yigitcanlar, T., Velibeyoglu, K., & Baum, S. (Eds.) (2008b). *Knowledge-based urban development: Planning and applications in the information era: Planning and applications in the information era.* Hersey: IGI Global.

Contents

Part I Theory of Innovation Geography

1   Introduction to Part I 3
    1.1   Rationale and Scope 8
    1.2   Methodology 8
    References 9

2   Conceptual Foundations of Innovation Geography 11
    2.1   Agglomerations and Creative Destruction 14
    2.2   Clusters 19
    2.3   Proximate Relatedness 20
    2.4   Summary 22
    References 22

3   Place Dynamics for Innovation 25
    3.1   Innovation in Urban and Regional Studies 25
    3.2   Differentiated Systems of Innovation 27
    3.3   Innovation Systems as Platforms for Knowledge-Based
           Urban Development 30
    3.4   Innovative Urban Concepts and Image Creation 32
    3.5   Summary 34
    References 35

4   Characteristics of Innovation Geography 39
    4.1   Production, Location and Economy 39
    4.2   Smart and Knowledge City Characteristics and Properties 41
    4.3   Policies, Governance and Locations 43
    4.4   Summary 48
    References 49

5   Conclusion to Part I 51

Part II Practice of Innovation Geography

6 Introduction to Part II 55
  6.1 Rationale and Scope 56
  6.2 Methodology 57
  References 58

7 Insights from Northern European Countries and Regions 59
  7.1 Helsinki, Finland 59
  7.2 Stockholm, Sweden 64
  7.3 Copenhagen, Denmark 69
  7.4 Tallinn, Estonia 74
  7.5 Comparison of the Best Practice Cases 79
  7.6 Summary 95
  References 98

8 Conclusion to Part II 101

Part III Place Making for Knowledge Cities

9 Introduction to Part III 105
  9.1 Rationale and Scope 107
  9.2 Methodology 108
  References 108

10 Theory and Practice of Knowledge Cities and Knowledge-Based Urban Development 109
  10.1 Evolution of Knowledge Cities 111
  10.2 Essential Assets of Prosperous Knowledge Cities 113
  10.3 A New Planning and Development Paradigm: Knowledge-Based Urban Development 118
  10.4 Prospects and Constraints in the Formation of Knowledge Cities 126
  10.5 Summary 128
  References 129

11 Global Knowledge City Best Practice 135
  11.1 A Best Practice Knowledge City from Europe: Barcelona 136
  11.2 A Best Practice Knowledge City from North America: Austin 139
  11.3 A Best Practice Knowledge City from South East Asia: Singapore 144
  11.4 A Best Practice Knowledge City from Oceania: Melbourne ... 148
  11.5 Lessons from the Global Knowledge City Best Practice 153
  11.6 Summary 156
  References 156

12	**Benchmarking City Performance**		159
	12.1	Knowledge-Based Urban Development Assessment	160
	12.2	Methodology of the Knowledge-Based Urban Development Assessment Model	162
	12.3	Global Standing of Brisbane as an Emerging Knowledge City	171
	12.4	Strengths, Weaknesses, Opportunities and Threats of Brisbane	189
	12.5	Summary	195
	References		196
13	**Recommendations and Strategic Directions for Knowledge City Place Making**		199
	13.1	Economic Aspects	200
	13.2	Societal Aspects	200
	13.3	Spatial Aspects	201
	13.4	Governance Aspects	202
	13.5	Recommendations for a Vision	204
	13.6	Summary	204
	References		205
14	**Conclusion to Part III**		207

Part IV Place Making for Knowledge Precincts

15	**Introduction to Part IV**		211
	15.1	Rationale and Scope	212
	15.2	Methodology	213
	References		213
16	**Theory and Practice of Knowledge Precincts**		215
	16.1	Evolution of Knowledge Precincts	216
	16.2	Challenges for Urban Planning and Development Practice	219
	16.3	Essential Success Elements of Knowledge Precincts	222
	16.4	Emergence and Development of Planned Knowledge Precincts	226
	16.5	Summary	233
	References		234
17	**Global Knowledge Precinct Best Practice**		239
	17.1	A Best Practice Knowledge Precinct from Europe: 22@Barcelona	240
	17.2	A Best Practice Knowledge Precinct from North America: Silicon Valley	243
	17.3	A Best Practice Knowledge Precinct from South East Asia: One-North	247

	17.4	A Best Practice Knowledge Precinct from Oceania: Macquarie Park	251
	17.5	Lessons from the Global Knowledge Precinct Best Practice	255
	17.6	Summary	257
	References		258

18 Benchmarking Knowledge Precincts 261
    18.1   Place Making Facilitators of Knowledge Precincts 262
    18.2   Comparison of a Knowledge Precinct from Brisbane with Global Best Practices 264
    18.3   Cambridge Science Park, United Kingdom 265
    18.4   22@Barcelona, Spain 267
    18.5   Arabianranta, Finland 269
    18.6   Strijp-S, The Netherlands 271
    18.7   Digital Hub, Ireland 272
    18.8   Macquarie Park, Australia 273
    18.9   Monash Employment Cluster, Australia 276
    18.10 Dutton Park Knowledge Precinct, Australia 278
    18.11 Findings from the Comparison Study 282
    18.12 Insights into Brisbane's Knowledge Precincts 282
    18.13 Summary 291
    References 292

19 Recommendations and Strategic Directions for Knowledge Precinct Place Making 295
    19.1   Context 296
    19.2   Space and Place 297
    19.3   Companies 298
    19.4   Users 299
    19.5   Prospective Investigations 301
    19.6   Summary 302
    References 302

20 Conclusion to Part IV 305
    Reference 306

Afterword 307

Definition of Terms 311

Index 319

About the Authors

Tan Yigitcanlar is an Associate Professor at the School of Civil Engineering and Built Environment, Queensland University of Technology (QUT). He also carries out the Executive Director role at the World Capital Institute, Monterrey, Mexico, and Adjunct Professor positions at the Federal University of Santa Catarina, Florianopolis, Brazil, and Open Innovation Academy, Daegu, Korea.

He has been responsible for research, teaching, training and capacity building programmes on the fields of urban and regional planning, development and management in esteemed Australian, Brazilian, Korean, Finnish, Japanese and Turkish universities. The main foci of his research include the following themes: 'Knowledge-Based Urban Development and Knowledge Cities', 'Sustainable Urban Development and Sustainable Cities' and 'Intelligent Urban Technologies and Smart Cities'.

He is the Editor-in-Chief of *Elsevier's Smart Cities Book Series*, Editor-in-Chief of *International Journal of Knowledge-Based Development*, Section Editor-in-Chief of *Sustainability*, Regional Editor of *Journal of Knowledge Management*, and Associate Editor of *International Journal of Environmental Science and Technology*, *Global Journal of Environmental Science and Management*, *Journal of Open Innovation: Technology, Market, and Complexity*, *Asia Pacific Journal of Innovation and Entrepreneurship*, and *Journal of Smart Cities*. He is also an editorial board member of the following five prestigious academic journals: *Journal of Urban Technology*, *Knowledge Management Research & Practice*, *Measuring Business Excellence*, *International Journal of Knowledge-Based Organizations*, and *Urban Science*.

He undertakes the Chairman role of the annual 'Knowledge Cities World Summit' series, and organised conferences in Monterrey (Mexico), Shenzhen (China), Melbourne (Australia), Bento Gonzalves (Brazil), Matera (Italy), Istanbul (Turkey), Tallinn (Estonia), Daegu (Korea), Vienna (Austria), Arequipa (Peru) and Tenerife (Spain).

He published extensively including over 130 high-impact journal articles and over 70 book chapters, and the following 11 books: 'Geographies of Disruption' (Springer, 2019), 'Urban Knowledge and Innovation Spaces (Routledge, 2017),

'Technology and the City (Routledge, 2016)', 'Knowledge and the City (Routledge, 2014)', 'Sustainable Urban Water Environment (Edward Elgar, 2014)', 'Building Prosperous Knowledge Cites (Edward Elgar, 2012)', 'Knowledge-Based Development for Cities and Societies (IGI Global, 2010)', 'Sustainable Urban and Regional Infrastructure Development (IGI Global, 2010)', 'Rethinking Sustainable Development (IGI Global, 2010)' 'Knowledge-Based Urban Development (IGI Global, 2008)' and 'Creative Urban Regions (IGI Global, 2008)'.

Tommi Inkinen is a Professor and a Research Director at the Centre for Maritime Studies. He is also the Vice Director at the Brahea Centre of the University of Turku, Finland. He has been senior staff member in the Universities of Turku and Helsinki for developing academic research as well as undergraduate and postgraduate teaching in the study programmes of geography on the fields of urban and economic development. He has worked as the Vice Chair of the Department of Geosciences and Geography at the University of Helsinki, Finland, where he worked as a Professor of Economic Geography and a Professor of Human Geography 2008 till 2016. He also holds Docentships of Regional Studies and Economic Geography. He has been a Visiting Professor at the Queensland University of Technology in 2014.

He has a multi-disciplinary background bringing together economic geography, urban technologies and logistical systems of maritime transportation. His research interests cover the issues of transport, logistical systems and port cities together with urban innovations and economic development of cities and regions. These include the regional and social structuring of innovation systems and technology integration. Currently, he is one of the key researchers in the University of Turku's strategic profiling area of Maritime Studies.

He oversees the organisation of the Doctoral Symposium as part of the World Capital Institute's Knowledge Cities World Summit academic conference series since 2014. He also serves as an international advisory board member to the World Capital Institute's executive board—providing input and directions in the innovation, economy, societal and geographical aspects of knowledge-based development.

He and his research team have published extensively on these topics in international journals and collections. He is a current Chairman of the International Geographical Union's (IGU) Innovation, Information and Technology Commission and previously Editor-In-Chief of peer-reviewed quarterly journal Terra, Finnish Geographical Journal. His latest projects include among the others 'Geography and Innovative Competitiveness? Finnish Knowledge Regions in European and Global Economy' funded by the Academy of Finland (2009–2012) and 'Innovative regions and economic development in the Helsinki Metropolitan Area' funded by the Helsinki Metropolitan Region Urban Research Program (2014–2016). He is currently an editorial board member in five international journals, which also include the *Journal of Urban Technology, Journal of Open Innovation* and *International Journal of Knowledge-Based Development*.

List of Figures

Fig. 4.1	Electronic government in practical context and relevant elements for service consideration.	47
Fig. 10.1	Foundations and activities of knowledge cities (derived from Van Den Berg et al. 2005)	117
Fig. 10.2	Conceptual framework of knowledge-based urban development (derived from Yigitcanlar 2014a, b)	126
Fig. 11.1	Symbol of Barcelona's knowledge-based urban development	138
Fig. 11.2	Snapshot from a vibrant social scene of Austin	140
Fig. 11.3	Snapshot from a knowledge and innovation space of Singapore	147
Fig. 11.4	Snapshot from one of Melbourne's knowledge precincts—Parkville	151
Fig. 12.1	Cluster analysis results (derived from Yigitcanlar 2014a, b)	184
Fig. 12.2	Brisbane's knowledge-based urban development standing	185
Fig. 12.3	Brisbane's comparative performance in knowledge-based urban development	185
Fig. 12.4	Brisbane's comparative performance in economic development	186
Fig. 12.5	Brisbane's comparative performance in societal development	186
Fig. 12.6	Brisbane's comparative performance in spatial development	187
Fig. 12.7	Brisbane's comparative performance in institutional development	187
Fig. 16.1	Framework of emergence of knowledge precincts (derived from de Van Winden et al. 2012)	227
Fig. 16.2	Key factors of knowledge precinct planning (derived from Wasim 2014)	230
Fig. 17.1	Snapshot of a creative industry at 22@Barcelona	241
Fig. 17.2	Google bikes at Mountain View headquarters	245
Fig. 17.3	Snapshot from One-North	249

Fig. 17.4	Snapshot from Macquarie Park	252
Fig. 18.1	Place making facilitators (derived from Pancholi et al. 2015)	264
Fig. 18.2	Cambridge Science Park (Google Earth)	267
Fig. 18.3	22@Barcelona (Google Earth)	268
Fig. 18.4	Arabianranta (Google Earth)	270
Fig. 18.5	Strijp-S (Google Earth)	272
Fig. 18.6	The Digital Hub (Google Earth)	274
Fig. 18.7	Macquarie Park (Google Earth)	276
Fig. 18.8	Monash Employment Cluster (Google Earth)	278
Fig. 18.9	Dutton Park Knowledge Precinct (Google Earth)	281

List of Tables

Table 7.1	GDP in metropolitan regions in million EUR (Eurostat 2018)	79
Table 7.2	Tertiary education (% of population) in NUTS 2 regions (Eurostat 2018)	81
Table 7.3	Patents per million people (Eurostat 2018)	82
Table 7.4	High-tech patent applications (per million) to EPO (Eurostat 2018)	83
Table 7.5	Ratio of high-technology patent applications to all patent applications (Eurostat 2018)	84
Table 7.6	Patents on recycling and secondary raw materials per million people (Eurostat 2018)	85
Table 7.7	R&D expenditure (EUR per people) in NUTS 2 regions (Eurostat 2018)	86
Table 7.8	Ratio of R&D personnel and researchers per total employment (Eurostat 2018)	87
Table 7.9	Employment in technology and knowledge-intensive sectors (%) (Eurostat 2018)	89
Table 7.10	Ratio of tertiary educated and employed science and technology workforce to all (Eurostat 2018)	89
Table 7.11	Ratio of tertiary educated and employed workforce to all (Eurostat 2018)	90
Table 7.12	Ratio of internet interaction of individuals with public authorities (Eurostat 2018)	91
Table 7.13	People (%) having done online purchase during the last 12 months (Eurostat 2018)	92
Table 7.14	Ratio of value of e-commerce sales to all sales (Eurostat 2018)	93
Table 7.15	Ratio of enterprises who have ERP software package to share information (Eurostat 2018)	94
Table 10.1	Recipients of the most admired knowledge city (MAKCi) awards	114

Table 10.2	Essential assets of a knowledge city (derived from Yigitcanlar and Dur 2013)	115
Table 10.3	Success factors of knowledge cities (derived from Ergazakis et al. 2006; Yigitcanlar 2009)	116
Table 10.4	Desired policy outcomes (derived from Van Geenhuizen and Nijkamp 2012)	117
Table 10.5	Evolution of the knowledge-based urban development concept (derived from Carrillo 2015)	121
Table 10.6	Common knowledge-based urban development policies (derived from Yigitcanlar 2014a, b)	123
Table 10.7	Prospects and constraints in the formation of knowledge cities	127
Table 12.1	Popular city performance indices	161
Table 12.2	Structure of KBUD-AM and alternative indicator weightings (derived from Yigitcanlar 2014a, b)	164
Table 12.3	Raw data (derived from Yigitcanlar 2014a, b)	173
Table 12.4	Equal weighted normalised index scores (derived from Yigitcanlar 2014a, b)	176
Table 12.5	Ranking of cities based on alternative weighting schemes (derived from Yigitcanlar 2014a, b)	182
Table 12.6	SWOT analysis of Brisbane	190
Table 16.1	Lifecycle of knowledge precincts (derived from Huggins 2008)	224
Table 18.1	Comparative findings (derived from Pancholi et al. 2015)	283

Part I
Theory of Innovation Geography

Chapter 1
Introduction to Part I

Abstract Technology and innovation driven information society research has experienced a great shift in the methodological mindset of how the technological developments affect the surrounding world—and cities in particular. Originally theories and theorisations concerning technological impacts of innovation and new media were literature-based and lacking a proper empirical evidence-base to support the conceptual aspects. In recent years empirical data sources have been developed and empirical investigations concerning knowledge cities, smart cities, technological embedding, and socioeconomic-technology relations have gained more explanative power and interest within the academia. This chapter provides an introduction to this part of the book.

Keywords Economic geography · Urban studies · Technological innovation Information society · Disruption

Information networks and innovations, which are based on a technological development, affect and change human practices in the contemporary world. The development of new technologies is manifested in all aspects of life including economy, politics, education, work and leisure. In the 1960s, McLuhan (1964) presented an idea of the 'global village' that is realised through economic globalisation and technological mediation of information. This process is today taking place through the meta-concept of digitalisation. Technological innovations and disruptions cause not only economic but also political and social changes. They are realised at changing observation scales starting from individuals and ending to the global scale. The first part of the book analyses those special characteristics and processes through which spatiality and economic geography shapes and rebuilds the essence of the information driven societies (e.g., Tuomi 2002; Scotchmer 2004).

The information society and its innovative development have been essential research subjects of social sciences for more than four decades. They have commonly been described through productivity, convenience and the growth of choice possibilities, provided by the technological innovations that are continuously developing and further improving the human life. Their origins lie in communication technologies and in a simple need to liberate the broadcaster (or sender) and the

© Springer Nature Switzerland AG 2019
T. Yigitcanlar and T. Inkinen, *Geographies of Disruption*,
https://doi.org/10.1007/978-3-030-03207-4_1

receiver from the limits of wires and physical locations (e.g., where phones locate). The digitalisation process, however, also has enabled the separation of the content and physical product properties. This has enabled new means of electronic service distribution and it has caused fundamental changes in the ways how content driven businesses work.

Generally, technologies should not be dealt as the truths given above. They are most often commercial products designed to generate profit for the developer. However, these systems also give roots for the routines and practices of communication. This differentiation (or converge) of service provider end-user platforms, content storing and security cause already a significant mixture of technology-society relations those final outcomes are yet to be seen. Therefore, the play-ground for globalised world is formed in relation to continuous technological development enabling growing data volumes and simultaneously smaller units of observation (e.g., in business intelligence). Quite often this is associated with targeted marketing and identification of single customers that may be considered as one of the most important business drivers for large social-media companies aiming to generate profit. National legislations and customs have to cope with the ever-increasing flows of goods (and people) creating an extensive field for legal studies in international law. These also include questions of intellectual property rights (IPRs) and the problematic of crowd-sourcing.

Innovations and technologies are therefore not only questions of the actual products themselves but rather a question of how their impacts are discussed, communicated, and disseminated (Nicolau and Santa-Maria 2013). For example, argumentation concerning alternative truths, the constitution of knowledge and what is verified and what is not are manifestations of the impacts caused by the internet and related social-media tools. There is a connection to a wider theoretical discussion between individualism and collective systems of belief (e.g., Eisenberg 1999). A crude example might be that for a social-media company dissemination of false information may prove to be highly profitable whereas the society as whole suffers from the impacts of this false information—particularly in the issues of health (e.g., vaccination and its importance) and other fundamentals of life in which public sectors, in most countries, have a significant role as service providers.

The relationship between a new technology and market economy is an essential in the development of information societies. Information (as well as other forms of capital) tends to cluster and accumulate in certain locations. Therefore, large cities are hosting the vast majority of information and communication technology (ICT) and media-related businesses. However, the economy is not the only point of view in the study of the innovation, smart cities and information society. Social sciences have argued for a long-time on the issues of equality, societal fracturing and citizen-government relations and their changes caused by digitalization (e.g., Mansell 2002; Borins 2008). Questions related to information reliability (true information vs. false information), data security and customer relations (freely given data concerning one's preferences or practices), and fragmentation of business orientation (particularly in the end-customer markets traditionally bound to physical locations and short distances).

The topic, therefore, may be approached from numerous angles. Actually, the problem field is similar to those defining 'what is a city?' Technological developments are impacting all the aspects of modern human life and they are also having significant impacts not only to the city employees (when cities are considered as organisations and administration structures) but also to the physical aspect of cities and their geography (e.g., daily traffic flows, smart integration of sensors and measurements devices into these structures). This information collection then provides more data, more accurate predictions and disseminated information both the vehicles and their drivers (or their 'managers' as self-driven cars are already reality and in the future decades implanted into practice on the streets).

The implications of technologies in (smart) city developments, therefore, include aspects of physical environment (urban geography of the location—architectural and planned space) and e-services adoptable to management (e-government, e.g., e-voting and interaction with the administration) together with commercial services, infrastructure, telematics, transportation, and flow management. Numerous e-city initiatives and measures tackle these issues. An approach applied by Yigitcanlar et al. (2015) used four main categories in order to assess knowledge-based development in second-tier cities. These categories include: (a) societal; (b) economic; (c) institutional; (d) spatial dimensions. All four categories are then further divided into two more detailed categories that each are composed of four single indicators. As an example, the 'spatial development' is divided into 'quality of life and space' including single indicators of 'quality of life', 'personal safety', 'housing affordability', and 'cost of living'; and secondly to the indicator category of 'sustainable urban development' that includes single indicators of 'urban form and density', 'sustainable transport use', 'greenhouse gas (GHG) emissions', and 'climate change adaptation/mitigation'.

Particularly, the environmental aspect is relevant as congestion has been an ever-increasing problem in, practically, all large cities. Regulative measures have been implemented (e.g., in Stockholm and Oslo) in order to reduce the daily traffic volumes in the city centres. Technological development has not been able to reduce the needs for actual movement but they have provided means to regulate and monitor traffic conditions with considerably higher efficiency than before. Therefore, questions connected to the capability and the potential possessed by the new innovations on ICTs and other forms of urban technologies (e.g., environmental solutions for urban environments) is targeted towards both the actual movement and flows taking place in urban environments as well as tools to monitor and possible to direct and guide these flows and movements (Romjin and Albaladejo 2002).

Especially within the sphere of the regional and urban studies attention has been paid to spatial and location properties to adopt, apply, and develop new forms of innovations and electronic solutions aiming to improve economy, environment and social condition. These are common aspects in technology foresight studies and evaluations (e.g., Georghiou and Keenan 2006; Veugelers 2012). A widely used term is regional resilience and regional adoptability to cover from impacts of global economy locally. These issues are fundamentally bound to national contexts in which urban and local units (such as municipalities) locate. Therefore, the geographical study of the information society is conducted and focusing on several starting points and designs.

Essentially, the subject can be approached with the tools offered by a classic network and innovation diffusion theories (e.g., Rogers 1995). The interest of the study has extensively focused on to the spatial structures, networks and their distributions. In addition to the quantitative approaching the phenomenon can be studied through the individuals, organisations and processes (Hemphälä and Magnusson 2012). Particularly, the human aspect in terms of ICT use, adoption and individual capabilities have been under extensive amount of research during the last decades (e.g., Inkinen et al. 2018). These problems are focusing on individuals and, in most cases, they produce unique and point-of-views that are difficult to generalise. In these cases, the data is normally collected by using qualitative methods such as interviews and surveys. Likewise, humanistic and psychological emphasises have become paramount topics within the research field of ICTs and the internet (Kellerman 2002).

Regional starting point in the information society studies brings forth the concept of spatial scaling. This means that there are steps and scales of analyses starting from single spaces (such as homes or offices or floor plans) that continue to neighbourhoods, postal code areas, sub-regions within cities, cities as a whole, functional urban areas, counties and regions and finally to nations, and finally international categories (e.g., alliances) and global issues. Information society studies have strongly concentrated on global questions. Some of the most central problems are estimates of the information divide between the industrial countries and the development countries, of the rearrangement of the labour and of the change in the processes of political economy.

An important aspect in these studies is the definition and conceptualisation of the information society and the following main questions have been addressed: What is the actual content of the information society? This question may be developed by asking the definitions of sub-categories defining the information society. These include concepts such as knowledge-based society, smart cities and communities, innovation systems, innovation society, electronic capital and digitalisation. Spatiality is important here as regional and spatial characteristics of a location have an impact on the ways how people trust, use and apply new innovative services. These are fundamentally connected to the cultures, education systems and other belief systems characterising social conditions and the ground to adopt new technologies. This takes place both on individual (micro) and aggregated levels of analyses and observations.

Technological development and innovations have been studied in urban studies for a long time. Spatial understanding of socio-technical phenomena such as smart city technologies provides a platform for understanding the current developments taking place in the contemporary world. As stated, geographical and spatial variations also exist between and within a national context. For example, urban-rural distinction and the question of technological adoption in densely populated cores are different to those areas with low populations and long distances. Graham and Marvin (1996) produced an influential book assessing the importance of telecommunications in cities. They emphasised the role of technological development particularly in the context of the internal development of cities. The development has two sides of the coin.

The physical and material aspect of production is the first. For example, locations of technology producing companies and businesses have their impact to the city's structure and design. Planning and design of the 'industrial' districts become more blurred when production requirements are e.g., based on start-up needs. Secondly, the products quite often themselves are immaterial information-based platforms or applications that do not require a physical distribution channel. This has been widely addressed in electronic commerce studies, which have pointed out the significant change in warehousing, customer support and customer interfacing. An example of this are the information commodities—as Lash and Urry (1994) name the consumer goods which are immaterial and related solely to information combination and distribution. They further create new interdependences between consumption, urban life and data networks.

The manifestations of the new and changing interdependencies are visible in the everyday functions and use of urban space. Internet mediated delivery platforms have created new markets e.g., for small restaurants and food services and they are also impacting traffic. Location-based services have enabled possibilities for closely targeted private car sharing systems lowering the need for car ownership and directing the demand towards collectively shared and on-the-need based car deliveries. Changes have been visible particularly through the uber-service that has impacted taxi-services greatly and is now testing national legislations particularly in Europe where taxi driving is a licensed business. Similar developments and examples are found in numerous other services targeted mainly for urban environments such as Airbnb enabling ordinary people to rent their accommodations and making challenges to traditional hotel industry.

From all these differentiations, developments and changes new business opportunities emerge and they have implications for cities from all aspects how to approach them: they create issues for governance and administration (e.g., electronic services and e-government), planning and design (e.g., citizen feedback, information distribution, transparency of planning process through digital media), urban design (participatory designing and planning of buildings and public structures), movement and traffic (intelligent transport systems, on-demand-based transit services, specified journey combinations merging together different forms of transit), consumption (online shopping, the use and role of physical stores, order logistics and transportation of goods), and the use of public spaces and online argumentation (real-time commentaries from locations and online information regarding happenings and events taking place in the city). In addition to these, citizen focused, changes the digitalization processes and applications of new media have a global reach. Significant changes are visible in the ways how governments and administrative offices disseminate information regarding their activities and decisions. Similarly, several companies are forwarding their product development increasingly towards 'internet of things' (IoT) or 'internet of everything' (IoE) applications. Cities are melting pots for all these integrative developments of economy, technology and urban form.

1.1 Rationale and Scope

The rationale and scope of the first part of the book include two major themes: (a) Innovation geography; (b) Urban economic development. Cities themselves are defined as large and intensive population concentrations that have established administrative structures for themselves. There are numerous categorisations concerning city sizes (e.g., hamlets, towns, cities, metropolises, megalopolises). The conceptual mix of urban development rhetoric is diverse. There are numerous terms that have been applied in order to illustrate technological and knowledge-based causes and effects in cities. Innovative and smart cities have been under a great deal of interest during the last decades. There have been numerous books and journal articles focusing on urban dimensions of innovation, technology and knowledge (e.g., Simmie 2001; Komninos 2002; Yigitcanlar 2016; Rodriquez-Bolivar 2018).

The main axes of the book include literature-based review of economic geography, innovation systems, knowledge-based development and smart solutions for governance and government. Cities are locations where these conceptual developments and categories are manifested. The first part of the book presents conceptual foundations and academic trajectories applied during the emergence of agglomeration concept and clusters. An important involves a discussion concerning innovation systems. This includes the distinction to national and regional scales, and economic spatial development. Part I focuses mainly on contemporary issues and literature but historical references and trajectories are included, for example, creative destruction, path dependency and evolutionary approach of spatial economy (see Martin and Sunley 2006, 2007; Musterd et al. 2007). These ideas are further discussed in the light of empirical data (see Part II) explaining selected four cases from the Northern European context, including best practice examples from Denmark, Estonia, Finland and Sweden.

1.2 Methodology

The first part of the book is based on a thorough literature review. The main information sources include internationally published influential books and articles. The starting point for Part I is in traditional agglomeration and cluster theories. In addition, diffusion of innovations, technological impacts on regions, and innovation systems are examined. Cities are the nodes that host and accommodate agglomeration and benefit most of their presence and of their interregional connectivity to other locations. It will be discussed that urban planning and design of new areas for innovation creates also new attractive residential areas. Technology and urban governance are enhanced and supported by digital solutions creating foundations for information and knowledge transfer in cities. Reflective discussion regarding these theories apply research results that have been collected in numerous research projects conducted by the authors since the early 2000s around the world. To summarise, Part I presents reflected and discussion-based approach for understanding urban changes and transformations in the light of innovations and technological progression.

References

Borins, S. (Ed.). (2008). *Innovation in government: Research, recognition and replication*. Washington, DC: The Brookings Institution.

Eisenberg, J. (1999). How individualism-collectivism moderates the effects of rewards on creativity and innovation: A comparative review of practices in Japan and the US. *Creativity and Innovation Management, 8*, 251–261.

Georghiou, L., & Keenan, M. (2006). Evaluation of national foresight activities: Assessing rationale, process and impact. *Technological Forecasting and Social Change, 73*(7), 561–777.

Graham, S., & Marvin, S. (1996). *Telecommunications and the city*. London: Routledge.

Hemphälä, J., & Magnusson, M. (2012). Networks for innovation: But what networks and what innovation? *Creativity and Innovation Management, 21*, 3–16.

Inkinen, T., Merisalo, M., & Makkonen, T. (2018). Variations in the adoption and willingness to use e-services in three differentiated urban areas. *European Planning Studies, 26*(5), 950–968.

Kellerman, A. (2002). *The Internet on earth. A geography of information*. London: Wiley.

Komninos, N. (2002). *Intelligent cities: Innovation, knowledge systems, and digital spaces*. New York: Taylor & Francis.

Lash, S., & Urry, J. (1994). *Economies of signs and space*. London: Sage.

Mansell, R. (Ed.). (2002). *Inside the communication revolution. Evolving patterns of social and technical interaction*. Oxford: Oxford University Press.

Martin, R., & Sunley, P. (2006). Path dependence and regional economic evolution. *Journal of Economic Geography, 6*(4), 395–437.

Martin, R., & Sunley, P. (2007). Complexity thinking and evolutionary economic geography. *Journal of Economic Geography, 7*(5), 573–601.

McLuhan, M. (1964). *Understanding media. the extension of man*. p. 397. New York: McGraw-Hill.

Musterd, S., Bontje, M., Chapain, C., Kovács, Z., & Murie, A. (2007). *Accommodating creative knowledge. A literature review from a european perspective. ACRE report 1*. Amsterdam: University of Amsterdam.

Nicolau, J. L., & Santa-María, M. J. (2013). Communicating excellence in innovation. *Economic Letters, 118*, 87–90.

Rodríguez Bolívar, M. P. (Ed.). (2018). *Smart technologies for smart governments. Transparency, efficiency and organizational issues*. New York: Springer.

Rogers, E. M. (1995). *Diffusion of innovations*. New York: Free Press.

Romjin, H., & Albaladejo, M. (2002). Determinants of innovation capability in small electronics and software firm in Southeast England. *Research Policy, 31*, 1053–1067.

Scotchmer, S. (2004). *Innovation and incentives*. Cambridge, MA: MIT Press.

Simmie, J. (Ed.). (2001). *Innovative cities*. New York: Spon Press.

Tuomi, I. (2002). *Networks of innovation. Change and meaning in the age of the internet*. Oxford: Oxford University Press.

Veugelers, R. (2012). Which policy instruments to induce clean innovating. *Research Policy, 41*(10), 1770–1778.

Yigitcanlar, T. (2016). *Technology and the city: Systems, applications and implications*. New York: Routledge.

Yigitcanlar, T., Inkinen, T., & Makkonen, T. (2015). Does size matter? Knowledge-based development of second-order city-regions in Finland. *disP-The Planning Review, 51*(3), 62–77.

Chapter 2
Conceptual Foundations of Innovation Geography

Abstract This chapter introduces theories of spatial economic development. There is an extensive amount of international research evidence that the main driving areas within nations are relatively large sized cities and their regions. The growth often concentrates on a limited number of key regions and cities of the world—mostly those are global cities of the world. Their number on the other hand is dependent on the population and geographical size of the nation. The relationship is not, however, linear or simple question of size. The location advantaged specific for each location have a significant impact and role on how it fits to the needs of varying markets. Industrial history and the advantage of the initiator are clearly visible in several key-hotspots in the technology industries producing innovations.

Keywords Agglomerations · Clusters · Proximate relatedness
Creative destruction · Spatiality · Innovation

Regional innovation activity is one the traditional research themes within economic geography and urban studies (e.g., Cooke and Morgan 2000; Cooke 2004; Crevoisier 2004, 2014). There is a long historical tradition concerning location decisions undertaken by economic actors (such as private companies) and they are traceable back to the birth of classical economic theories. The study of spatial economy and agglomerations was strongly developed at the end of the 19th century. Theories concerning economic agglomeration of businesses, and therefore clustering, emerged in the beginning of the 20th century. Marshall (1919) is one of the first developers of spatial agglomeration theory. His work was one of the first to consider economic activity in spatial context where distances, topographies, location advantages and market potentials either exist or do not. Company location and relocation theories started to evolve after the initial models in economic geography and regional science including the seminal works of Weber, Lösch, and Isard.

The first task of this Chapter is to define and understand innovation in a rigid way. Traditionally it is defined as a new idea resulting product, service or practice development that has an impact on market demand. Innovations may be incremental (thus small improvements accumulating in time) or radical (or disruptive) changing quickly the product scenario or production mindset. Therefore, the concept refers

© Springer Nature Switzerland AG 2019
T. Yigitcanlar and T. Inkinen, *Geographies of Disruption*,
https://doi.org/10.1007/978-3-030-03207-4_2

both to the process and to the outcome of that process. Innovation is therefore an idea or an activity that improves (incremental) or presents new (radical) service, product, or enhancement in production process. These ideas and improvements should also be in accordance with market demand and therefore have an economic impact on the performance of the innovation producing company. There are five identifiable functions that innovation process needs that are discussed below.

Firstly, considering the initial foundations of innovation there are elements of personal creativity, communication among colleagues, and a creation of an idea. These aspects are related to personal qualifications of each individual and therefore are highly difficult to be generalised other way than stating that innovation usually requires capable, resourceful, and skilful persons with intellect. Innovations the most often created in teams where knowledge transfer between persons and personal skills in communication play significant role.

Secondly, individuals and teams that create innovations are often working in the fields of research and development (R&D), science or other knowledge-intensive professions. Therefore, collaboration between education and skills are connected to pooling-up resources. This means that local universities commonly work with local companies and intermediaries in order to produce a platform or an innovation system (Howells 2006; Inkinen and Suorsa 2010). Individual and team capability on the topic of each innovation in question therefore require also skills and tools of R&D management, technological forecasting, IPR management, and marketing of innovations. Thus, innovation broadens from individuals to organisations that may benefit themselves of good practices or process innovations created in themselves or in other organisations. Therefore, the adoption of innovation and technology transfer is significant factor in the macro-level development of innovative region and cities.

Thirdly, technology transfer takes place in physical environments as well as in digitalised networks. In the case of cluster development, a frequently observed synergic gain of close proximities and co-location is the easier and more instant interactivity with the stakeholders. This is tightly linked to the transfer of tacit knowledge and weak signals obtainable from those places where the buzz is on (e.g., Malmberg and Power 2005; Gertler 2008). The transfer of these resources, signals or technologies are essentially locations of business and technology parks. Therefore, they are also places of planning and investments for innovative growth. In several occasions organisational mixes are associated to these locations and the presence of innovation system brokers such as various types of intermediaries (from consultancy to public support offices). There are easily identifiable properties available to increase operations efficiency in these locations, including technology auditions and benchmarking exercises (science park level), and the existence of supportive services such as law (e.g., IPR licensing, copyrights) and finance.

Fourthly, the factoring of innovation involves a significant role of finance. Successful locations of innovation, whether science parks, innovation districts or clusters, attract external finance. The question is about creating a positive cycle that creates a pooling effect of resources. The optimum situation also requires a critical mass of relevant actors. The attractiveness of an innovation centre creates a continuous emergence of start-up companies that is needed as majority of start-ups will fail. Silicon Valley is still the best example of this as it provides a great variety of companies from global giants to numerous micro-scale companies. The support services and availability of risk-funding creates a potential for the development of disruptive technologies and radical innovations. On the other hand, tools of quality management and assessment are essential in the incremental product development. Product and process development concerns particularly traditional industry fields and manufacturing companies. Main tools include product development and design improvements, product cycle management, and business process re-engineering.

Finally, networking requires further attention. Collaboration modes and optimally functioning inter-firm relations require relevant knowledge of supply-chain management and related vertical and horizontal linkages between firms, particularly on regional level (e.g., Ala-Rämi and Inkinen 2008). Digitalisation has provided efficiency gains in stock and inventory management relevant for companies producing physical products. The old just-in-time (JIT) principle is relevant as timely flow operations and the use of right amount of resources to produce right amount of goods according to market demand is the key-characteristics of all efficiently operating companies.

In contemporary setting innovation and smart city concepts are among the most widely used terms applied in regional economic development studies. They are also widely applied in public discussions and forums among politicians and professionals. An important conceptual development has been the integration of network theory into the analyses of local businesses. Issues of knowledge-transfer, knowledge-assets, knowledge bases, and knowledge-based development have gained an extensive amount of interest within the academia resulting into numerous single papers, special issues and books dealing with the topic (e.g., Asheim and Gertler 2005; Manniche 2012).

In the literature of socio-technical relations Castells (1996) provided a widely accepted classification into the spaces of places and flows. This distinction highlights the difference between global and non-place specific actions done in the digitalized environments and its counterpart physical places where the production and economic geography are realised. Additionally, Webster (2002) produced a widely read book focusing on the concept of information society as a broad overall concept to describe the increasing significance of information and its impacts on societies. The roots of his approach may be found in the earlier works of Toffler (1971) and the future oriented studies of technological change (e.g., Philips 1971).

There have been numerous conceptual applications such as social construction of technology (SCOT) applied mainly in the 1980s and the relating terminology such as technological determinism and digital divide that were popular in the 1990s and early 2000s. An interesting critical view on this topic was produced by May (2002) arguing and pointing out several problems associated with technological progress and the conceptual idea behind the 'information society'. Spatial examinations of the internet and spatial impacts both on economic production as well as society have been produced among the others by Graham and Marvin (1996) regarding cities, Dodge and Kitchin (2001) regarding virtual environments, and Kellerman (2002) regarding spatiality of the internet. A common denominator in their works is the recognition of geographical specifics found in their various study locations and its significance in how digital technologies and information channels are being used and how they impact the society.

The applied view on innovation (and innovative locations) may be summarised as follows: The starting point are the individuals and their groups capable of innovating. They work and collaborate in networks of organisations that are parts of each national innovation systems. They require market intelligence information as well as services often provided by innovation intermediaries or technology brokers (e.g., Millar et al. 2012). These organisations sometimes co-locate in same places and these places and locations may be either carefully planned or organically emerged. They may be branded as science, innovation, technology, or creativity parks, cities or districts. The critical success factor is the presence of adequate number of organisations that are needed in order to establish a critical mass of activities to keep these locations attractive for growth. This requires both finance as well as educated, creative or innovative workforce.

2.1 Agglomerations and Creative Destruction

In the theory base of innovative locations and cities commonly two main question types may be defined. The first question is: Why do other areas grow and prosper (in economic terms) faster than the others? The second one is: Why does the growth agglomerate (or cluster) in those areas that it does? The former question deals with and applies relations between locations and areas. It is traditionally approached via macroeconomic indicators (such as gross value-adding (GVA) or regional employment rate). A common application of this research type is to do comparisons with the selected spatial categories, and it aims to bring out causal-relations of spatial characteristics depicting current and potential growth areas. The latter question on the other hand concerns more specific locations and their properties themselves. This view may be considered as micro-economic approach that highlights internal properties of the region or area, in which the agglomeration is realised. This approach is most often characterised by case study designs and focuses on understanding processes. Study subjects may concern one industry field (e.g., ICT) or specific geographic or network cluster formations.

In the case of innovation and spatial development one influential theory tradition is traceable to the seminal works of Schumpeter (1939). He is most well-known for his analyses on innovation and creative destruction—one of the earlier and most eminent scholars emphasised the importance of innovation. However, in Schumpeter's theories the main focus was not exactly on spatial concentrations. His main contributions concerned firm properties and their relationship to innovation and creation of new markets. Simmie (2005) points out that there are two main eras in Schumpeter's views on how he has viewed firm-level innovation (see Phillips 1971).

Schumpeter's work on creative destruction and path dependency form foundations for the studies of evolutionary economic geography that has also been influenced by evolutionary economics. The work of Schumpeter had a clear and identifiable impact on regional scientists who were interested in the ways how he combined the market types into the realisation of innovations in specific locations. The concept of path dependency is exceedingly applied in current literature to describe the formations of trajectories on certain paths dependable both on public decision-making (e.g., planning, taxation or economic development expenditure) and private sector decisions (e.g., constructions, location decisions, and outsourcing).

The first point-of-view in Schumpeterian perspective is the highlighting of small and medium sized companies (SMEs) that are operating in a competition market (or a monopolistic market). The companies have marketing power at least some extent but they are experiencing hard competition (e.g., Larsen and Lewis 2007). In this view, the innovation activity is high due to the competitiveness and there is a need for constant improvement in volatile market condition. Schumpeter's first view emphasises the swaying development process of the market economy in which the uncertainty factors of competition, market demand and global resource prices are met and dealt with by small and agile companies able to modify their production and cost structures within a short amount of time. Firm's technological and organisational agility and ability to modify to market impacts is strongly present and creates a platform for innovation that is considered as the central motor for the dynamic market economy (Konsti-Laakso et al. 2012).

The second point-of-view that was raised under a loop of investigation later in Schumpeter's work instead highlights quite the opposite: it stresses the role and importance of large oligopolies and market conditions with limited number of players in the innovation creation and product development: Large companies have larger resources and therefore it is easier for them to experiment with designs and products that do not necessarily lead to an end-product. In this regard, focusing on R&D investments of firms and corresponding regional R&D expenditure gives motivation justified analyses in understanding of creative destruction and openings of new markets. R&D (and resulting innovation or R&D&I) activity of large companies also creates ground for smaller companies through subcontracting and networked production chains. The internal technology processes and market recognition of large companies tend also to cause spatial concentration and clustering. Schumpeterian thinking also takes temporal variations into consideration. Short term changes may be significant and fast but they are contrasted with small incremental changes and continuous development according to path-dependency argument. The long-term view on the development of economic activity is founded on the macro-perspective build on the actions of shorter time-frames.

The significance of Schumpeter's work may be considered in the ways how his thinking about innovation creation through market conditions took place. The combination of innovative production, product development, R&D and location produces interesting openings in the understanding of economic conditions and related growth goals set in cities. As such regional economics provide econometric background. In the most successful situations econometrics are supported with suitable and closely targeted qualitative data sets enabling cross-disciplinary analyses of economic activity (e.g., Storper 1985, 1991, 1997; Scott 2000, 2001; Krugman 1995, 2000).

The multidimensionality required for analysing industrial or service clusters is extensive (e.g., Bathelt et al. 2004; Simmie 2004; Inkinen and Kaakinen 2016). Clusters refer here either to geographical concentrations (agglomerations) or they may refer to networks combining different business under the roof of same brand in a specific geographical context (e.g., national). This distinction is similar to ones applied in innovation systems literature where commonly a distinction is made to national systems of innovation (NIS) and regional systems of innovation (RIS). In some cases, even local levels or specific cities have been identified to have their own innovation systems or highly similar arrangements supporting innovation creation (Asheim and Isaksen 2002). This often requires recognition of start-ups and related business incubators, often associated with a well-functioning business eco-system capable of producing successful start-ups (Aernoudt 2004).

In the growth theory of enterprises, the forms and modes of business co-operation external factors (external economies) is seen to have an affect also in the internal processes (internal economies) of the firm. Additionally, companies are able to produce pools of resources, in other words combinations, in which case they are able to utilise similar production facilities, labour resources, other forms of infrastructure, and capital. The larger the pools become, the better opportunities single companies have in order to pursue specialisation and resulting production efficiencies. This is achieved through lower internal production costs obtainable through external pooling. The central idea is that commonly these pooling opportunities and thus economies of scope locate in the same areas with short proximities. The main principles of Marshallian agglomerations contain the fundamentals still observable in firm-level analyses of geographies of production.

Conceptual clarity is important as there are generally three types of innovations separated: Technical, organisational and marketing innovations are commonly separated in the innovation literature. A widely used division is also the separation into radical (fast, unpredictable, market changing) and incremental (small improvements, enhancements, market supporting) innovations. Innovation impact scales are also very different from each other. On the smallest scale an incremental technical innovation may have small impact in product performance and on the largest scales innovations (such as social-medias) change the way of how societies function and communicate.

Simmie (2001) argues that Schumpeter's first point-of-view highlighting the role of smaller and agile companies together with the Marshallian agglomeration theory have provided strong explanative power in the models of urban economics all the way till the end of 1970s. Since then the emergence of multi- or transnational corporations that have a global reach and impact have changed the mode of innovation towards the second phase of Schumpeterian thinking. One may argue that the aspect of social innovation (and network) is perhaps the most pronounced in the cases of oligopoly markets: there are only a limited number of companies and they are aware of each other's existence and role in that market including their competitive position. Oligopoly markets are prone to cartel formations due to their competition knowledge.

As stated above, the second main phase and point-of-view of Schumpeterian approach towards innovation is fundamentally very different from the first one. They

2.1 Agglomerations and Creative Destruction

represent two opposing ends in a spectrum in terms of actors, size and resources to conduct science and engineering-based product (or service) development. The process may be seen as a cyclic process where positive experiences and feedbacks of innovative enhancements (mainly incremental after initial launch of radical innovation) great impetus for higher R&D expenditure. This creates a positive development circle, in which successful solutions and decisions fuel the future inputs for either incremental enhancements or radical alternatives. In other words, innovations create innovations, and large established companies have better resources to ensure the continuation of the innovation cycle. This is also a result of the financial system. Innovation creation and thus R&D efforts often require extensive and continuous investments. In other words, success requires innovation that requires success. This causal relation has been studied e.g., by Makkonen and Inkinen (2013) in the European context and further developed by Makkonen and Mitze (2017) stating that the innovation triad (finance, education and innovation) are linked in the order where education comes first, innovation second and finance third.

The work of Schumpeter has influenced numerous researchers and therefore also gained to be addressed as a creator of a research tradition. Evolutionary economics and innovation studies are always founded to the key-ideas of Schumpeter (e.g., Ianmario 2005). There are other key persons who continued the work of Schumpeter and they have been extensively accounted by Simmie (2001). Among them are French Perroux (1950) and American Raymond Vernon. Their particular contribution in the mid-1900s concerned the connection between innovation dynamics (firms) as a part of regional accumulation. One of the main arguments according to Perroux is that an innovative company more likely operates with higher efficiency compared to similar companies that are less innovative or do not conduct innovative activities.

This line of thought then leads to an idea that business volumes and growth are taking place with higher phase in innovative industries (and companies). In turn, the higher growth rates attract more investments and expectations towards these industries resulting into higher activity rates and thus also lowering relative costs through accumulation. The locations those are able to attract (or produce/introduce) innovation prone companies then enable higher growth rates for the regional economy in question. The presence of supporting industries in close or neighbouring proximity supports the spatial accumulation and agglomeration. Thus, innovation activity tends to support and create economic activity. The process is also cyclic and self-supporting. A traditional view of positive multiplier effects is identifiable within this view of spatial development.

The agglomeration effect can also be viewed in the light of networks. Traditional concepts of economic geography include horizontal and vertical connectivity in the production chains. The connectivity (e.g., subcontracting networks) brings forth the essence of geographies of production, particularly in the case of multinational corporations, which are able to distribute and manage their production chains in global level. Thus, the geographies of production also involve the understanding of product life-cycles that are constant targets of incremental innovations. An excellent example of this are contemporary mobile telephones that were formed in their current standard state (touch pads) during the years 2007 (launch of the first Apple iPhone) till 2009 (emergence of competing models from other producers). Considering the life span of one radical innovation (well-functioning touch pad operating system and interface) incremental innovations have supported the product lifespan to exceed more than a decade. The product life-span is also affected and resulted through after sales services, support and maintenance offered by the technology companies.

Innovative solutions are therefore deeply rooted into the production phases. These include the product development phase (that may or may not include radical or incremental innovations); the testing phase and quality insurance before market launch; the actual production phase and search for cost efficiency according the product standards; logistics phase and optimization of product availability in different (geographical) markets; and market response after the launch. These phases are then monitored and followed by key-indicators such as total revenues, number of sales, market responses and competition responses, and (customer/retailer) feedback. A guiding principle is that products portfolios are in continuous change. Product differentiation (incremental change) is a common tool to achieve these goals expanding the product life-span. In the case of software dependent products, such as mobile phones, the operating system (OS) support is perhaps one of the best examples expanding the product lifespan.

Short time-periods and intervals for product innovation and development are seen as a means to allow broader scope for resource distribution (Lundvall 1985). Vernon (1966) continued this line of thought and highlighted that the innovation phases should be quite short (shorter the better) as the new products and innovative solutions tend to lose their cutting-edge value relatively shortly after the public launch. Reasons for this include product cloning and imitation. The aim to shorten the elapsed times in production cycle is supported by spatial agglomeration. Therefore, it is probable that the most innovative, cost-efficient and economically efficient companies locate in metropolises where support services and economies of scope are most likely to be present for different types of industries. There is also a recognised differentiation in terms of in-house production and networked subcontracting production chains. The argument has been made that innovative product developers are more often reliant on highly specialised capabilities and they are also doing more intra-firm collaborations than producers of standard products.

The availability and cost-efficiency (through competition) of support services and external resources is generally better in large (and diverse) economic locations. Larger economic volumes also enable wider opportunities and flexibility in terms of experimentation and testing. Economies of scale are present particularly in the operations of multinational corporations operating on global oligopoly markets. On the other hand, economies of scope are realised in cluster formations. Perhaps one of the most widely studied clusters of economic activity and innovation is the Silicon Valley located close to the city of San Francisco in California. Locations attractiveness has developed since the 1970s and it is a text book example of regional context (e.g. 'the mindset of California') through openness, presence of the top-class universities and a long history of garage start-ups that have developed to the global conglomerate companies particularly in the field of ICTs.

Economic geographers have developed and applied different modelling techniques for understanding the co-variations between population statistics, socio-economic factors and economic performances of firms. Perhaps one of the most traditional means to assess their interdependencies is the gravitation modelling aiming to produce limits between functional areas surrounding neighbouring cities (or nodes). Geographical information system (GIS) tools have enabled more and more detailed analyses on clusters and their micro-geographies. The development of analytical tools has also integrated more robust statistical methods into the arsenal of spatial analysis. Today the use spatial autocorrelation models have gained empirical interest, as the data sources and analyses tools available in software are enabling their use.

2.2 Clusters

Specialisation, flexibility, adoptability, innovation capability and cluster formations are widely applied concepts in the understanding of economic clusters. Michael Porter is the most well-known for his diamond classification and cluster definition that pointed out that clusters are may be considered through networks. The networking takes place between and within companies and between similar and dissimilar industries or business fields.

Clusters are commonly defined as networked organisations (network clusters) operating on the same field or geographically agglomerated units (spatial clusters). There are numerous interest points in the analysis of spatial clusters. These include the decisions made by the enterprises to locate themselves into certain areas. In addition, the decisions of urban planners, taxation and other public-sector decisions and services have a role in the cluster formation. Employer satisfaction has become an important factor in the attraction of highly-skilled employees. Therefore, also the properties and qualities of life are significant.

Spatial clustering may be regarded through objective measures of macroeconomics and therefore analysed according to distance decay, gravity models, and spatial autocorrelation. These methods have been extensively applied in quantitative research of economic geography. Another way of looking and collecting data concerning clustering is to use qualitative methods and process analysis tools. Tacit knowledge, social networks, and expectations that are not based on rationality, are examples of factors that are very difficult to quantify for causal analysis. As such, innovation, information society and cluster indicators are also statistically related. New innovations are changing the traditional parameters according to which companies have traditionally considered their locations. Recent studies have indicated that even tough software and information-based companies that have the largest freedom of choice to locate themselves still, however, follow rather traditional routes in their location decisions. The tradition and history of the location as well as the reputation have an impact and location decision is as much an issue of image as it is functionality.

Clusters may be considered also as locations where common interests and ideas meet. This sets requirements to single office spaces and premises but continues also outside office buildings to other services (restaurants, sport-facilities, markets, shops) located in cluster core areas. Those areas locating high level of businesses and also residential apartments and houses are interesting as they transcend the traditional planning doctrines separating business districts from residential areas. This type of mixture development is easily identifiable in several cases applied in Part II of the book.

These ways how synergies are gained and produced through co-location is gaining further interest among urban researchers and economic geographers. Numerous empirical works have confirmed both in quantitative and qualitative ways that short proximities and nearness to other fields of industries creates synergies. This is connected to transformation process of tacit knowledge towards codified knowledge and knowledge transfer within and between organisations (Nonaka and Takeuchi 1995). These transfers require conjunctions between people from different organisations and fields. Close proximities and distances also enable easier learning processes both on individual and organisational levels (e.g., Rutten and Boekema 2013).

Innovative production and economic efficiency in urban context is tightly connected to the digitalisation process. The concept of digitalisation is widely used in current literature but its roots are identifiable back to the classic concepts of automation, innovation diffusion and micro-electronics. On the one hand, digitalisation may be approach as a means to an end that is commonly referred to cost reductions and minimisation of logistic costs. Digital services increase reliability, it enhances stock management and warehousing needs, and it helps monitoring the whole logistic process. On the other, digitalisation also produces new services and products that provide new business potentials for operations management. The third and perhaps the most significant impact of digitalisation is that how it changes (or transforms) use patterns or practices in broader societal contexts. For example, customer relations and feedback systems are not dependent on fields of industries—and they have extensively been digitalised already.

Innovation in urban context is therefore all encompassing. It concerns urban technologies and integration of smart technologies into urban infrastructure such as roads and rails. It also is exceedingly evident in numerous applications providing information regarding services and locations in urban space. Sometimes these services are called 'augmented' spaces referring to digital information integration into physical objects.

2.3 Proximate Relatedness

Proximate relatedness refers to similarities and dissimilarities of locations with either narrowly or broadly focused industrial profiles. There are number of different proximity conceptualisations such as geographical (distance), cognitive, organisational, institutional and social proximities. The narrowly focused areas are those that are relying on few similar industry fields and thus supporting similar subcontracting networks making the economic profile of the location biased towards these few industrial branches. The broadly (or diverse) profiled areas on the other have broader industrial profile but they are not necessarily as successful or dominant in terms of global market shares or profit making.

Proximate relatedness is connected to the concept of regional resilience and regional capability to survive a market shock in relation to other areas with similar or dissimilar economic profiles (proximity metrics). Industrial mixes and broadness in industrial profile is often considered to be more robust condition to survive severe economic downfalls often caused by external sources. Specialisation and decentralisation of production, marketing and resourcing are common in contemporary global business. Geography matters as there are clear distinct distance benefits in terms of logistics and marketing. In general, the global markets follow distinctions to continents.

A concept of flexible specialisation in production was developed particularly in order to answer demand driven market feedback. On the other hand, flexibility and specialisation may be thought two options or pathways towards innovative enterprising. They are strategic choices how to position one's product or service within the production chain. The strategic opposing ends may be defined as 'connecting/collaborative networks' (assembly resulting from an extensive outsourcing and use of intermediate products) and 'in-house' production (all parts of production done under one roof). Outsourcing is tightly intertwined with spatial economies as large multinational companies often have extensive subcontractor networks, often located close to the production facilities. This causes a multiplier effect in the case the production facility is shut down as these supportive functions lose their income source. This is related to the spatial trickledown effect caused by the existence of large economic actors and is also connected to the concept of regional resilience and industrial profile of the region. If the profile is too narrow or reliant on one industry branch recovery potential is limited in the actualisation of a crises.

One important milestone in innovation and flexible specialisation research was the continuation of the life cycle model development in the innovative production (e.g., Piore and Sabel 1984). The lifetime of a product is a result or is at least affected by complexities, diversities and uncertainties of the market. The demand pressure coming from market drives companies to specialise. Companies producing mass products usually have to outsource the production to low-cost countries as specialisation requires higher production qualifications and quality control. This often means also a relatively narrow product portfolio that is easier to manage and control. There are examples in the ICT business of how too large product portfolio sets may cause severe problems for highly successful companies, such was the case with the mobile phone manufacturer Nokia in the early 2010s.

2.4 Summary

The theoretical perspectives in economic geography require the contextual knowledge on nodes and their connecting networks. Urban centres are excellent examples of nodes, in which socioeconomic activity concentrates. Proximities are important as they refer to absolute, relative and relational distances between objects within spatial categories and constructs. For example, market proximity may be considered in terms of costs, distances or potentials. The internal structure of location based economic activity is also important. As these structures always exist in local, regional and national context they experience a national policy bound starting point. These starting points have different positions according to proximity concepts in relation to external (international) markets. Some proximities are political (e.g., restrictions in trade, trade alliances, internal markets, and customs), some are economic (e.g., transportation costs, market distance, customer potential), and some are legislative (e.g., differentiations in legislation concerning environment, security or quality).

References

Aernoudt, R. (2004). Incubators: Tool for entrepreneurship? *Small Business Economics, 23*(2), 127–135.
Ala-Rämi, K., & Inkinen, T. (2008). Information technology, communication and networking in small and medium size software companies. The Case of Northern Finland. *International Journal of Knowledge Management Studies, 2*(3), 320–334.
Asheim, B., & Gertler, M. (2005). The geography of innovation: Regional innovation systems. In J. Fagerberg, D. Mowery, & R. Nelson (Eds.), *The oxford handbook of innovation* (pp. 291–317). Oxford: Oxford University Press.
Asheim, B. T., & Isaksen, A. (2002). Regional innovation systems: The integration of local "sticky" and global "ubiquitous" knowledge. *Journal of Technology Transfer, 27*(1), 77–89.
Bathelt, H., Malmberg, A., & Maskell, P. (2004). Clusters and knowledge: Local buzz, global pipelines and the process of knowledge creation. *Progress in Human Geography, 28*(1), 31–56.
Castells, M. (1996). The rise of the network society. (Vol. I, p. 556). Oxford: Blackwell.
Cooke, P. (2004). Evolution of regional innovation systems—Emergence, theory, challenge for action. In P. Cooke, M. Heidenreich & H. -J. Braczyk (Eds.) *Regional Innovation Systems* (2nd ed., pp. 1–18). London: Routledge.
Cooke, P., & Morgan, K. (2000). *The associational economy. Enterprises, regions, and innovation.* Oxford: Oxford University Press.
Crevoisier, O. (2004). The innovative milieus approach: Toward a territorialized understanding of the economy. *Economic Geography, 80*(4), 367–379.
Crevoisier, O. (2014). Beyond territorial innovation models: The pertinence of the territorial approach. *Regional Studies, 48*(3), 551–561.
Dodge, M., & Kitchin, R. (2001). *Mapping cyberspace.* London: Routledge.
Gertler, M. (2008). Buzz without being there? Communities in of practice in context. In A. Amin & J. Roberts (Eds.), *Community, economic creativity and organization* (pp. 203–226). Oxford: Oxford University Press.
Graham, S., & Marvin, S. (1996). *Telecommunications and the city.* London: Routledge.
Howells, J. (2006). Intermediation and the role of intermediaries in innovation. *Research Policy, 35*(5), 715–728.

References

Iammarino, S. (2005). An evolutionary integrated view of regional systems of innovation: Concepts, measures and historical perspectives. *European Planning Studies, 13*(4), 497–519.

Inkinen, T., & Kaakinen, I. (2016). Economic geography of knowledge intensive technology clusters: Lessons from the Helsinki Metropolitan area. *Journal of Urban Technology, 23*(1), 95–114.

Inkinen, T., & Suorsa, K. (2010). Intermediaries in regional innovation systems: High-technology enterprise survey from northern Finland. *European Planning Studies, 18*(2), 169–187.

Kellerman, A. (2002). *The internet on earth. A geography of information*. London: Wiley.

Konsti-Laakso, S., Pihkala, T., & Kraus, S. (2012). Facilitating SME innovation capability through business networking. *Creativity and Innovation Management, 21,* 93–105.

Krugman, P. (1995). *Development, geography, and economic theory*. Cambridge: MIT Press.

Krugman, P. (2000). *Geography and trade*. Cambridge: MIT Press.

Larsen, P., & Lewis, A. (2007). How award-winning SMEs manage the barriers to innovation. *Creativity and Innovation Management, 16,* 142–151.

Lundvall, B. Å. (1985). *Product innovation and user-producer interaction*. Aalborg: Aalborg University Press.

Makkonen, T., & Inkinen, T. (2013). Innovative capacity, educational attainment and economic development in the European Union: Causal relations and geographical variations. *European Planning Studies, 21*(12), 1958–1976.

Makkonen, T., & Mitze, T. (2017). Deconstructing the education-innovation-development nexus in the EU-28 using panel causality and probability tests. *Journal of the Knowledge Economy*. https://doi.org/10.1007/s13132-017-0454-4.

Malmberg, M., & Power, D. (2005). (How) do (enterprises in) cluster create knowledge? *Industry and Innovation, 12*(4), 409–431.

Manniche, J. (2012). Combinatorial knowledge dynamics: On the usefulness of the differentiated knowledge bases model. *European Planning Studies, 20*(11), 1823–1841.

Marshall, A. (1919). *Industry and trade*. London: Macmillan.

May, C. (2002). *Information society. A sceptical view*. London: Polity.

Millar, C., Udalov, Y., & Millar, H. (2012). The ethical dilemma of information asymmetry in innovation: Reputation, investors and noise in the innovation channel. *Creativity and Innovation Management, 21,* 225–237.

Nonaka, I., & Takeuchi, H. (1995). *The knowledge-creating company: How Japanese companies create the dynamics of innovation*. New York, NY: Oxford University Press.

Perroux, F. (1950). Economic space: Theory and applications. *Quarterly Journal of Economics, 64*(1), 89–104.

Philips, A. (1971). *Technology and market structure*. Lexington: Heath Lexington books.

Piore, M., & Sabel, C. (1984). *The second industrial divide: Possibilities for prosperity*. New York: Basic Books.

Rutten, R., & Boekema, F. (2013). Beyond the learning region: A new direction for conceptualizing the relation between space and learning. *European Planning Studies, 21*(5), 722–734.

Schumpter, J. (1939). *Business cycles. A theoretical, historical and statistical analysis of the capitalist process*. New York: McGraw-Hill.

Scott, A. J. (2000). *The cultural economy of cities. Essays on the geography of image-producing industries*. London: Sage.

Scott, A. J. (Ed.). (2001). *Global city-regions: Trends, theory, policy*. Oxford: Oxford University Press.

Simmie, J. (Ed.). (2001). *Innovative cities*. New York: Spon Press.

Simmie, J. (2004). Innovation clusters and competitive cities in the UK and Europe. In M. Boddy & M. Parkinson (Eds.), *City matters: Competitiveness, cohesion and urban governance* (pp. 171–196). Bristol, UK: The Policy Press.

Simmie, J. (2005). Innovation and space: A critical review of the literature. *Regional Studies, 39*(6), 789–804.

Storper, M. (1985). Technology and spatial production relations. Disequilibrium, interindustry relations, and industrial development. In M. Castells (Ed.) *High technology, space, and society* (pp. 265–283). New York: Sage.

Storper, M. (1991). *Industrialization, economic development and the regional question in the third world. From import substitution to flexible production*. London: Pion.

Storper, M. (1997). *The regional world: Territorial development in a global economy*. New York: The Guilford Press.

Toffler, A. (1971). *Future shock*. London: Pan Books.

Vernon, R. (1966). International investments and international trade in the product cycle. *Quarterly Journal of Economics, 80*(2), 190–207.

Webster, F. (2002). *Theories of information society*. London: Routledge.

Chapter 3
Place Dynamics for Innovation

Abstract Innovative urban development commonly requires the recognition of technological development directing the physical infrastructure. This is related to current digitalisation trend taking place in all societal domains including economy, infrastructure, and policy. Leading cities in innovative development have actively participated the development and enablement of services and solutions aiding visitors, inhabitants and businesses on their vicinities. Traditional examples include open access urban networks (enabling access) and different mobile applications helping the use of urban space (e.g., public transport systems and route planners). The availability and low use cost of digital services contributes to urban image and cities potential to advertise themselves as knowledge-intensive locations. This chapter discusses and reflects on innovation system approaches and urban development.

Keywords Innovation systems · Place dynamics
Knowledge-based urban development · Urban competition · Urban marketing

3.1 Innovation in Urban and Regional Studies

Eminent smart city researcher Komninos (2002) named the first chapter in his book *Intelligent Cities* as 'innovation is an island' discussing this dualistic definition of innovation and the problematic of cause and effect. This aspect has been widely acknowledged in empirical innovation studies since in the form that innovation inputs (e.g., R&D spending and researcher employment) are treated separate from innovation outputs (e.g., applied or granted patents). The two components are naturally interlinked and in one sense innovation may also be regarded to be conceptually very close to 'product development'. The main characteristic of innovation is that the new idea or development is fundamentally developing the process or product forward. Therefore, product differentiation (a new product) is not an innovation as such if it lacks the development process behind it.

Cites are verifiably deemed as the motors of economic growth as they are nodes of human inhabitancy. Urban and regional innovation geographies have applied several conceptual frames (e.g., de Bruijn and Lagendijk 2005). Knowledge bases approach has been among the most applied ones. It recognises the importance of tacit and codified knowledge but considers them to be a too narrow starting point for innovation analyses since it commonly involves both forms of knowledge. The separation between the two is highly difficult as they are intertwined into the roots of creativity and skill. To overcome this division knowledge bases approach applies distinction by focusing on applications and industry fields. This has led to a formation of three conceptual categories, namely analytical, synthetic and symbolic knowledge bases (Asheim and Gertler 2005; Asheim et al. 2011a).

These three dimensions involve the recognition of the interlinkage and change from codified knowledge, which primary function is to be transferable, towards more subjective and individual forms of knowledge. Symbolic knowledge is therefore considered to be most 'creative' and its main use purposes are directed towards artistic and creative endeavours of innovation (e.g., Florida 2002; Boschma and Fritsch 2009). This logic may be also described as quantifiable (or measurable) forms of knowledge and the transition towards qualitative (process) knowledge. Knowledge bases have been applied in empirical innovative city analyses (e.g., Inkinen 2015) and they provide one potential point-of-departure.

Knowledge base approach is connected to the triple-helix (Etzkowitz and Leydesdorff 2001; Etzkowitz and Klofsten 2005) concept that highlights the co-existence and collaboration between the universities, innovative businesses, and public sector (government). This classification has also been described as B-U-G model—business-university-government model (Anttiroiko and Kasvio 2006). The main line of thought goes that universities are the main producers of analytical knowledge that impacts mainly medicine production or biotechnology. Universities are also producers of the highly educated workforce impacting all fields. The producers of synthetic knowledge base innovations, for example different fields of engineering, have a varying demand for tertiary educated workforce. Symbolic knowledge base has the most difficult relationship directly linked to university education as in several cases arts, creative professions and culture professions do not necessarily require similar formal education as engineering, marketing or analytics. The university-business linkage often benefits from smart (or innovative) government services and solutions (e.g., Altshuler and Behn 1997; Lambooy 2007). Together the three fields form the starting point for understanding regional development.

The conceptual frames applied in urban and regional innovation studies (see Fisher 2001; Fagerberg and Verspagen 2009) require also other dimensions than technology and digitalisation. Environmental attractiveness and well-functioning service provision have become key-characteristics in the development of desirable location—not only for inhabitants but also for attracting business. There are extensive varieties within large cities in terms of their attractiveness. Several global cities have, on one hand, highly desirable locations and, on the other, very difficult locations for attracting workforce or businesses. The internal variations of urban space present an increasingly popular study field in economic geography. The locations desirability is also tightly connected to the functionality of governance and public-sector policies towards innovative businesses. Practically all industrialised countries have established their development policies to attract international business and investments. Public-private-partnerships (PPPs) have become a widely used strategy to pursue economic growth and new innovative services, products and practices in regional and local development.

3.2 Differentiated Systems of Innovation

Innovation systems consist of knowledge producers including higher education institutions along with research organisations, and of knowledge users including private companies as well as government agencies. Innovation system can be understood broadly to include all factors that have an impact on innovation activities (i.e., the whole economic sector) or narrowly to include only actors that affect directly to innovation (Lundvall 1992; Clarke and Tracey 2004). Innovation system frameworks commonly recognise the importance of cooperation and interlinkages together with the conditions of organisational culture, social capital, trust and interactive learning between actors (Asheim et al. 2011b; Asheim and Isaksen 2002; Asheim and Gertler 2005).

The traditional way of separating innovation systems is to consider national policies and national systems of innovation (NIS) as the main doctrine of innovative development (Lundvall 1992; Nelson 1993; Lundvall and Maskell 2000). The second and particularly favoured by the economic geographers is the regional approach that recognises that different regions and areas within a nation may have highly diverged conditions or opportunities to execute or benefit from the national level decisions. Therefore, the concept of regional innovation system (RIS) has been developed. There are numerous studies and research conducted around this concept (Oughton et al. 2002; Asheim et al. 2007, 2011a). In general, it may be defined to include regionally specific tools and objective setting recognising regional strengths and conditions aiming to make the best of them in terms of innovation development (Ahlqvist and Inkinen 2007).

In the case of NIS, the main question boils down to consider the relationships and marching orders between industrial and economic policies, tools that are executing these policies and development organisations commonly implementing these tools. NIS considers national resources and potentials according to the networks and competition advantages obtainable from one country. NIS and RIS are fundamentally combined because in majority of countries national legislation and policies overcome those of the local level. In several cases, municipalities or regions have their taxation rights in order to obtain resources to execute the tasks identified for them (according to governmental structure of each country). However, municipal governments are always secondary to the national legislation. Therefore, RIS may also been as a local implementation of NIS in several cases.

National actors are in the driving position in the development and focusing of innovation, technology and science policies. Their decisions and policy documents are clearly visible in the strategies of regional actors (Suorsa 2007; Makkonen and

Inkinen 2014). The similarities of national and regional strategies are to be expected but too extensive similarities cause that the regional specialisations and specifications are easily neglected. Typically, those regions that have developed systemically their own RIS often are dependent on one or few distinct special industries making their case special. These special needs commonly require more detailed and focused tools (in networking, collaboration, and funding) that may be under-presented in national strategies. It is common that NIS is targeted to provide broad national guidelines and then regional organisations (or regional offices of the national bureau) make their own specific need assessment and further target the available support systems towards industries and research organisations. There are extensive variations in how much targeting power regions have depending on their national contexts. For example, the NIS applied in South Korea is highly top-down oriented and monotonous whereas the US is fragmented according to state legislations.

There are some key-elements observable from both NIS and RIS frameworks. The starting point is that innovations are fundamentally produced in companies. This makes the separation to innovation (and/or technology) policies and traditional industrial policies a blurred line (Freeman 1987; Lemola 2003). Innovation is, however, also adopted in the public-sector organisations. These adoptions concern mainly the service provision development and enhancement of efficiency through the adoption of new service platforms and delivery systems. Digitalisation plays a key-role in this service development. There are also studies focusing solely on the public-sector innovation activity among city organisations (e.g., Makkonen et al. 2018). The main obstacles for city organisations and officials to produce innovations are strongly linked to hierarchical structures of administration and low level of flexibility. Still, there are some good examples of patent applications and suggestions to improve service provision also from the public sector.

The efficient development of the NIS and RIS requires the recognition of a limited number of focal points and spearheads. These should function as the main elements to where the resources (services and financial support) are targeted. Economic activity is always located in places with specific characteristics and draws on the resources of the regional context—including the use of it as a platform to make connections elsewhere. The related micro-level activities of individuals, organisations and their local-scale interactions need to be understood. In the analysis of the regional development process, focus has to be on the connections between different spatial scales; the local scale is integrated to the regional, national and international economic and institutional processes by networked structures. The ability of a regional economy to succeed in and trigger changes in the broader economic system depends on the innovation ability and successful market strategies of its economic agents.

Innovation policies (e.g., in Finland) are often based on customer driven understanding of national and regional growth (Lemola 2003, 2004). This highlights the traditional definition of innovation that requires the presence of market demand. New services or products are less likely to succeed in consumer markets if their usability or functionality is too complicated. The user driven approach is handled better in some countries than in others as the question is connected to specific cultural contexts of different countries (e.g., Scott 2000, 2001). A good example here is the differences between Sweden and Finland in the ways how marketing and user experience are integrated to the end-product and how Swedes have been better in achieving this.

Other fundamental concerns the market impact expected from innovative companies. As stated, the demand driven expectation may lead more towards incremental innovation activity than radical. There are also distinctive spatial scale differentiations identifiable in NIS and RIS agendas. NIS policies and strategies quite often have their market focus on the international or global level innovations where as RIS focuses are relatively often focused on their specific regional vicinities. Therefore, there have been arguments that RIS produces mainly small-scale products and services that are not able to break through to national or international markets.

RIS are fundamentally connected to the concept of functional urban areas (FUAs). They may be conceptualised according to a definition applied by the OECD (2013, p. 2): "The definition of urban areas in OECD countries uses population density to identify urban cores and travel-to-work flows to identify the hinterlands whose labour market is highly integrated with the cores". In the context of urban areas, an important aspect concerns urban regeneration that refers to actions and developments concerning structural and social changes taking place in urban areas. These include deindustrialisation, demographic changes, labour market conditions (unemployment and employment levels) as well as social segregation and gentrification.

Intelligent and smart city solutions are crucial part of any innovation system strategies. Large cities and particularly capital (either national or regional) areas are those most actively aiming to improve their attractiveness and competition advantage with the aid of innovation support and growth. The most common elements to which smart city solutions are target for are grid information systems (electronic), security and identification for service use and geographical information system (GIS) services and platforms. In addition to these, location-based services (LBSs) and their enablers (such as digital platforms) have been under interest for several years (e.g., Inkinen 2010).

Based on the existing literature and conducted empirical investigations it is possible to identify some trajectories and development paths that are probably the most relevant ones in the upcoming years of smart city development. It is likely that cluster-based development will continue to be a key-asset in urban economies. The interconnection between geographical agglomerations and networked fields of industries will be supported by value creation and ecosystem development of immaterial products, ideas and new interdisciplinary integrations (e.g., combination of art and design into end-product development). There are also several other societal aspects to which smart city development tackles. These include the environment, wellbeing and health services, and education.

Firstly, environmental aspects of smart cities are important. Particularly, the use and treatment of renewable energy in the city infrastructure and the potentials provided by the ICT in pursuit of lowering the amount of waste are essential.

Secondly, when wellbeing services are being considered, the development of efficient and well-functioning electronic patient databases and platforms are important. The development of such services is highly demanding task as the privacy legislation and data management issues differ considerably between different organisations and operators of the health industry. The health-related ICT development has been one of the most important areas in terms of the amount of public sector expenditure spent on ICTs in cities in the Nordic countries where health services are produced and provided by the public sector to a very high degree.

Thirdly, issues of smart learning platforms, cities potential to acquire them to their local schools and teachers' know-how to use them present a significant development field. One of the goals designated to ICTs is that technologies should aid and help the lifelong learning. However, the implementation and applying the new technologies have also caused quite opposite end-results, particularly for those doing office work.

3.3 Innovation Systems as Platforms for Knowledge-Based Urban Development

Knowledge-based urban development is a relatively new term and is strongly associated with the wider concept of knowledge-based development (KBD). Yigitcanlar and Lönnqvist (2013), Yigitcanlar et al. (2015, 2017) have studied this concept extensively with empirical evidences. In the approaches of KBUD studies it is common to make distinctions to sub-categories applicable for empirical designs and classifications. These are often defined to concern meta-aspects of society including dimensions of economy, the environment, social condition or wellbeing, and urban governance and administration. All these categories are treatable with statistics available from cities and therefore making the KBUD a suitable analysis framework. There are several interesting points of variations that need to be considered if innovation system concepts and KBUD are wanted to be integrated.

Firstly, the variations of specific cities in their relative locations compared to other cities (either in the same country or in other countries). This is important as KBUD itself may be considered as an assessment framework where as RIS is a concept describing support tools and structures for innovation in specific locations. Thus, they are similarities but fundamentally they look at different things taking place in urban contexts.

Secondly, cities are evolving through the guidance and decisions made by the planning authorities. A traditional doctrine in planning has been to separate residential areas, business areas and recreational areas to their own segments. These distinctions vary extensively according to each country. However, the distinctiveness of planning traditions is traceable through the history. Healey (2007) produced a sophisticated recount of the fragmented and diverse planning philosophies applied in contemporary societies. One integrative view is to consider physical properties and 'knowledge-intensive business locations' such as science parks and other high technology clusters in urban space. They provide an interesting venue for research as they often combine high-end architecture (part of image building), presence of highly educated and knowledge-intensive workers, and technology driven product or service development. Thus, they portray location properties that are crucial both for KBUD and RIS. These locations are also interesting from the smart city point-of-view because with the help of social-media and other media these locations also advertise and tell about themselves to the world.

Consequently, the essence of smart city development may be seen through KBUD and RIS conceptualisations. They are tools that enable classifications and segmentation of the urban development into more precise and understandable (smaller) parts. The content defining of these parts is then the task of the researcher. The presence of formal high education, which is easily classified according to education levels, is perhaps the most distinctive indicator applied in smart and innovative literature. In terms of causality, education precedes other innovation and smart city indicators such as economic measures from relevant industries, patents, and R&D expenditures. Thus, the spatial characteristics of KBUD development and their combinations bring forth new empirical and theoretical openings to consider cities as producers and providers of education, urban renewal, social change and urban image creation.

Florida (2002) studied the role of creativity and diversity of urban locations in relation to their economic condition and success. His approach has been met a good deal of criticism and other empirical studies have indicated that creative professions do not straightforwardly lead to better economic outcomes. Hence, the relationship between creative occupations (particularly concerning arts and crafts) and higher spatial economy performance is a complex one. In empirical sense, economic performance of innovative regions is the best explained with education and research related variables. Furthermore, the clustering of successful companies (particularly in the field of ICTs) is empirically verified in numerous studies conducted on varying spatial scales.

A current challenge in the innovative and knowledge-based development is that how to transform education and R&D measures to economically viable activities. This concerns the transformation process from learned information towards knowledge (signified and processed information) and then finally to economic results. The distinction between formal education and operational skills (or know-how) is important as education as such functions as the background for innovative growth but it requires implementation skills and practice that are obtainable only trough actual work and experience. The commitment and self-motivation is as important as formal education. Individual differences are high in this respect: some people are better suited to apply and make benefit of their university degrees than the others. Thus, the process requires a personal commitment and capability as well as an advanced educational system.

Kellerman (2002) examined information and innovation processes and he presents two main drivers (or motives) combining the interest of two main sectors of innovation systems, namely enterprises and universities. The first may be defined as the intellectual motive that is the driver for the universities. University presence has been deemed as one of the main indicators of economically better of regions and cities (compared to similar cases that do not have a university presence). However, the short-term measurement traditions of academic work have caused a challenge for the intellectual motive as research project funders and other interest groups demand faster results. There are significant amount of research indicating that this short-term and market-oriented phase of basic research is causing problems for a long term demanding research. This is directly linked to the second main motive that may be defined as the market motive that combines university level formal education to labour production and business driven product development. Thus, the distinction between the two main drivers have blurred during the last decades alongside with the extensive expansion of innovative production. The main impact for regions comes from the education and the labour force they produce. This is strongly supported by the university start-up activities have gained momentum during the last decade.

3.4 Innovative Urban Concepts and Image Creation

There are numerous terms and characteristics of how cities have been described since 1980 until today in terms of knowledge production and technology integration. These terms commonly include separations to 'digital', 'intelligent', 'innovative', 'smart', 'knowledge-based', and 'creative' (e.g., Simmie 2001; Komninos 2002; Polenske 2007; Yigitcanlar 2016) . The content lines of the definitions are overreaching and mixed. For example, Yigitcanlar (2016) applied a classification of 'smart urban information technologies' and 'smart urban systems' of which the former includes infrastructure related technologies (location and sensing technologies, ubiquitous computing, and augmented reality technologies), and the latter includes larger conceptual entities of transport, grid, water and waste processing, emergency and safety, and participatory decision-making systems. These technological categories lie in the heart of 'smart' city concept. As stated, the smart city focus mainly concerns particularly technological progress taking place in cities through physical structure development and digitalisation of administration. Innovative city concept is the most often associated with businesses and economic development. These conceptual differences are related to the following characteristics:

- Economy (companies, producers, commerce, investors, investments, market segmentation)—'innovative city'
- Society (governance and government services, front-office, back-office, public sector data management)—'digital city'
- Knowledge (education, know-how, skilled workers, creativity, awareness, responsibility, functionality)—'knowledge-based city'
- Technology (infrastructure, software, hardware, services, products, platforms, integration)—'smart city'.

The fourfold classification is a crude and simple way of depicting the most common characteristics concerning 'high-end' urban developments. As a reference, Komninos (2002, p. 201) considers a concept of an 'intelligent' city and its three basic components: "(a) The island of innovation; (b) The virtual innovation system; (c) The connection between real and virtual innovation systems". The first one refers to clusters and their manifestations such as technology parks, industrial districts and innovation centres that exist in physical (real) world and are formed through communities of researchers and scientist creating new knowledge. The virtual innovation system, on the other hand, refers to digitalised operating environments where knowledge management and innovation processing (e.g., design and architecture development) takes place. It is a technological domain within technological development itself. The final component combines the first two together. This concerns the essence of transferable information (or knowledge) as digitalisation process requires codified form of information (data stream) even though this codified knowledge may represent also tacit forms of knowledge (e.g., streams of behaviour patterns) (e.g., Capello 1999; Chesbrough 2003; Chesbrough et al. 2014).

Komninos (2002) developed the intelligent city concept by identifying five structuring functions identifiable in the interplay of the three main elements (real-virtual-combination). These functions are R&D, technology transfer, finance of innovation, product development, and networking. All these functions are present in daily real-world interactions in innovation creation as well as digitalised forms of interactions. These five functions have been widely discussed in the previous chapters as the fundamentals underlying the conceptual foundation of innovation and cities.

The 'islands of innovation' i.e., location bound realisations of economic activity are a local scale way of looking at innovation dynamics. This is because innovative firms may be (and quite often are) connected to global networks. These networks may concern ownership relations (e.g., subsidiaries), main markets, or essential partnerships in production and supply-chains. Localised clusters of course may have a great impact on global markets, as is the case of the company profile of Silicon Valley, where significant number of global market leaders, particularly in ICT industries, is located.

One of the contemporary and persistent claims in urban development has been the doctrine of global competition (e.g., Clarke and Tracey 2004). The idea of competition has caused several cities to promote and market themselves as potential locations for international investors and corporations. These marketing efforts are most often combined with subventions and taxation benefits for foreign companies to locate themselves into the target cities. Information society, innovation and progress in the high-technology sectors have been fondly applied in locations development strategies and documents across Europe. An influential aspect in this respect has been the combination of urban characteristics and KBUD. These dimensions include, for example, the following aspects of brand creation:

- The verifiable properties of cities that are often measurable and identifiable from statistics or other data sources (physical properties):
  - Availability of services (commercial, cultural, others)
  - Availability of housing (costs, locations)
  - Socioeconomic variables (profile)
  - Number and characteristics of business (economic activity and success)
  - Number of administrative units and services (public sector presence)
  - R&D&I indicators
  - Education indicators (number of universities, graduates, publications)

- The networks that organisations locating in these urban locations have and how they are used and applied (network properties):
  - Business networks
  - Public-Private-Partnerships
  - Innovation intermediaries
  - Reputation and accessibility to employment market after graduation
  - Start-up traditions (e.g., garage companies, peer-support, investors)
  - Potential to move to other locations
- The images and ideas (including stereotypes) observed and created concerning cities. Image creation and urban marketing are easier if the national context and earlier imaginary supports the desired image projection (marketing properties):
  - Environmental cleanliness
  - Life style and variety of choices (services, happenings, attractions)
  - Quality of life (wellbeing, availability and accessibility to health services)
  - Security (low crime rates)
  - Climate and environmental benefits.

Traditional location marketing has been mainly targeted to tourists and short-term visitors. There has been however a constant and clear change, mainly caused by more efficient financial markets, towards investor marketing and location promotion for business purposes. Different countries and locations have different types of location-based attractiveness. These may include specific historical traditions that have created a competition advantage to certain regions (such as ICT industries in Silicon Valley), specific expertise or availability of highly educated professionals (e.g., Boston) or climate related benefits. Traditional location factors such as relative prices of real-estate, labour costs (in relation to education and skills) and logistical efficiency and market proximities still have also a significant role on how to construct the most attractive urban image.

Urban marketing connects also to the current trend of recruiting international students to universities. University rankings have been existing already more than a decade and they are examples of the information needs of global education markets. Some ranking providers have also started to rank student friendly cities. This joint ranking (cities themselves and universities they host) is an excellent example of the elements in current global urban competition. The most attractive global universities (and via them cities) are mainly located in the US and the UK. Due to the dominance of English language in science publishing this is an expected result and the annual variations among the world leading universities are quite small.

3.5 Summary

This chapter focused on elaborating the place dynamics for innovation. Urban innovations are studied according to innovation producing organisations. These include

3.5 Summary

technology firms as well as universities and research institutions. Theoretical considerations concerning innovations, locations and economic growth are tightly connected to the location's properties, relative and relational distances between other locations and their connectivity with similar and non-similar locations nationally and internationally. Cities are the key-nodes where clusters physically manifest themselves. They also function as locations where new innovative urban technologies are being created, implemented, and applied (e.g., Simmie 2002). The following main points may be drawn based on the studied literature:

- Urban and regional development theories highlight and require understanding of agglomeration, clustering and diffusions (flows). Urban networks and interactions are in key-roles in studying relative and relational spatiality of economic activity.
- Technological progression has been the key in opening-up new research themes and developments observable in economic landscapes of cities.
- Conceptualisations related to innovative and technological growth are numerous and they are often applied as synonyms. However, different concepts have their fundamental root in different elements of urban development. Some focus on education and skilled people (knowledge-based), some on technologies (smart), and some on new products, processes, and practices (innovations).

References

Ahlqvist, T., & Inkinen, T. (2007). Technology foresight in multiscalar innovation systems. A spatiotemporal process perspective. *Fennia, 185*(1), 3–14.

Altshuler, A., & Behn, R. (Eds.). (1997). *Innovation in American government: Challenges, opportunities and dilemmas*. Washington, DC: The Brookings Institution.

Anttiroiko, A. V., & Kasvio, A. (Eds.). (2006). *e-City: Analyzing the efforts to generate local dynamism in the city of Tampere*. Tampere: Tampere University Press.

Asheim, B., Coenen, L., Moodysson, J., & Vang, J. (2007). Constructing knowledge-based regional advantage: Implications for regional innovation policy. *International Journal of Entrepreneurship and Innovation Management, 7*(2–5), 140–155.

Asheim, B., & Gertler, M. (2005). The geography of innovation: Regional innovation systems. In J. Fagerberg, D. Mowery, & R. Nelson (Eds.), *The Oxford handbook of innovation* (pp. 291–317). Oxford: Oxford University Press.

Asheim, B. T., & Isaksen, A. (2002). Regional innovation systems: The integration of local "sticky" and global "ubiquitous" knowledge. *Journal of Technology Transfer, 27*(1), 77–89.

Asheim, B. T., Moodysson, J., & Tödtling, F. (2011a). Constructing regional advantage: Towards state-of-the-art regional innovation system policies in Europe? *European Planning Studies, 19*(7), 1133–1139.

Asheim, B., Boschma, R., & Cooke, P. (2011b). Constructing regional advantage: Platform policies based on related variety and differentiated knowledge bases. *Regional Studies, 45*(7), 893–904.

Boschma, R., & Fritsch, M. (2009). Creative class and regional growth: Empirical evidence from seven European countries. *Economic Geography, 85*, 391–423.

Capello, R. (1999). Spatial transfer of knowledge in high-technology milieu: Learning versus collective learning processes. *Regional Studies, 33*(4), 353–365.

Chesbrough, H. (2003). *Open innovation: The new imperative for creating and profiting from technology*. Boston, MA: Harvard Business Review Press.

Chesbrough, H., Vanhaverbeke, W., & West, J. (2014). *New frontiers in open innovation*. Oxford: Oxford University Press.

Clarke, G., & Tracey, P. (2004). *Global competitiveness and innovation. An agent-centred perspective*. London: Palgrave Macmillan.

de Bruijn, P., & Lagendijk, A. (2005). Regional innovation systems in the lisbon strategy. *European Planning Studies, 13,* 1153–1172.

Etzkowitz, H., & Klofsten, M. (2005). The innovative region: Toward a theory of knowledge-based regional development. *R&D Management, 35*(3), 243–255.

Etzkowitz, H., & Leydesdorff, L. (Eds.). (2001). *Universities and the global knowledge economy. A triple helix of university–industry–government relations*. London: Continuum.

Fagerberg, J., & Verspagen, B. (2009). Innovation studies: The emerging structure of a new scientific field. *Research Policy, 38*(2), 218–233.

Fisher, M. (2001). Innovation, knowledge creation and systems of innovation. *The Annals of Regional Science, 35*(2), 199–216.

Florida, R. (2002). *The rise of the creative class: And how it's transforming work, leisure, community and everyday life*. New York: Basic Books.

Freeman, C. (1987). *Technology and economic performance: Lessons from Japan*. London: Pinter.

Healey, P. (2007). *Urban complexity and spatial strategies: Towards a relational planning for our times*. New York: Routledge.

Inkinen, T. (2010). Urban travel information and wireless technologies in Helsinki, Finland. *Journal of Urban Technology, 17*(2), 57–75.

Inkinen, T. (2015). Reflections on the innovative city: Examining three innovative locations in a knowledge bases framework. *Journal of Open Innovation, 1*(8), 1–23.

Kellerman, A. (2002). *The internet on earth. A geography of information*. London: Wiley.

Komninos, N. (2002). *Intelligent cities: Innovation, knowledge systems, and digital spaces*. New York: Taylor & Francis.

Lambooy, J. (2007). The transmission of knowledge, emerging networks, and the role of the universities: An evolutionary approach. *European Planning Studies, 12*(5), 643–657.

Lemola, T. (2003). Innovation policy in Finland. In: P. Biegelbauer & S. Borra´s (Eds.) *Innovation policies in Europe and the US* (pp. 77–92). Aldershot: Ashgate.

Lemola, T. (2004). Finnish science and technology policy. In: G. Schienstock (Ed.) *Embracing the knowledge economy. The dynamic transformation of the finnish innovation system* (pp. 268–284). Cheltenham: Edward Elgar.

Lundvall, B. Å. (Ed.). (1992). *National systems of innovation: Towards a theory of innovation and interactive learning*. London: Pinter.

Lundvall, B. Å., & Maskell, P. (2000). Nation states and economic development: From national systems of production to national systems of knowledge creation and learning. In G. L. Clark, M. P. Feldmann, & M. S. Gertler (Eds.), *The Oxford handbook of economic geography* (pp. 353–372). Oxford: Oxford University Press.

Makkonen, T., & Inkinen, T. (2014). Innovation quality in knowledge cities: Empirical evidence of innovation award competitions in Finland. *Expert Systems with Applications, 41*(12), 5597–5604.

Makkonen, T., Merisalo, M., & Inkinen, T. (2018). Containers, facilitators, innovators? The role of cities and city employees in innovative activities. In: *European Urban and Regional Studies*, iFirst.

Nelson, R. R. (Ed.). (1993). *National innovation systems: A comparative analysis*. Oxford: Oxford University Press.

OECD. (2013). *Innovation-driven growth in regions: the role of smart specialisation*. Paris: OECD Publications.

Oughton, C., Landbaso, M., & Morgan, K. (2002). The regional innovation paradox: Innovation policy and industrial policy. *Journal of Technology Transfer, 27*(1), 97–110.

Polenske, K. (Ed.). (2007). *The economic geography of innovation*. Cambridge: Cambridge University Press.

References

Scott, A. J. (2000). *The cultural economy of cities. Essays on the geography of image-producing industries*. London: Sage.

Scott, A. J. (Ed.). (2001). *Global city-regions: Trends, theory, policy*. Oxford: Oxford University Press.

Simmie, J. (Ed.). (2001). *Innovative cities*. New York: Spon Press.

Simmie, J. (2002). Knowledge spillovers and the reasons from the concentration of innovative SMEs. *Urban Studies, 39*, 885–902.

Suorsa, K. (2007). Regionality, innovation policy and peripheral regions in Finland, Sweden and Norway. *Fennia, 185*(1), 15–29.

Yigitcanlar, T., & Lönnqvist, A. (2013). Benchmarking knowledge-based urban development performance: Results from the international comparison of Helsinki. *Cities, 31*(1), 357–369.

Yigitcanlar, T. (2016). *Technology and the city: Systems, applications and implications*. New York: Routledge.

Yigitcanlar, T., Edvardsson, I. R., Johannesson, H., Kamruzzaman, M., Ioppolo, G., & Pancholi, S. (2017). Knowledge-based development dynamics in less favoured regions: Insights from Australian and Icelandic university towns. *European Planning Studies, 25*(12), 2272–2292.

Yigitcanlar, T., Inkinen, T., & Makkonen, T. (2015). Does size matter? Knowledge-based development of second-order city-regions in Finland. *disP-The Planning Review, 51*(3), 62–77.

Chapter 4
Characteristics of Innovation Geography

Abstract Global geography and the organisation of human activities on earth are fundamentally managed by governments of different nations. Each sovereign country has their own legislations and interest governing economic activity within their boundaries. Governments have numerous roles in the economic activity. Conceptually 'government' refers here to a set of public sector organisations that are responsible for governing the specific areas and spatial scales. As an example, municipalities are interested on the local scale and activities that impact their specific vicinities where as regional or national governments or bodies are focusing on larger spatial scales. The nation state may be considered as the fundamental level of regulation as the states have the right to conduct taxation of its inhabitants. This is the reason why international economy is governed with trade and commercial treaties and is based on international agreements. This chapter elaborates the key characteristics of innovation geography.

Keywords Innovation geography · Nations · Planning · Governance
Government
E-government · Smart city · Knowledge city

4.1 Production, Location and Economy

Global economy functions trough circulation of capital and resources. Immaterial flows of finance and codified information, and material flows of raw materials, machinery, products, and labour force (e.g., Dicken and Lloyd 1990; Dicken 2007). Majority of the innovations addressed in popular discussions concerning technological drive of development are focusing on the immaterial aspects of economy. The location-based analyses on innovation and geography includes recognition of space-time situation (context), interrelations between different types of networks (physical, social and economic), and locations relation to other locations (similar or dissimilar) (Asheim and Gertler 2005). These dimensions may be regarded as a means to understand the complex wholeness in which companies operate. The following elements in firm-level activity require different types of approaches to understand their spatial linkages:

- Marketing (and business intelligence):
  - Identification of customer/target groups (i.e., B2B, B2C, B2G)
  - The use of analytical marketing tools and collection of data
  - Scale of product viability (local, national or global)
  - Type of product (end- or intermediate-product)
- Production (and product development):
  - In-house (location bound) or subcontracted (diversified)
  - Vertical and horizontal relations (combination and integration)
  - Logistics and market proximity (cost, reliability and efficiency)
  - Quality control and assessment (particularly in subcontracting)
  - Centralisation versus network mindsets.

The abovementioned classification is an example of understanding classical division of marketing (sales) and product development (innovation). There has been a long tendency towards more networked-based production solutions particularly in post-industrial societies that are commonly highlighting high quality and both economic and environmental effectiveness in their national production strategies and economic policies. An interesting distinction is to look at the export driven companies and companies that are focusing on national markets. For example, in Finland the small size of local markets has created a situation in which majority of industrial policies are directed to support export companies or companies aiming to international markets. Therefore, the product viability is an important starting point in the consideration of economic geography and market scale.

Target markets and segmenting are in the core of analytical customer relations management. In this business intelligence the new potentials provided by the Big Data and data mining have gained extensive amount of interest applied particularly by social media companies and smaller subcontractors focusing on customer databases. The customer segmenting and provision of individually identified marketing products will become continuously focused in the near future challenging privacy legislation and national laws.

In terms of spatiality of production (either material or immaterial) the key-question are intertwined with the horizontal and vertical production chain relations that themselves are fundamentally connected to the processes of convergence and divergence. In other words, where different parts and segments are produces and how they are linked together. For a long time, cost-minimisation has been one guiding principle in global economy. This is often manifested as transfer of production to low-cost countries such as China and India. This is a classic example of diverged production model where corporate main offices, main markets and physical production have been dispersed. However, there are counter developments taking place. The retrieval of physical production closer to main markets or home-offices results from quality issues and more environmentally aware consumerism. Short distances from manufacturing to assemblies and finally to markets mean shorter logistics and lower environmental stress. The doctrine of close-proximity production is also gaining more attention and significance in corporate responsibility strategies (Oerlemans and Meeus 2005). The most common responsibility areas include workforce rights and labour conditions together with environmental sustainability and transparency of production waste management.

The solutions for complex production networks require an understanding of several political and legislative aspects of economy that are not only financial. Environment and social rights are the clearest example. These dimensions are then needed to be streamlined with the innovative production and the market demand. In practice this is realised through efforts in: (a) production processes that is aided by new technologies and intelligent systems; (b) product development and creation of new products and ideas together with more efficient production cycles and quality; (c) organisational adoption to new market conditions and changes caused by disruptive radical innovations including more efficient logistical systems and supply-chain management (Gemünden et al. 2007; Nemet 2009; Pullen et al. 2009, 2012).

4.2 Smart and Knowledge City Characteristics and Properties

There are numerous ways to identify elements and parts that are significant denominators for smart and knowledge-intensive cities in relation to their innovative development. The following segmentation is applied in order to demonstrate selected view-points that have been widely applied in recent literature and development discussions. The first aspect is technology. This includes smart data solutions covering sensor networks and databases; data mining based on these sensor networks and other Big Data sources; LBSs and geographic urban modelling and engineering; other urban data analytics including statistics and GIS data; and urban data management (e.g., relational databases and SQL and alternatively NoSQL).

These listed factors focus on data science and management of urban technologies. Sensor networks prove an important development area in urban studies combining maintenance and improvement of physical infrastructures into digital interactive intelligent systems. Therefore, systems of spatial intelligence and environment have an important role in the future. Perhaps one of the clearest examples of this field are intelligent waste collection units and litterbins. They are able to inform collectors if they are either full before common scheduled emptying time or if they are empty enough for to be skipped in scheduled route. This is also an example of real-time surveillance technology focusing on waste management. There are already a number of cities implementing these types of urban technology solutions including cities of Copenhagen, Stockholm and Helsinki that are discussed as examples.

Systems for city intelligent management also include building infrastructures, energy distribution solutions, and transport. From the technological view-point the development of urban transport may be divided into two main problem fields: (a) the development of vehicles; (b) public transports physical attributes (rails, smart traffic lights, integration of cables and information sensors into the roads and rails). Smart cities have also invested and developed different types of telecommunication systems in their spaces. Perhaps the most widely studied topics of urban telecommunication integration is the expansion and availability of wireless local area networks (WLANs/WI-FI) for inhabitants across the city space. The distribution and service provision solutions of WI-FIs have been under an investigation by Inkinen (2010), who proposes a triad framework in order to understand logics behind outdoor service provision including elements of access, affordability and transparency. These three dimensions illustrate essential societal values is open and transparent digitalisation of e-governance.

Mobility and transport services are significantly affected by digitalisation. There are numerous service solutions based on mobile applications that provide help and information about public transport. These include location-based services (real-time traffic monitoring), online ticketing, and multimodal transport solutions (e.g., combined ferry, local bus, train, and taxi fees). The future goals of digitalisation in (public) transport have been set high. For example, Nordic countries have their own transport digitalisation policies implemented. They commonly slogan the future movement in cities as services enabled by real-time data enabling flexible changes and orderings of different transport modes. Cities are commonly the responsible public organs that arrange public transport and therefore urban technology dimension is pronounced in this topic.

The development work focusing on transport is extensive and global. For example, traffic timetable services are, in several cases, offered by multinational corporations such as Google maps, and by local traffic authorities such as Helsinki Region Transport Authority (HSL) that has done extensive development effort in the journey planner of the Helsinki capital area. Transport monitoring, congestion information dissemination, and accident alert systems are therefore in the very heart of urban technology developments. These services also have become embedded parts of daily lives and urban mobility patterns. Mobility also includes some of the most advanced technological development fields in automated and autonomous vehicle development and sensor technology.

The second aspect involves people and inhabitant related demographic and skill related properties of knowledge-based cities (or knowledge cities as referred to more commonly). These include employment levels in high-technology sectors, the absolute and relative numbers of R&D personnel and researchers, and employment in other knowledge-intensive sectors. For example, employment levels in knowledge-intensive business services (KIBS) has been one widely used categorisation in knowledge management studies. One of the main challenges in current research has been the limited amount of data availability particularly on regional and urban scales. In several cases these statistics are available only on national level. This problem is still relevant even though the data availability has improved during to some extent during the last decade.

Another approach to include population characteristics into a regional analysis is the use of education statistics. They also have a better regional availability and they provide a good starting point for classifying and assessing spatial properties of knowledge-intensiveness underlying innovations and innovation indicators. Particularly the significance of highly educated people (tertiary education) has been deemed as a significant contributor (and causally proceeding) of economic and innovation indicators. The significance of education has mainly been debated concerning creative industries (such as musical talent, crafts and professions of praxis, and in some cases even coding) but importance of formal education particularly concerning analytical and synthetic knowledge base industries is undeniable.

Finally, skilled people aspect can be connected to democracy and execution of electronic governance and government. Digitalisation has been one of the key-development areas in public administration and systems for public participation defining an elemental part of knowledge-based city (transparent and efficient governance). The role of feedback and grass-root notions concerning public spaces, parks and local transport have had a great impact on the development of the human-centred urban planning mind-set. In several cases, urban technologies and human aspect are realised in the practices and outcomes of planning. Digital feedback and interaction channels have enabled stronger voices for local inhabitants and visitors concerning urban renewal and restoration of developing urban areas. These processes also include gentrification and segregation effects observable all around growing cities.

Thus, education, employment and digital (online, mobile and IoT/IoE) interaction statistics provide a widely applied starting point for urban innovation analysis. The main questions concern fundamentals such as where in urban space innovations are being produced; where innovative people work and live; how these locations interact with each other in terms of measurable flows (e.g., goods, finance, people); and how digitalisation or other forms of innovations change or have an impact on the location quality (such as environment, quality of real-estate, cleanliness, low-crime rates).

4.3 Policies, Governance and Locations

The importance of urban policies in innovative development has been recognised a long time ago and several initiatives have been established around the world and in Nordic countries. The current urban development actions commonly involve the recognition of urban cores and their regions. The two standard scenarios for national development include either convergence or divergence in the regional economic performance. The former commonly emphasises the importance of regionally balancing policies that are applied in Nordic countries, particularly in Finland and Sweden. This is due to their large land areas and relatively small populations requiring government intervention to support remote and peripheral areas. However, there are successful examples, particularly from the northern parts of Finland (Oulu region) that have been highly successful in the past (early 2000s) in creating innovation hub for high-technology (e.g., Inkinen and Suorsa 2010). The latter option (divergence) is associated with the presence of economically highly successful urban cores that are influencing their respective regions and simultaneous presence of slow progressing regions and cities that experience considerably lower levels of economic activity and performance.

In terms of urban and regional policies this question concerns also internal urban structures and locations that are experiencing divides within urban space (Laranja 2004). These divides are manifested in economic and socio-demographic variables such as income per household, education and employment. There are also other significant differentiating indicators such as crime rates, number of businesses and unemployment rate. A widely applied concept in urban studies is segregation referring to processes causing neighbourhood differentiation to 'highly wanted' and 'highly avoided' areas. The larger the urban core becomes it is more likely that these divergent tendencies become stronger. The segregation processes are observed and they tend to become more pronounced in the Nordic countries. For example, there are already significant problems with segregation in the largest Swedish cities such as Stockholm and Gothenburg, and similar developments are observable also in Finnish cities such as Helsinki and Turku.

Urban policies are therefore connected to economic development but they are interlinked with social condition of urban development. There are solutions for urban policies that have been applied with good results. For example, coordinative and collaborative cross-administrative work overcoming organisational borders (e.g., offices of economic development and welfare division) have produced encouraging results in Helsinki where the eastern parts of the city have been lacking behind in economic development. One way of balancing this is to locate educational units and campuses to these areas that support socioeconomic mix. These decisions also activate the local businesses and buzz as the number of daily users of the urban spaces increases. These tasks are, however, tightly connected to the political context and mind-set of each country. In this example from Finland, municipalities have significant influence in providing funding and making decisions for polytechnics in their vicinities, and therefore are able to have decision making power over the location of these new establishments.

On the broadest level national governments are involved in economic activities through the maintenance of property rights and broad governmental frameworks. These activities include legislation issues and also maintenance of law and order. The trade and commercial legislations also direct the whole functioning of economic system as they regulate questions of property rights for individuals and organisations together with definitions of ownership and property. These form the fundamental platform for trade-based societies. Agreements of employment and labour rights are also fundamentally connected to nation state legislation.

If the relationship with the nation state and economic activity is considered more narrowly, the approach may be limited to issues of economic policy. Traditionally, economies are governed with macroeconomic tools including monetary and fiscal policies. Monetary policy concerns tasks and decisions affecting the total amount of circulating money and assets in that specific state. They also include decisions affecting interest rates and other measures dealing with money and currency. Fiscal policy on the other hand concerns income transfers and public sector subsides done between different policy fields and socioeconomic groups. Different countries differ considerably from each other on their traditions and histories in conducting fiscal policies.

Innovation activities have been strongly supported in several national growth strategies of nation states staring from the early 1990s. A concept of electronic government (or e-government in short) has been developed and applied to describe and include tasks and exercises through which governments aim to make their processes more efficient and cost-effective. Evidently, it is a broad concept that intertwines information technologies and governance (and government). E-government studies commonly separate three elements according to which the research emphasis is focused on: citizens (as users of services); administration itself (back-office functionality resulting into cost-savings); and technologies themselves (what services and to what extent digitalisation is implemented into governmental functions). Electronic government combines these elements from all fields of technology-society integration including an efficient physical network structure (hardware), user-friendly service design (software developments), and also non-technological dimensions, for example, networking tools and other collaboration models that aim to bring together professionals and developers in order to collaborate and work together (see Bryson 2004).

The identified three focal areas include a number of cross-cutting issues of importance. These include for example data security and privacy (also including internal and external threats and weaknesses towards system reliability), usability and inclusion (mainly citizen focused services and ability of all citizens to use them), and reliability and interoperability (system functionality with other systems and the minimisation of errors). E-government service studies also identify aspects of organisational provision responsibility (what governmental organisation provides the service and the question of administrative borders within governmental structure) together with the development path and decision related to e-government (is it dependent on each sector of government or is it jointly managed). These dimensions are essential in order to analyse electronic services and thus answering the following questions: Who does provide the service? How secure and reliable is the service? Are the services provided in accordance with the concept and idea of inclusive good governance?

E-government may be approached also with broad or narrow definitions. The broad definition refers to a mindset of development in government. The adoption and efficient use of digitalisation and immaterial services provide undeniable cost savings. However, their implementation is not as straightforward as is the case in commerce and business. This is due to fact that governmental processes commonly involve transactions of private and user sensitive information (such as social security numbers, taxation information or health related issues). The data security and reliability issues are paramount in the development of e-government.

The narrow approach refers here to a limited number of processes or outcomes that e-government implementation brings forth. There might be only one process that is under the study or a set of interconnected processes. The main point is to have identified and understandably defined framework for the study. An example of this type of an approach is an end-user citizen survey concerning the adoption of online website(s) providing e-government services. This approach would provide information concerning the user levels, views and satisfaction towards the service. Another easily identifiable example is an analysis focusing on automated traffic control or any other monitoring system of the use of public space. The narrow perspective thus raises issues of single services and applications in relation to their functionality, operability and security.

The fundamental distinction in e-government research concerns the separation to government and governance. Bovaird (2003) applied this distinction to make a separation between the e-government and e-governance. According to him (2003, p. 38): "e-government will be used to denote electronic enablement of all the services provided or commissioned by the public sector" and "e-governance will be used to denote electronic enablement of all the other activities of government (e.g., management of democratic activity, ensuring fairness and transparency of decision-making in public bodies)." This distinction is a bit problematic as commonly government is referred to include organisational structure of public administration whereas governance is used to describe the actual conduction and implementation of government decisions. According to Bovaird (2003), definition of a 'service' is separated from 'other activities', and this separation is a difficult one to do in empirical classifications.

Another way of seeing e-government and e-governance is to define the first to include electronic services and tools that aim to improve internal data, information and knowledge management within government organisations and the latter to include those services and tools that are targeted to external customers (citizens or businesses) in order to help their transactions and communication with the government(s) and public-sector authorities. City administrations belong into the category of 'government' in this text as they conduct decision-making having an effect on the physical environment (urban planning) and they also in several national cases have a right for taxation and therefore also responsibility for public service provision (such as police or fire-departments). Moreover, quite traditional view is that government represents top-down management ideology whereas governance is related to grass-root level and bottom-up management. There are numerous excellent e-government implementations done around the world those have significantly changed the customer interface for citizen-to-government interaction. Figure 4.1 presents a schematic that indicates the relevant dimensions that may be considered when e-government services are implemented.

The following segments are considered in the implementation of a successful e-government service. The customer interface requires a consideration of citizens' online skills and citizens overall capabilities to adopt new user interfaces. Even though user interfaces and usability has improved substantially, there are still large portions of population who are not able or willing to use or willing to prioritise e-services (Inkinen et al. 2018). E-government tools and services also require a specific attention as the contents of the interaction commonly involve personal data

4.3 Policies, Governance and Locations

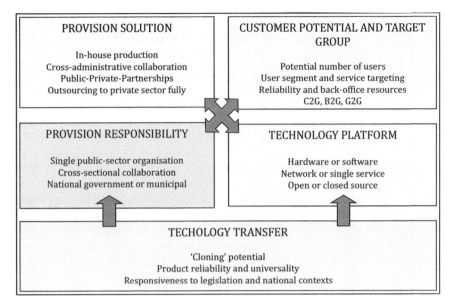

Fig. 4.1 Electronic government in practical context and relevant elements for service consideration

and information resulting to the needs of higher security standards and software interoperability compared to commercial or business services.

The four main components of Fig. 4.1 illustrate that the e-service provision needs to recognise the internal structure and the way how the service is provided (provision solution). This is directly linked to the customer target group and to the simple distinction of citizen services and to business services. Provision responsibility determines the drivers for the development work. A combination of these dimensions then results to different technology needs that are then either produced in-house or acquired from the market if an existing solution already exists. Figure 4.1 also indicates the importance of technology solutions in how electronic government services are developed. E-government services have, therefore, a significance role in the information society development. The e-government services may also be considered as manifestations of technological expansion in digitalised societies.

Arguably, the politics of building a service provision structure have relevance for electronic government and public-sector online service provision. National legislation matters as electronic services are always produced in a spatial context. They are designed to meet customer needs together with certain quality requirements. Therefore, they have a predetermined purpose within the administrative process, which fundamentally aims to increase efficiency and reduce costs. The automation of data management is targeted at enhancing data security (both in terms of individuals as well as organizations), reliability, and efficiency (time) in services provision (Wimmer and Traunmuller 2000; Kellerman 2002; Gauld et al. 2009).

It is highly common that development strategies highlight different types of collaboration models. Triple-helix is widely used as it is a generic framework of key-stakeholder groups and their interactions (Jauhiainen and Suorsa 2008). In majority of identifiable regional innovation systems, the role of intermediaries is highlighted as the coordination units. It goes without saying that inter-regional coordination in the smart city development is essential. Best practice duplication and modifications are examples of knowledge-transfers taking place between regions. However, there is still a need for maintaining enough flexibility in the operational decision making so that each adopting region is able to modify the selected approach in accordance with the strength areas of the region. This is one core ideas in the recognition of historical trajectories and path-dependency in spatial development analyses.

A significant aspect in e-service provision is the possibility for technology transfer and service 'cloning'. This concerns particularly numerous municipalities that perform and provide similar services. It has become more common that municipalities collaborate in their service acquisitions but the overall picture is still very fragmented. This problem concerns particularly those countries that have large land areas that are segmented into numerous municipalities. The efficient public-sector management requires collaborations among national subunits. In addition, there are several cases in which public sector IT procurements have been targeted too narrowly to technologies that are dependent on one or two producers of the services. This leads into problems when technologies are needed to be updated and maintained as service providers may have monopoly status.

4.4 Summary

The chapter has indicated key-properties underlying urban development and, thus far, the definitions of applied hierarchies to dissect the large problematic of spatial economic growth, innovation, smart cities and knowledge-based development requires at least a careful classification of the following elements (or main categories of observation):

- Physical location attributes:
    - Physical geography (proximity to water, topography, climate, hazards)
    - Population statistics (demography, age profile, density, total population)
    - Physical properties of land and possibilities for agriculture and other forms of production
    - Distances to other locations
      Similar of dissimilar locations
      Absolute (kilometres/miles), relative (time), relational (factor relations dependent)
    - Planning and design (architecture, planning form, historical tradition)
- Economic profile and economic condition of location:

4.4 Summary

- Value-adding and gross-domestic/national product
- Number of present industries
- Number of present companies
- Economic key-performance indicators (KPIs) of companies
- Distances to main/secondary/tertiary markets
- Production profile (agriculture, industry, services)

• Societal condition of the location:

- Security (crime rates, other measures of violence/crime)
- War and conflict
- Differentiation of wellbeing (Gini-coefficients)
- Unemployment/employment rates
- Expenditure of social care, health and wellbeing services
- Universal or restricted accessibility to services
- Share of public services/and public sector

• Environmental properties of location and quality of space:

- Functionality of recycling
- Amount of airborne waste
- Amount of other waste types
- Amount of diseases.

The above list is not intended to be an all covering listing of location properties that are societal but have also an impact to the concept of smart cities as several of the technological solutions aim to help different elements that are fundamental several characteristics on the list. For example, innovative e-health system may provide additional information to the patient and through that reduce costs and time in healthcare system. Additionally, some smart city solutions are targeted particularly to deal with physical material flows (such as garbage and waste) and smart solutions in the collection and processing the waste may contribute extensively on smart city development.

There are a number of requirements commonly needed before well-functioning smart, knowledge-based or innovative urban services may be initiated. The first is that societal condition needs to be in order. This means that societies having too extensive differences in e.g., wellbeing or income are less likely to succeed in implementing innovative solutions in public space. This is also reflected to the business presence and locational decision-making as it is likely that the most innovative companies will not locate themselves to those areas that have significant social or welfare problems.

References

Asheim, B., & Gertler, M. (2005). The geography of innovation: Regional innovation systems. In J. Fagerberg, D. Mowery, & R. Nelson (Eds.), *The Oxford handbook of innovation* (pp. 291–317). Oxford: Oxford University Press.

Bovaird, T. (2003). E-Government and e-governance: Organisational implications, options and dilemmas. *Public Policy and Administration, 18*(2), 37–56.

Bryson, J. M. (2004). What to do when stakeholders matter: Stakeholder identification and analysis techniques. *Public Management Review, 6*(1), 21–53.

Dicken, P. (2007). *Global shift. Mapping the changing contours of the world economy* (5th ed.). London: Sage.

Dicken, P., & Lloyd, P. F. (1990). *Location in space. Theoretical perspectives in economic geography* (3rd ed.). New York: Harper and Row.

Gauld, R., Graya, A., & McComba, S. (2009). How responsive is e-government? Evidence from Australia and New Zealand. *Government Information Quarterly, 26*(1), 69–74.

Gemünden, H. G., Salomo, S., & Hölze, K. (2007). Role models for radical innovations in times of open innovation. *Creativity and Innovation Management, 16,* 408–420.

Inkinen, T. (2010). Urban travel information and wireless technologies in Helsinki, Finland. *Journal of Urban Technology, 17*(2), 57–75.

Inkinen, T., Merisalo, M., & Makkonen, T. (2018). Variations in the adoption and willingness to use e-services in three differentiated urban areas. *European Planning Studies, 26*(5), 950–968.

Inkinen, T., & Suorsa, K. (2010). Intermediaries in regional innovation systems: high-technology enterprise survey from Northern Finland. *European Planning Studies, 18*(2), 169–187.

Jauhiainen, J. S., & Suorsa, K. (2008). Triple helix in the periphery: The case of multipolis in northern Finland. *Cambridge Journal of Regions, Economy and Society, 1*(2), 285–301.

Kellerman, A. (2002). *The internet on earth. A geography of information.* London: Wiley.

Laranja, M. (2004). Innovation systems as regional policy frameworks: The case of Lisbon and Tagus Valley. *Science and Public Policy, 31*(4), 313–327.

Nemet, G. (2009). Demand-pull, technology-push and government-led incentives for non-incremental technical change. *Research Policy, 38,* 700–709.

Oerlemans, L. A. G., & Meeus, M. T. H. (2005). Do organizational and spatial proximity impact on firm performance? *Regional Studies, 39*(1), 89–104.

Pullen, A., de Weerd-Nederhof, P. C., Groen, A. J., & Fischer, O. (2012). SME network characteristics versus product innovativeness: How to achieve high innovation performance. *Creativity and Innovation Management, 21,* 130–146.

Pullen, A., de Weerd-Nederhof, P. C., Groen, A. J., Song, M., & Fischer, O. (2009). Successful patterns of internal SME characteristics leading to high overall innovation performance. *Creativity and Innovation Management, 18,* 209–223.

Wimmer, M., & Traunmuller, R. (2000). *Trends in electronic government: Managing distributed knowledge.* New York: Springer.

Chapter 5
Conclusion to Part I

Abstract In order to establish a disruptive geography through innovation one should have a clear understanding on the conceptual and theoretical aspects of innovation and geography symbiosis. This part of the book focused on expanding our conceptual and theoretical understanding and relevant aspects associated with innovative growth, cities and economic geography. The old trajectories traceable back to Marshall and Schumpeter are still evident in the current classifications and view-points on how to look at the economic landscape in cities. The latest trajectories include knowledge-based urban development paradigm and knowledge and smart city notions. This chapter provides a concluding remark to this part of the book that focuses on theory of innovation geography.

Keywords Theory of innovation geography · Technological innovation Economic geography · Reflections

Urban technologies, innovation and smart development are most often concepts that are directly associated with the economic development or social aspects and the use of space in cities. Planning, urban design and the usability of different spatial types (or categories) within cities have a significant impact on the ways how these locations are perceived, what kind of ideas and location related stereotypes are associated with them. This brings the historical continuum and traditional locations of certain activities into the picture. The chapters thus far have indicated that economic production and place-based creation of innovations always require an understanding of the complex interlinks between local resources, proximity concepts, concepts of nodes, cores, clusters, networks, and vicinities. Traditional information needs also include the availability of resources and their costs in relation to other locations, spaces, and places. One may approach economic development through technological progression. This path leads quickly to the debates of innovation, properties of capability, and knowledge-based applications, thus to the intertwinement and juxtapositions of economic geography, regional and national innovation systems, urban and regional planning and regional development and resilience to the management sciences, industrial management, and civil engineering.

The Part I has presented that technological development involves all societal sectors. Private companies and businesses are the drivers of market economy and in the heart of the innovation systems approaches. However, numerous classifications have been developed in order to take into an account that also public sector and the third sector or non-governmental organisations may have an important role in the innovative growth. Commonly many things are joint efforts including different types of PPPs or triple/quadruple-helix arrangements between cities, companies, universities and other research units, as well as civil society and crowd-sourcing. Similarly, efforts conducted in the field of administrative development of city governments, e-government platforms and public administration management are of importance. Particularly, enhanced communication channels developed to improve the transparency and civil participation in urban planning and design are clearest examples of innovative smart city applications.

The summarising conclusions drawn from the Part I may be presented as follows:

- Theoretical foundations of the innovation and smart city development may be drawn to the agglomeration formation, cluster emergence, and path dependency;
- Concepts of proximity, distance, and size/volume (e.g., population of a city) are significant basic components through which economic geographies (e.g., transportation costs between points A and B) are realised;
- Innovation systems require understanding of both scales, national and regional, when they are applied in empirical studies concerning smart or knowledge-based growth;
- Innovative and smart cities are often aware themselves regarding the economic activities taking place in their vicinities and therefore also taking into an account these issues in their own strategy development and growth planning;
- Technological proficiency is identifiable in the (urban) innovation literature as the traditional elements of production, location, and economy are studied through smart city characteristics and properties. These have also impact on the measurement and indicator creation done in empirical studies and measurement campaigns (e.g., Part II);
- Finally, economic fundamentals such as market demand, supply, cost structures and relative advantages form the foundations for successful economic activity in a city. If these conditions are not favourable it is likely that development efforts will not lead to successful long-term results.

Part II
Practice of Innovation Geography

Chapter 6
Introduction to Part II

Abstract As much as having an in-depth understanding on the conceptual aspects of innovative geographies, it is also important to have empirical evidence on how these geographies are formed or being formed. This part of the book focuses on practical examples from the Nordic context identifying the most well-known and successful urban locations that may be deemed as forerunners in smart development. Nordic countries and Estonia are easily selectable locations as they all have gained interest as highly adopting ICT countries. All the other cities are from the Nordic countries except the Baltic state of Estonia. The chapter provides an introduction to this part of the book that searches for empirical evidence from the practice.

Keywords Practice of innovation geography · Smart development
Best practice cases · Nordic countries · Baltic Sea capitals

The following focuses on the practices of innovation geography and empirical insights from highly advanced countries and cities. The 'practice' is here defined as empirically observable and identifiable characteristics that define the macro-condition of the specific region and city in question. Practice also recognises the proactive understanding of urban and regional development. This means that observable and measurable indicators are products of activities done by participants within the innovation system. In other words, firms develop and fine-tune their products in accordance with market demand—new products or improved ones have better markets. Regional intermediaries aim mainly to support the product development, create networks and help local businesses of their regions (Lichtenthaler and Ernst 2008). Public sector then regulates and supports these activities quite often through the intermediaries, in which they are commonly strongly involved through ownerships and funding. Therefore, innovation practice may be considered as implemented designs of actions aiming to impact the target regions.

In order to clearly understand urban technologies and smart city developments the four main axes are needed to be considered: Firstly, the economy (innovation systems, research and development, economic profile, and regional resilience); secondly, the governance and government (electronic services for citizens and businesses, back-office solutions for governance and data management, applications for

C2G interaction, and spatial planning); thirdly, the environment and clean technology (waste management, development of infrastructure, recycling); fourthly, the wellbeing and social condition in urban space (low crime rates, well-functioning daily services, recreational spaces such as parks and green areas). Practice of innovation geography in cities may be seen as a layered structure combining urban administration and public policy, the economic characteristics within the city and locations of enterprises, the environment, and well-being. Similar approach has been applied in international (Yigitcanlar and Lönnqvist 2013) and intra-national (Salles-Filho et al. 2011; Yigitcanlar et al. 2015) comparison studies. These layers are empirically investigated in the following chapters concerning cases of Denmark (Copenhagen), Estonia (Tallinn), Finland (Helsinki) and Sweden (Stockholm). The examination of the case locations is done in accordance with empirical evidences. All cases are capital cities of their countries and all of them have been recognized among the leaders in e-society development.

6.1 Rationale and Scope

The rational and scope for the Part II derives from the Nordic experiences. All presented countries and their capitals have a long tradition of smart city applications, urban and location-based marketing, and they all have verified strong commitment and reliance on innovative growth and knowledge-based development. The four case studies provide a comprehensive and comparative view of the theoretical ideas and issues raised in Part I. The focus is on innovation policies, innovation measures and regional economic performance. In terms geographical properties some remarks are needed. Firstly, the locations are Nordic capitals, except the capital city of Estonia, Tallinn that is a Baltic state country. Secondly, all capitals are relatively close to each other, i.e., the longest distance is between case cities (Helsinki and Copenhagen) measures approximately up to 900 km.

The scope of the Part II is empirical and materials include qualitative documents, statistics and also personal experiences from all case locations. The selected cases also have interesting empirical similarities and dissimilarities that concern proximity concepts, policy options, and economic development trajectories resulting into diverging and converging path-dependencies. As stated, the cases are relatively close to each other. This results into quite similar logistical transport costs of goods and services (also during the winter when the Baltic Sea is extensively frozen). This holds also true in the case of exports and, thus, international trade costs of the cases. All the cases cities also host international airports and they function as main hubs for their respective countries creating similar positions to national second-tier cities.

There are historical differences among the cases, particularly concerning Tallinn. The Nordic capitals also have practices quite similar solution frames for their social-policies, often referred as universal model of social policy. Due to the short distances also climate is quite similar among the cases. This is a significant issue particularly concerning the attraction of skilled human capital. The recruitment of highly specialised professionals has been a bottleneck in several fields of programming, machine learning, and ICTs in general. All these factors contribute to the attractiveness of locations for international workforce. For this reason, policy options have also been considered and particular attention has been paid to the environmentally attractive location marketing. The main question of Part II asks that how and to what extent case cities take innovations and smart solutions into an account in their strategies; and what actual conditions they do in the light of comparative statistics?

6.2 Methodology

The cases are presented through available statistics and literature. The main statistical resource is the Eurostat (2018) database and the extensive regional statistics domain. There are some region-specific issues concerning the data. Firstly, a majority of regional statistics are available on NUTS2 or NUTS3 levels that are commonly applied in regional studies. The case locations have some similarities and dissimilarities. Secondly, Finland and Sweden are relatively large countries in respect to their land area. Denmark and Estonia are, on the other hand, quite the opposite. This is also reflected to the availability of regional statistics. This is a problem particularly in the case of Estonia that belongs to the NUTS3 category as a nation. Sweden and Finland are instead composed on numerous NUTS3 regions. Therefore, the presented statistics are used to give the national context (average) in which their capital cities locate. Capital region then are applied to indicate that whether or not they perform better in respect to statistics. This is possible in all the other cases expect Estonia.

Standard descriptive statistics are used to illustrate differentiation and the extent of variation of each indicator. Eurostat provides annual time-series till the recent (commonly 2016) figures starting from early 2010s. For the consistency, year 2012 is used as a starting point to indicate the recent macro developments. Descriptive indicators have been calculated from the data such as variation, standard error, average, and median. Specific development activities are presented according to the information provided online (webpages and downloadable documents).

Cases are also examined in accordance to selected smart city/knowledge city or innovation strategies (Suorsa 2007). Smart city strategies are the preferred ones but there are some cases in which there is no explicit smart city strategy available. There are also spatial differences in the strategies. In some cases, the capital region has their own strategy and the city has its own. In these cases, strategies are selected on the basis of relevance and scope. At the end, all cases have relevant strategies publicly available online and their presentation is based on these documents to a large extent. All the case locations are members of the EU and therefore it should be recognised that the EU strategy for smart city (and knowledge city) development gives the broad guidelines to all cases.

References

Eurostat. (2018). Database. Available at https://ec.europa.eu/eurostat/data/database.

Lichtenthaler, U., & Ernst, H. (2008). Innovation intermediaries: Why Internet marketplaces for technology have not yet met the expectations. *Creativity and Innovation Management, 17,* 14–25.

Salles-Filho, S., Bonacelli, M. B., Carneiro, A. M., Castro, P. D., & Santos, F. O. (2011). Evaluation of ST&I Programs: A methodological approach to the Brazilian small business program and some comparisons with the SBIR program. *Research Evaluation, 20,* 159–171.

Suorsa, K. (2007). Regionality, innovation policy and peripheral regions in Finland, Sweden and Norway. *Fennia, 185*(1), 15–29.

Yigitcanlar, T., & Lönnqvist, A. (2013). Benchmarking knowledge-based urban development performance: Results from the international comparison of Helsinki. *Cities, 31*(1), 357–369.

Yigitcanlar, T., Inkinen, T., & Makkonen, T. (2015). Does size matter? Knowledge-based development of second-order city-regions in Finland. *disP-The Planning Review, 51*(3), 62–77.

Chapter 7
Insights from Northern European Countries and Regions

Abstract Cities and regions have a number of different strategies aiming to improve the urban environment and economic activities. The concepts of innovation and smart city development are among the key phrases or catchwords. These strategies apply to similar topics and the goal setting the present is very alike. The cloning idea is visible, even though similarities are probably caused by the European Union level strategies that function as guidelines for national policies and strategies. As stated, environment and sustainability are strongly present in smart city agenda. Therefore, studied strategies expand the smart city definition to include traditional problems of urban growth. This chapter focuses on generating insights from Northern European countries and regions.

Keywords Innovation policy · Statistics · Strategies · European Union · Estonia Denmark · Finland · Sweden

7.1 Helsinki, Finland

Finland has a long national tradition in innovation policies and technology driven growth. The countries first documents that may be counted as 'innovation policies' were published in the 1980s mainly focusing on industrial development and ICT engineering needs (Sotarauta and Kautonen 2007). Since then there have been numerous different targeted policy programmes and development plans in order to support and develop innovation activities and technology planning and production in the country. One of the largest efforts has been the implementation of specific 'information society policy programme' that was implemented during the years 2003 till 2007. Information society programme was one of the four programmes aiming to support growth and development on their specific fields (Government of Finland 2006). The significance of this particular programme was that it was directed by the prime minister whereas the other three programmes were directed by their specific ministers. Innovation and information society were therefore raised to one of the national government priorities.

The City of Helsinki is the Finnish capital and is neighboured by the second and fourth largest cities of Espoo and Vantaa. These three cities form the main core of Helsinki capital area with an approximate population of 1.2 million people. Together they are also a text book example of a single FUA that are defined by the daily commuting of people to-and-from one city to another. Helsinki itself has a population of 616,000. There have been an extensive number of empirical studies conducted in Helsinki and its surrounding areas. In some cases, the concept of Helsinki Metropolitan Area (HMA) has been applied in order to include the functionality of the larger geographical area into the analyses. The capital area has several sub-locations that have been identified as the most important innovative locations within the city space. These include in the city of Helsinki the Centre, Ruohalahti and Pitäjämäki (in detail see Inkinen 2015; Inkinen and Kaakinen 2016; Kiuru and Inkinen 2017), and in the city of Espoo Keilaniemi and Otaniemi areas that host significant amount of ICT companies and the main campus of the Aalto University together with extensive presence of research institutions.

In a broader context, the Helsinki region (Uusimaa) produces approximately one-third of the total Finnish GDP. The regions specific importance for national economy has called for urban policies that have traditionally been considered under the theme of regional policy (Inkinen and Vaattovaara 2007, 2010). The question is between regional imbalances (peripheral areas) and supporting urban cores (nodes). The balancing between national regions and their economic condition has been contrasted in numerous arguments from the point-view of global (or international) competition. The logic is founded on the idea that the most successful cities and regions should be compared to international benchmark (cities of similar sizes and with similar social-conditions) in order to attract foreign investments and business to locate on. This attraction goal and the expanding need to attract international visitors and businesses are connected to urban marketing and branding. All the smart case cities discussed in this book are examples of locations that have strongly emphasised digitalisation, ICT, and other high-technology images of their locations.

There are two main levels of strategic guidance for smart cities in Helsinki. Firstly, there is the development strategy of the city of Helsinki that was implemented in the late 2017. The strategy period continuous till the end of 2021. Secondly, the Helsinki region (Uusimaa) council has produced its own smart city strategy that was updated also in the late 2017. Both strategy papers are inherently intertwined with digitalisation and innovative development but the regional strategy is the one that is explicitly dealing with smart city (knowledge city as its predecessor) concept. Therefore, it is discussed first as follows.

The Helsinki region strategy states that it is a part of the innovation policy of the capital area and recognises the need for regional specialisation, thus each region and city should find and enhance their specific strength areas in their search for economic prosperity. Overall the strategy focuses on four main thematic topics that are widely discussed in earlier chapters. These include: (a) urban clean tech (environmental and waste management with intelligent systems); (b) health and wellness (including e.g., digital patient data, digital prescriptions and other e-health applications); (c) digitalising industry (strong emphasises on robotics and artificial intelligence applications);

(d) citizen's city (e-government applications and citizen-to-government interaction tools). The strategy aims to support these four main themes through a common set of action tasks. They include the identification of spearhead industries. After the identification there is a need for enabling knowledge transfer with technology architectures that are realised through innovation platforms and piloting environments. These actions are needed to be supported by purposeful innovation policies and related funding instruments (about Finnish national support systems see Georghiou et al. 2003).

The regional strategy defines the spearhead industries as sectors with a distinctive association to the Helsinki region. They are also expected to have a rapid growth in the near future. This definition is contradictory as commonly the established industries having a longer existence and tradition behind them are experiencing lower growth rates than newly found industries. This is a typical characteristic of contemporary public-sector strategies as they consider regional growth through the business language. However, regional industrial mix and establishment of internationally successful clusters is a long-term process. However, these spearhead industries are argued to relate to innovation, digitalisation, and security. The strategy also states that the whole Finnish national economy is dependent on the development of Helsinki and its region therefore setting arguments concerning regional policy and urban hierarchy.

The strategy highlights triple-helix as a source for knowledge creation as it sees this to be a joint process taking place in universities, public organisations and private companies. This is further supported by the acknowledgement of the third-sector (non-governmental organisations and civic groups). The emphasis is given to researchers and professionals who should give birth to strong science and research base combining technological proficiency and human know-how skills. An interesting additional point is made in the strategy that emphasises the participatory aspect and civic engagement in knowledge creation. This emphasis may also be seen as a typical tendency of crowd sourcing ideology in urban and spatial development policies.

Innovation platforms are highlighted as key-role players in the development of regional innovation systems. The Helsinki region strategy applies common metaphors from natural sciences and it argues on the behalf of 'innovation ecosystems' that is often defined as the overall context in which organisations from different sectors interact. The role of intermediaries (supportive organisations) is commonly highlighted in regional strategies—as is the case also here. Secondly, innovation platforms are also defined to be important for the sake of international competition supporting the economic foundations and consideration that cities 'compete' with each other the similar way as private companies. Research has clearly indicated that this is not the case but contemporary use of corporate language is so widely adopted into current public-sector management that these references are visible in strategy papers around the world. Third, innovation platforms are considered to offer channels for collaboration and technology validation needed for international market breakthroughs.

The regional strategies named innovation policies and funding tools as the two final steps in the search of smart city competence. This is an interesting choice to be included into the strategy as it defines innovative growth and urban development through public policies. There are numerous examples literature, in which the public-sector guidance of innovation is strongly disputed. Again, this highlights the strong public-sector involvement in the process of making smart cities visible: public sector (such as the regional council responsible for the strategy) emphases roles and activities within innovation frameworks that are closely linked to its own functions. This is clearly observable as the strategy highlights the need of smart city strategies for gaining access to the EU structural funds. Thus, regional and urban strategies in the EU member countries are often inherently connected to the broader spatial scales—the EU strategies in this case. This is an example of strategic cloning and connectivity of spatial scales responding to funding opportunities. The EU connection also indicates that the funding requirements (of the public sector) often require larger expenditures than small national economies, such as Finland, are willing to put in innovation and R&D (or R&D&I).

The city of Helsinki strategy 2017–2021 highlights the functionality of the city as the title of the strategy is 'the most functional city in the world' (City of Helsinki 2017). The title impresses that the main goal of public service provision (including traffic and daily use of urban space) is their timeliness and quality. Digitalisation is deemed as one of the key-drivers ensuring the quality-based service development. The strategy focuses strongly on environmental quality, sustainability and cleanliness. The strategy states that the most essential task of the city is to ensure sustainable growth. In the strategy language this means growth (or improvement) in economy, environment, and social conditions. Each of them has their own monitored measures.

The strategy points out the steady population growth in Helsinki. The growth is caused mainly by international migration and it has been around 1% annually during the last decade. The whole development strategy for the city relies on the future prognosis that the population growth will continue in the future. Helsinki and the surrounding region are small in international perspective and there is a general agreement that the growth is clearly a positive phenomenon. The level of urbanisation is still relatively modest and the negative (such as congestion, segregation, crime, limited waste management) effects associated with urban growth are still minor. However, they are needed to be dealt with proper resourcing as addressed in the strategy. Particularly the growing ethnic diversity and increasing income differences between households create stress on housing policy. Helsinki have practiced so called balancing policy in urban planning for decades now referring to a residential mix that includes all types of housing solutions in newly planned residential areas. Therefore, social cohesion is seen as a strategic competitive advantage.

Knowledge-based urban development is recognised through education and human capital. Finland is among the highest educated nations and as education has been verified to be the fundamental factor in innovation causality, it is also deemed as the fundamental source of social development and reducer of inequalities within population. Particular attention is paid to the inclusion of education as the preventer of deprivation. The strategy links this to broad societal goals and obligations of the cities (such as providing elementary school, day care services, libraries, and health services). In this regard, identification of knowledge and skill sets is deemed as central issue in the integration for those entering the labour markets.

In the heart of the strategy paper is urban planning. The strategy stresses social issues and social cohesion as continues doctrine of residential area planning and design. Numerous international comparisons have indicated that spatial differentiation (segregation) is more limited than in reference cities (e.g., Stockholm and Gothenburg in Sweden). Housing policy and planning require well-functioning housing markets. In contemporary situation, in which Euro-zone interest rates have been negative for several years, housing prices have increased with steady pace in Helsinki. However, the increase has been lower than in Denmark (Copenhagen) and Sweden (Stockholm) that are suffering from overheating of housing markets and average prices have been stagnant or even on decline at the early 2018.

The City of Helsinki strategy arguably focuses on broad societal questions as addressed. The strategy, however, has a specific segment considering service development and means of achieving presented societal goals. This is an interesting point as the 'smart city' or 'digitalisation' is seen as a tool and not as a changer or disruptor. The rhetoric of the strategy is therefore rather different compared to regional council's strategy goals of spatial development.

The smart city content is focused on citizen services and therefore deals explicitly with e-government and e-services. The emphasis is laid on accessibility of e-services that are founded on three principal development areas: digitalisation, artificial intelligence, and robotics. The strategy states that e-services are the primary development area and therefore other means of provision (e.g., face-to-face offices, phone services, paper interaction) are inferior in development priorities. The strategy also points out the development work needed to make the service use available to population segments with differentiated user skills of technologies. In order to achieve this, the city states it actively collects electronic feedback when appropriate regarding the service functionality. The city has designated idea-boxes, where service users are able to give their feedback.

The stratum of education from elementary school to Ph.D. degrees creates foundations for smart and intelligent specialisation. Finland and Helsinki have promoted the educational system for a decade now after a good success in international learning benchmark studies such as PISA-study (OECD 2018). Following the ideas of urban branding the Helsinki strategy sets a goal for itself being the world's leading city in learning. Digitalisation and technological learning platform development enables personal choices for accessing study resources and contents.

The strategy emphasises that Helsinki will be continuously fine-tuned as an innovative and experimental city of education and lifelong learning. The strategy explicitly states that the city will invest in physical learning environments (buildings and locations) that are able to promote and enhance learning skills required in contemporary working life and information society. These investments include new campuses built for polytechnics (called as universities of applied sciences in Finland) and high-schools. The attractiveness of the University of Helsinki and Aalto University (including the units of technical university, business school, and the school of art and design) are crucial in this goal setting. The number of international students and teaching staff are considered to be key-performance indicators for this.

To summarise, the strategy highlights quadruple-helix networking involving universities, companies, government and NGOs in pursuit of better urban environment. A practical and interesting development platform is the 'Smart School' initiative. It is an operating model where new pedagogical learning methods and solutions are developed and implemented. The strategy combines data analytics as tools to enable learning progress than is easily monitored on the individual level providing an extensive array of potentials for individual specifications. Therefore, the case of Helsinki emphasises education, learning and individual skills in smart city development that is supported by digital tools.

7.2 Stockholm, Sweden

Stockholm is the capital of Sweden and is the largest city in Scandinavia. Metropolitan area has approximately a population of 2.2 million. Stockholm has been perhaps one the globally most recognised smart or innovative cities since the emergence of the concept. Swedish government has invested extensively on research and development during the last decades and Stockholm hosts the most important development offices of the country.

Like all the case locations, Stockholm has specific areas that may be considered innovation and technology hot-spots. Kista is one of the most well-known concentrations of ICT activity in Stockholm. It is mainly marketed through Kista Science City concept that is a brand of a development company. The location is considered to be among the top three ICT clusters in Europe and the largest in Northern Europe. The beginning for Kista's cluster development was started in the mid 1980s as typical triple-helix (business-government-university) collaboration that resulted into the establishing of Electrum Foundation (EF). The purpose of this collaboration and the newly founded foundation was to establish a world-class research centre for electronics including radio and telecommunications comprising the foundations for the ICTs. The triple-helix idea is still strongly present in the current undertakings in Kista.

Currently, EF has two functionalities carried out by the Kista Science City AB that is responsible for Kista Science City concept and related functions; and Stockholm Innovation and Growth AB that is a business incubator supporting and helping start-up companies for faster growth and market expansion. EF defines its own activities to concern ICT education, economic growth, and innovation. The triple-helix framework is strongly identifiable as the foundation aims to support and develop businesses with strong connection towards science and research. Urban infrastructure and environment are taken into the development agenda together with the provision of residential housing and an extensive provision of commercial, cultural, and leisure services.

7.2 Stockholm, Sweden

The collaborative tendency of Kista Science City is identifiable from the key-organisations functioning in the management of EF. The foundation's board includes representatives from large global businesses (such as IBM), a strong presence of Stockholm's universities including the Royal Institute of Technology (KTH) and Stockholm University, and a number of regional authorities including the city of Stockholm, Stockholm County Council and County Administrative Board.

Stockholm has implemented a new strategy specifically focusing on smart city development (City of Stockholm 2017). The strategy reaches up till the year 2025. The strategy is founded on the same conceptual and societal principles discussed earlier and explicitly highlights the four dimensions of sustainability crucial for smart city development in Stockholm (and elsewhere). These dimensions are ecological, financial, social and democratic forms of sustainability that require openness, connectivity and innovativeness as their overreaching principles.

The strategy sets a clearly defined target vision for Stockholm and its digitalization efforts. The first goal is to produce the best life quality for the local inhabitants, and the second one, is to offer the best business conditions for entrepreneurs. Thus, the goal setting includes the recognition of economic performance and citizen services. The strategy outlines that these two main goals are achieved through innovative solutions, transparency and connectivity.

The strategy paper of Stockholm is interesting as it applies the principles studied in regional innovation systems from its foundation. The strategy has been developed by following principles of quadruple-helix model (added form of traditional triple-helix) so that in addition to business-government-university interaction also citizens have been included in the strategy process. According to the strategy an important emphasis has been put on the analysis and consideration of the global developments in key-areas. In this regard, the strategy production follows close examples and outcomes suggested in innovation policy studies and therefore provide an interesting example, of how conceptual and theoretical interaction modelling may be implemented in practice. The development process is well documented in the strategy paper itself.

There are several steps worthy of investigating in the strategy development and creation (see Ramstadt 2009). The inclusion of local inhabitants has applied both traditional means (e.g., direct dialogue meeting at the city hall of Stockholm) and applications of social media. Through these main channels of participation more than 3300 people were involved in the strategy creation. The participants expressed views regarding the initial vision and goal setting of the strategy. They also provided feedback concerning Stockholm's current digital citizen services and interfaces. This provides also a good example of crowd sourcing as it is likely that development suggestions thousands of people will include excellent ideas. This work also requires skills in content analysis that may even approach tools of qualitative data mining.

The inputs and suggestions from the citizens were integrated to the traditional collaboration of triple-helix partners in development meetings of the strategy in order to get the overall acceptance and consensus of the content. The business-university-government collaboration resulted into a substantiated result when the City of Stockholm, Royal Institute of Technology (KTH), and four large Swedish companies from the fields of electronics, energy, construction, and transport established *Digital Demo Stockholm* innovation arena. It is a platform for innovation intermediation and project development aiming to enhance and provide digital solutions for citizens. The strategy also recognized the Kista Science City and the comparable efforts (e.g., ICT arena) conducted there in innovation testing and service development.

The international aspect and the analyses of global best practices has been conducted in the strategy process. The strategy has observed and monitored experiences from other countries and cities through the use of key-performance indicators applied in a benchmarking exercise of Citykeys (2017). The Stockholm strategy also indicates active interaction with other cities and their best practices on the smart city development.

The strategy contains two main chapters describing the principles for enablement and implementation. These following steps are good examples of systematic development planning in urban technology integration. In the case of enablement there are seven points highlighted and they represent key-issues of the urban studies literature. In practice, the principles of enablement are:

- Common digital solution platforms
- The use of central platforms as the nodes of system exchanges
- Open standards are the foundation for technical solutions
- Modularity of technological solutions
- Agreements and collaboration are the enablers of innovations
- Security and data protection is secure
- Openness of internal and external data.

There are several best practices of technology implementation identifiable in the enablement list. Firstly, the goal of using common solution platforms is desirable, particularly in the public administration. Far too often single silos of administration develop and deploy their specific solutions that are not interoperable and experience problems in service maintenance, particularly if the service provider is a small company applying its own system designs. The second main point is the emphasis on applying open standards for technical solutions. This will be interesting development area to monitor and observe that to what extent open and closed system architectures are being applied in smart city development. It also entails the roles of multinational corporations (e.g., Google, Microsoft) as their role in local system development. The third main observation from the enablement list is the emphasis on data security and privacy issues. The question of data-base combining and more efficient data management will evitable be contrasted with the privacy legislation, particularly concerning public data-bases (e.g., medical data, taxation registries) and their cross-referencing.

The technological foundations of enablement principles provide the pathway to implementation. The Stockholm strategy implementation principles are the practical guidelines for strategy execution. The strategy names eight main points (for its own execution) as follows:

- Citizen needs are the foundation of initiatives
- The continuation of existing development activities (accumulation)

7.2 Stockholm, Sweden

- Prioritizing is conducted in line with the overall goals
- Internal and external collaboration in development
- Long-term principle in all investments
- Information collection is contextualised and visible
- Digitalisation integration into urban planning processes
- Communication is the key-driver in internal and external development.

The identified eight points have visible similarities with the other case location strategies. The first point identifies citizen needs as the foundation of the strategy. There are clear similarities to Helsinki city strategy. Citizen focus evitable brings forth e-government services (citizen targeted public services) and questions of urban functionality as is the case in Helsinki. This approach may be connected to the broad societal goals of functionalities in transport, environment and welfare services. The citizen focus also makes an interesting distinction between, and even divergence, of smart city development goals (and strategy rhetoric) compared to the business development strategies and their common rhetoric and emphases in relation to smart city development.

The second point highlights the continuation of existing development activities and in other words accumulation and beneficial applications of the already conducted work and efforts. This point recognises the need to overcome common problems related to project and time-frame bound development activities. The question is most often about the resources and funding as still too many of the initiated (and funding period bound) projects focusing on urban development are terminated after the initial funding period or they wither away after a while. A common phenomenon is also that the project ideas are repackaged or remarketed with highly similar contents but with other key slogan depending on the pitch words hot at the moment of applications. The emphasis of continuation is therefore an important strategic decision aiming to overcome the project tendency troubling smart city initiatives.

The Stockholm strategy makes a worthy point in addressing that development project prioritising is done in accordance with the overall goals of the strategy. This is often easier stated than realised in the project portfolios. Smart cities projects are most often technology projects. The overall strategic guidance of the smart development integration focusing on societal functions is tightly interwoven with politics and local power relations in city parliaments. Therefore, the integrative approach of combining societal goals with technology and innovation is a fresh and welcome approach. The plurality of the conceptual foundations of smart cities is clearly visible here. An interesting research task appears in the longitudinal analyses of smart city development. It is also likely that the most important indicators will change in time. For example, in the early 2000s the relative number of personal computers, tablets and other forms of technical devices were important measures of information society. Currently, they are uninteresting or even banal indicators, as their penetration levels have reached the saturation threshold (approximately 90%) of fully embedded technologies.

The fourth point brings another important element of successful creation of smart cities. Thus, internal and external collaboration in development efforts is a necessity.

There networking approach is one manifestation of new ideas and innovations in the creation of sustainable and time enduring solutions in urban environments. The networking directions (outwards and inwards) are defined from the view-point of public administration (government). Internal directions are linked to concern silo-type of administrative organisations and potentials for cross-sectional or cross-policy collaboration. External collaboration instead refers to quadruple-helix approach commonly applied in the Nordic smart city strategies. Collaborations are essential as essentially all successful disruptive innovations are created in private companies. Collaborations take place in different arrangements of PPPs and quite often they create beneficial conditions for urban innovations. Clean tech area is one of the most current topics today as waste management and efficient waste disposal handling impacts several goals of good living environment.

The fifth point of the strategy states that long-term principle is applied in all investments done by the city. This theme is also thematically interlinked to the second point of accumulation and making benefit of the already conducted project works. This thematic is also highly crucial as long-term investments often dictate the technological premises for the development path. There are too many examples, in which city organisations have chosen technology providers whose solutions are not standardised. These providers are often also rather small in terms of turn-over and employees making them more vulnerable in market turmoil and recessions. As discussed, the long-term approach requires visionary approach to understand the development pathways of technology combined with realism and reserves for changes taking place during the development path.

Openness and open principles are always connected to public sector operations as they are based on taxation and collectively collected resources. The sixth point in the Stockholm strategy makes a statement concerning information and data visibility and openness. The point also stresses the contextual consideration of information collection. The basic principle commonly recognises the line of thought that no information should be collected unless it is required. This may be connected to a philosophy of data mining and big data that commonly are interested on all possible data sources and also of that information that is 'not there'. Thus, advanced data mining and data analytics operate in recognition of the existing data and non-existing data of which the latter is in several cases more interesting than the former. The visibility principle and open data supporting is clearly in line with good administrative practices.

The two final points connect the Stockholm strategy explicitly to digitalisation and technology development. Digitalisation integration into urban planning processes is a necessity in contemporary information societies. The citizen-government interaction needs various kinds of digital platforms to be realised in an efficient manner. Different technology user segments use different types of technologies and for example tools of social media should be integrated into the public sector official online services and tools. It goes without saying that the more extensive social media coverage the city has the broader and wider the potential audience and feedback pool of inhabitants is in the city's disposal. Technologies enable communication between different actors and also the final point of the Stockholm strategy list stating that communication is the key-driver in internal and external development. Again, these communication directions refer to external stakeholders and internal offices within the city administration.

Arguably, all presented points (enablement and implementation) are connected to projects that are the grass-root level tools to achieve the main goals of strategy. Therefore, the selection and prioritising of the project conducted within the strategy framework require prioritising as stated in the point three of the implementation. The case of Stockholm has clearly brought up the fundamental integration of citizens into the all steps of strategy. In this regard, Stockholm initiatives are strongly focused on social and democratic dimensions of the smart city frameworks. Stockholm has also an explicit innovation strategy published by the local centre of commerce highlighting the economic conditions and needed improvements in the business environment development for investors and entrepreneurs.

Considering the locations of innovation in Stockholm, the Kista area is the most recognised single location associated with innovation development. Overall the strong strategic guidance and explicit implementation steps on smart city development are indications of systematic and seriously taken approach towards integrating new and more efficient channels for citizen interaction and economic development in Stockholm. Particularly, the development process of the strategy itself applies smart city technologies of citizen-government interaction and there is strong emphasis on open and transparent platform development.

7.3 Copenhagen, Denmark

Copenhagen is the capital of Denmark and has population of approximately 600,000 (city) and the broader urban area has more than 1.2 million inhabitants. In terms of population, Copenhagen, Helsinki and Tallinn are quite close to each other. Copenhagen has been considered one of the leading smart cities in the world and it has steadily ranked among the top cities of the world.

The EU ranked and evaluated the 28-member states (EU-28) in 2017 in 'Digital Economy and Society Index' (DESI). Denmark received the highest marks for digital development in the EU report indicating the strong national position in the smart city development. The same report ranked Finland the second and Sweden the third in e-society development. The report uses a relevant indicator set for all smart city studies and research. The indicators included five main categories listed as follows:

- Connectivity
- Digital skills
- Use of internet
- Integration of digital technology
- Digital public services.

Connectivity part includes indicators of broadband properties and speed (bandwidth). In total the category holds nine single measures. The second main indicator category is 'Digital skills' that includes four indicators. These include the number of internet users, the basic digital skill level holders, number of employees in ICT industry, and number of graduates from technological fields of science including engineering and mathematics. The third main category the 'Use of internet' is operationalised through seven variables including the use of social-media, online conferencing, e-banking, and e-shopping. The fourth category of 'Integration of digital technology' refers to commerce and business dimensions of ICT applications. This section is studied with eight variables divided into business digitalisation and e-commerce provision.

The final set of 'Digital public services' is also named as e-government having four individual indicators. The first indicator is the number of e-government users referring to those people who have sent filled forms to public sector officials via online tools during the last 12 months. The second one is 'pre-filled forms' that means the amount of documentation (data) that is pre-filled in administration. The third is 'online service completion' that is a relative number of administrative steps of major changes in persons record including, for example, changes of addresses, birth of a child, and other events that are doable online. The final fourth indicator is 'open data' and it is a composite indictor. It includes information regarding existence of open data policy, estimations of open data impact on society and other data characteristics such as the use amounts of national data portals, and so on. Considering the above indicators, Denmark and the other Nordic countries, rank constantly above the EU averages that are provided in single country reports.

Copenhagen, like all the other Nordic capitals, has strategic guidelines for innovation and smart city development implemented in 2015 (City of Copenhagen 2017). The strategy identifies two major poles as context: global view and user involvement. The former signifies the global benchmarking and state-of-the-art smart city solutions and technologies, and the latter the grassroot level of local inhabitants and customer-based approach in service creation and development.

There are three main components through which the smart city is realised in accordance with the two major poles. These are: (a) quality of life; (b) growth; (c) sustainability. These keywords are easily connected to the earlier cases of Helsinki and Stockholm further supporting the similarities of strategies emphasising the points raised in the innovation literature concerning meta-level policy cloning. This notion is critical as it considers that overall societal goals are too general to be able to get a grasp of local contexts and conditions needed in order to create best possible solutions for improving an urban condition. In other words, one solution fits-for-all does not provide enough precession for strategy actualisation. However, the rather similar contexts of Nordic countries and their approach towards social and welfare policies probably establish them as the best possible group of countries in which similar strategy guidance may be expected to work.

The dimension of quality of life focuses on safety, diversity, leisure and wellbeing and convenience. These four characteristics are clear indicators of human focused urban functionalities needed in smart city. Safety relates to indicators such as crime rates, low corruption, and resources of the safety organisations such as the policy and fire departments. Diversity is an interesting characteristic as it is applied extensively in the studies of 'creative' regions and cities. Florida's work on ethnic and social diversities has brought into an analysis of the issues of tolerance as a factor in urban and regional growth. However, empirical results have verified that diversity as an isolated indicator does not have a significant role in regional value-adding. The significance rather comes from the overall impact that diversity and tolerance measures bring. Therefore, they may be considered as secondary factors in growth enablement. Finally, wellbeing and convenience are dimensions that are derived from the inhabitant and urban service use experiences. Wellbeing may also be considered as a final goal of development actions. The monitoring of these dimensions needs reliable survey methodologies and analyses. They also bring forth the need of end-user feedback resulting improvements done on the basis of that feedback.

The second dimension is named as 'growth' including elements of knowledge production, innovation, employment, and investments. These keywords are widely applied earlier in this book and they are essentially the economy-sided elements of the Copenhagen smart city strategy. The strategy considers that the smart city is founded on innovative partnerships that is supported by continues technological development. Similarly, the strategy applies technology driven rhetoric raising elements of big data and data mining. This is due to the explicit acknowledgment that the smart city applies and needs data generated within the city (and surrounding areas) and these data sources are needed in the creation of efficient and resource optimised solutions and alternatives for the citizens and businesses locating in the city.

The strategy highlights sector collaborations the same way as the earlier examples of Helsinki and Stockholm. The Danish strategy considers that innovative solutions for different user segments concerning the city as an organisation and as a physical structure span to cover all city services. Their creation, maintenance, and development involve all stakeholder groups (NGOs, private sector, and research and education units). Copenhagen and its locations are considered as test-beds or living-labs for testing innovative solutions (Kusiak 2007). The life quality, sustainability and technological proficiency then should attract highly educated workforce, new businesses, and investments to the capital area.

The third and final component in the Copenhagen way of developing smart city is sustainability referring to carbon neutrality and clean air and water. Copenhagen has defined its vision for green city and the main goal is defined as Copenhagen should be the world's best urban environment that is able to produce experience of a unique urban life. There are four targeted vision goals set in the Copenhagen strategy through which the main goal is tried to attain. They are as follows:

- Green and blue capital
- World's best city for cyclists
- Carbon neutral capital
- Clean and health city.

The four pillars separately associated for the sustainability development are actualised in the strategy. The first point (Green and blue capital) relies on the goal that 90% of Copenhagen's residents should be able to reach parks, natural areas or sea/water areas in less than quarter of an hour by walking. The walkability paradigm in urban planning has gained also academic interest, particularly in Denmark, where significant contributions on this field have been done. The strategy also gives motivational goals for the development as there is a set goal of doubling the use of recreational spaces in the end of the strategy period—year 2020.

The second specific sustainability pillar singles out one technology for movement—the bicycle. This is highly context specific issue as cycling is very popular sport and general means of urban movement. The only comparative country in the adoption of urban cycling is found in Amsterdam and other cities of the Netherlands. The Copenhagen strategy sets a goal of 50% of residents to use bicycles in their travelling to work or study locations at the end of the strategy period. The number is significant as for example only approximately 10% of trips are done by bicycles in Helsinki. However, the traditions and climate between Copenhagen and Helsinki are considerably different. Cycling as a means to improve urban sustainability is also connected to injury risks and planning of safe cycling routes within urban structures. Cycling is esteemed to have a reductive impact of 20% on the total air CO_2 emissions in Copenhagen. This is substantial reduction goal.

The two final goals are carbon neutrality and clean and healthy city are tightly intertwined to the cycling city theme as all the carbon emissions to air are expected to be reduced by motivating inhabitants to use bicycles. The strategy states that the goal of Copenhagen is to be carbon neutral city by the year 2025. The goal is ambitious and is likely not to be realised but the goal setting indicates political motivation and mind-set needed in order to make changes in behavioural patterns resulting into smaller emissions and environmental friendliness.

On practical level the Copenhagen strategy states the need for co-operation across the city's administrative structure (a total of seven offices) that are involved in the project coordination activities and they all follow the defined strategy. There is a number of spearhead projects initiated and conducted including open data, intelligent map development, and big data platforms. Open data developments are significant in Copenhagen and they are connected to the development of big data and map interfacing. The defined strategic goals are to improve the administrative decision-making by enabling holistic and extensive overview on the urban planning.

There are four significant application targets designated for open data projects. The first is the goal of creating a data market place that is deemed to promote innovation, creativity, inclusion, and PPP collaborations (see Laforet 2008, 2009). Through these general keywords the strategy sees potentials for coordinated urban development. This requires improvements in infrastructure and optimal use of available resources. Open data helps to identify and make use these resources. The final goal appreciated in Copenhagen is the potential support obtainable through open data and its applications. These are realised in strategic initiatives, and they focus strongly on sustainability and green growth. The political benefits and initiatives are also expected to have an impact on employment level and occupational growth. Information and technology systems are defined as the foundation for these goals and overall smart city development. The strategic guidance of open data follows similar keywords as it has been the cases with Helsinki and Stockholm.

7.3 Copenhagen, Denmark

Open data solutions are linked and recognised but treated separately from the big data elements of smart city development. Copenhagen aims to create a big data platform for data management concerning databases and registries. There are three main objectives and desires associated with big data and data mining development: The first is to construct a digital platform, on which centralised data from public and private sources would be stored. The goal of combining public and private data sources will make an interesting study topic. Currently majority of companies applying or collecting big data are doing so for marketing purposes and customer segmenting needs. This will enable more individually targeted marketing efforts and campaigns. The second, the big data is deemed to create new and innovative business markets that are based on digital platforms. This goal is a contradictory one as there are limitations on how extensively public organisations (such as city administrations) are able to operate on business processes. The goal is mainly targeted to concern private companies and the city role is defined through PPP collaboration. Finally, the third goal is focusing on application side and practical improvement of urban planning. This should be streamlined with sustainability targets in order to promote urban structures for wellbeing and good living. This is widely discussed already also in the cases of Helsinki and Stockholm that have similar initiations aiming to increase participation and interaction between inhabitants and planning officers.

There are several ongoing smart city projects that define the most significant efforts conducted in Copenhagen. They constitute fundamental characteristics of the urban technologies. These are as follows:

- City Wi-Fi
- Smart parking
- Traffic optimisation
- Smart water defence
- Smart waste
- Establishment of the Copenhagen Solution Lab (CSL).

The first project concerns traditional development of open and publicly available free-of-charge city Wi-Fi. The aim is to establish an extensive coverage common infrastructure through which smart services and applications (software) may be used. The question is a traditional combination of hardware provision that targeted to enable software interfacing. The Copenhagen strategy does not directly state the desired extensiveness of the network coverage as it prioritises the use content directly to the services provided by the city. Only after securing the bandwidth for the primary services, additional connectivity will be established for tourist and visitor services.

Smart parking is a practical and well understandable extension of smart services on which digitalisation may provide value-adding. The foundations in this development task are in hardware technologies as sensors and meters are required to calculate the available parking spaces. This information is then forwarded to end users via software platforms. This helps the daily use of urban space and makes it easy for a driver to identify empty parking spaces in the centre areas. There are also planned extensions to this information service such as pre-ordering of specific spaces with dynamic and changing prices that are dependent on the level of the congestion.

Parking development is connected to the third theme of general traffic optimisation. Real-time route planners and traffic monitoring enable benefits to reduce travel times. Traffic optimisation produces also longitudinal time-series data of the traffic volumes and the collection of this data helps in the long-term urban planning and street volume design.

Environmental elements of the smart city development are managed and targeted in the fourth and fifth projects. The former concerns physical challenges that Copenhagen has as a sea-side city: the smart water defence against floods and over-reaching water disasters. For example, sensors and IoT solutions integrated to pumping stations and monitoring centres are able to identify the capacity and potential over volumes of water. These systems are also connectable to information sharing systems providing real-time information to persons locating close to flood areas. The system also enables new alternatives to tread rainwater directions in sewage system. The latter (fifth) project deals with smart waste management and is based on real-time data provided by sensor technologies from garbage bins. They send alerts to waste trucks when filled up and also affecting the truck routes of waste collection. This creates extensive efficiency gains in waste collection. These technologies also enable new pricing models of waste management—e.g., for households on the basis of real use volumes instead of weekly/monthly pricing.

The final part of the Copenhagen story concerns establishment of the 'Copenhagen Solution Lab' (CSL) that is a location-based idea of providing a test place for new smart city solutions where companies of different sizes and profiles are able to test and implement creative and innovative solutions for sustainable cities. It may be considered as an intermediary focusing on combining local knowledge resources. The collaborative work is seen to lead to the implementation of innovation and smart city development. The strategy points out that CSL is intended to function as a common meeting place for quadruple-helix stakeholders that are dealing with and have an interest in the development of data-driven urban solutions. Overall, the Copenhagen case shows that the goals are deeply connected to sustainability and green values.

7.4 Tallinn, Estonia

The e-society development of Estonia started already in the late 1990s less than a decade after gaining the independence from the collapsed Soviet Union in 1991. The steps identified by the e-Estonia online service (e-estonia.com) identify nine significant products and developments achieved on the societal level. An important notion is that e-government was decided to be a strategic choice already two decades ago and was considered to be the best way to organise and manage governance. It is obvious that the renovation of the governance and the non-existence of earlier western traditions in governance allowed a great deal of potentials for modern governance development for Estonia.

7.4 Tallinn, Estonia

The current information society strategy in Estonia reaches till the year 2020. The future strategy directions will be interesting to see. The current development mindset is focused towards applications on e-education, intelligent traffic and transport, cyber-security, and real-time economy (for conceptual definitions see Dodge and Kitchin 2001).

The case of Estonia is interesting as it is the first country in the world that has arranged national voting for public officials via online platform (i-voting) in 2005. The voting process takes place with privately owned devices and thus do not require physical voting locations commonly adopted in other countries. The solution is efficient from the end-user perspective, but it leaves open several questions concerning the confidentiality of the voting (e.g., presence of other persons and possibility for affecting the selection). These insecurities that are not technology related are also the main reason why only a hand-full of nations have implemented corresponding systems. The technology adoption is also a cultural question connected to national contexts.

As pointed out, the most well-known intelligent solutions for e-governance in Estonia concern taxation, voting and residency. The data management is also forerunning as Estonian services are extensively based on block-chain solutions that enables highly secure data processing and transparency. Currently, approximately 95% of Estonian tax declarations are processed and managed via online tools. Estonia has also implemented e-identity card system for interactivity for government affairs that also enables the digitalised voting system.

From the technical side, the use and application of block-chain technology in e-government is highly advanced. The Estonian government has tested the technology already since the 2008 and after 2012 it has been implemented in practice. The system is applied is various registries such as national health, legislation and law, and security. A recently implemented e-service is an 'e-resident' concept that is targeted to foreign individuals interested in establishing business in Estonia. The service allows residents of other countries to obtain government issued digital identification number and access to Estonia's public e-services. The fundamental idea is that entrepreneurs have a potential to conduct their businesses in other countries and have their main office in Estonia. This is particularly directed towards Finnish and Swedish entrepreneurs as the taxation and other costs are significantly above the Estonian ones.

The city of Tallinn has implemented its current innovation and growth strategy 2014–2018 (City of Tallinn 2013). The strategy focuses on innovation and knowledge-city development but it has several relevant dimensions concerning smart cities. The innovation strategy is based on broader city strategy 'Tallinn 2030'. It also relies on the local development plan 2014–2020. These three main documents envision Tallinn as an internationally attractive destination that is driven by innovation with a safe urban environment.

The strategy focuses more explicitly on economy and finance than the other cases. This is understandable, as the Tallinn strategy is more an innovation strategy than smart city strategy. Again, the conceptual differences come into play. However, the Tallinn case explicitly indicates the intertwinement of innovation and smart cities as they may be considered to be in deeply rooted relation with each other. Smart solutions are in several cases innovations and smart solutions function as platforms for innovation. They are also substrate for economic growth and sources of new business models and opportunities.

In this regard, the strategy highlights two key concepts as the guiding principles: entrepreneurship and innovation. The strategy actually also defines the contents for these two that is not always the case in current strategies. According the Tallinn strategy, (City of Tallinn 2013), "entrepreneurship is the process of creating something new with value by devoting the necessary time and effort … and receiving the resulting rewards of monetary and personal satisfaction and independence". This definition is taken from Hisrich et al. (2005), it emphasises the process-view of the business making it more useful to start-up companies. This definition brings entrepreneurship close to innovation definition as it argues on the behalf of 'something new with value' that is exceedingly much applied phrase in innovation definitions. Innovation on the other has been defined loosely as something useful and new with a significant improvement aiming at development (Bernier and Hafsi 2007).

The strategy sets two goals for itself in accordance with these definitions: Firstly, the strategy has to be based on strategic position. This is interesting as the strategy highlights the geographical location, demography, and current level of development, in the context of problems connected to entrepreneurship and innovation. Secondly, it has to be harmonised with the national goals, and with the goals of the private sector. The city is seen here as the context that offers the effective environment for pursuing these goals.

The Tallinn strategy considers also spatial layers of public administration as it separates different levels of activities in accordance to stakeholder potential to operate on each layer. The layers are the EU, state (Estonia), city (Tallinn) and private sector. The dimensions, on the other hand, are partnership, capital, knowledge and skill, and support systems (infrastructure). The strategy considers that partnerships together with infrastructure and support systems are those that the city has the most significant impact on. Interestingly, capital dimensions are estimated to be managed the largest extent by private sector together with the EU. This is not surprising as Estonia has been received a significant amount of income transfer from the EU in comparison to other cases. Estonia is the only nation among the study cases that receives more income that it pays membership fees to the EU.

The Tallinn strategy is written in an analytical way and it applies references as an academic paper. This tendency is also recognisable as it sets two aspects, argued to be based on analysis, that are needed to be assessed before moving to practical solutions of implantation. The first issue is to consider Tallinn's international competitive situation that will enable the definition of focal areas of the strategy. The second part is the practical solutions and information on the innovative ideas that should aid the achievement of the objectives.

Strategy considers strengths, weaknesses, opportunities, and threats (SWOT) as there is an emphasis on discussion of challenges and problems that exists in the way of successful urban development in Tallinn. These include, according to the strategy, insufficient knowledge and skill-bases for implementing innovations. This problem is connected to the education. There is therefore an increasing need to expand the relative proportion of the highly educated (and skilled) professionals. The limited knowledge sources are also connected to the second identified problem that is too low information exchange and cooperation between entrepreneurs and existing innovation systems (support structures) nationally. This problem is continued to include also international networks and limited international competence.

The use of this critical language in an innovation strategy is different if compared to assessed Nordic capital strategies. Interestingly, the strategy also is critical towards the city itself stating that public service provision is rarely innovative. The final criticism in strategy concerns limited resources of Estonia and Tallinn in comparison to other cases. Thus, the lack of resources in infrastructural development is considered as an obstacle for entrepreneurial growth. This is a significant difference in comparison to e.g., Helsinki strategy that strongly emphasises the need for innovative cooperation and service provision models. Overall, the Tallinn strategy emphasises the innovation of businesses instead of public sector. Based on this critical assessment, the Tallinn strategy identifies main goals of mission and vision for the city as follows (City of Tallinn 2013, p. 5):

- Mission: To create better opportunities for the rise of enterprise development and innovation in Tallinn and thereby to increase the competitiveness of Tallinn as a city and a region.
- Vision: By 2018, Tallinn is an internationally open city that stimulates entrepreneurship and creativity, which promotes cooperation among the enterprise stakeholders and the development of knowledge and skills.

The vision is supported by general objectives aiming to increase general economic indicators in Tallinn such as employment rate, fixed assets of companies and labour productivity per worker. The set goals for increase are substantial (approximately a double compared to the 2010 levels). The strategy proposes two main development sectors: (a) service industries; (b) future technology sectors. These are then interlinked through ICT sector that is defined to be the interlinking segment of services and future technologies, and it is also considered to be one of the most potential development industries itself in Tallinn. The development and prioritisation of ICTs in the strategy is founded on the notion that the sector provides foundation for the other sector to beneficially apply opportunities of digitalisation including efficiency and quality gains.

Service industries produce approximately 80% of the value adding in Tallinn's GDP. The prioritisation of the service sector further enhances the existing strengths within the city and therefore presents a traditional approach towards innovative growth (sticking with those fields that are traditionally important and strengthening them). Thus, service field includes five important segments in Estonian economy that are strongly focused on Tallinn area: Creative industries; tourism; transport and logistics; health services; financial services. The future technologies, on the other hand, include hardware and software development that may be considered as a new opening and a potential field for improving urban life quality. The future technologies field has three main components: health technologies; mechatronics; environmental technologies.

The Tallinn strategy provides systematic schematic presenting tools to achieve the set goals. These include four main domains that are interconnected. They are: (a) knowledge and skills; (b) international openness; (c) cooperation; (d) motivational urban space. The first pillar focuses on knowledge and skills resources. Increasing awareness of entrepreneurship through business counselling and consultation are important tools here. They are connected to the provision of business incubator services and innovation system. The strategy sees also that knowledge resources are dependent on availability of capital, particularly for start-up companies, and the city aims to support this through direct funding (e.g., grants and competitions for funds) and with networking events. An interesting suggestion presented in the strategy is innovation course (education) provision for not only university students and starting entrepreneurs but also in elementary and basic schools. Additional tools include scholarships for researchers and competitions for applied sciences. Majority of the presented toolbox therefore focuses on either organising networking events or providing funding opportunities. These tools may also use innovation awards as motivators for excelling in product development (Azadegan and Pai 2008; Wessner 2009; Makkonen and Inkinen 2014).

International openness is an important dimension for small economies such are all the studied cases. The Tallinn strategy sees that there is a need for support services for internationalisation that may be realised as market consultations, partner searches, and networking services. This is an example of outward directed approach. The strategy also considers incoming migrants and suggests that there is a need of creating an urban environment that promotes immigration. This involves also the development of urban brand and tools of urban marketing for foreign investors, tourists and potential immigrants. Again, traditional networking solutions (organising events, meetings and conferences) is presented in several occasions. The most interesting point in the openness section is perhaps the recognition of the Helsinki-Tallinn collaboration that is supported by politicians on the both sides of the Gulf of Finland. The physical distance between Helsinki and Tallinn is only some 80 km and there are several development plans done how to improve the connectivity between the cities. Some of these plans are more realistic than the others. For example, a railway tunnel between the cities has been examined for several years now and initial plans have been developed. However, the cost-benefit ratio of the project is not economically viable.

The third domain of cooperation is promising from the viewpoint of research. The Tallinn strategy emphasises mapping and surveying tasks of the current situation of businesses and innovation within the capital region. The understanding of up-to-date condition of business sector provides opportunities for additional support for the development plans and actions for new geographical clusters. There are also goals for promoting interdisciplinary cooperation between companies and universities.

The identification and monitoring views are expanded to concern also the final fourth domain of the strategy implementation, which is the creation of motivational urban space. The strategy calls forth regular identification of citizen and business expectations of urban space. This is connected to the development goals of the 'smart city' concept. It refers to the city's idea of creating a specific strategy for smart city that

7.4 Tallinn, Estonia

Table 7.1 GDP in metropolitan regions in million EUR (Eurostat 2018)

	2011	2012	2013	2014	2015	Average	Growth 2011–2015 (%)
Copenhagen	99,491	103,493	106,985	111,788	116,395	107,630	16.99
Tallinn	10,101	11,018	11,740	12,307	12,822	11,598	26.94
Helsinki	74,378	75,033	77,725	78,476	n/a	76,403	5.51
Stockholm	124,462	131,475	135,684	136,915	n/a	132,134	10.01

includes 10-year planning period for improving industrial, educational and residential areas and their relational proximities. This tightly connected to procurements made by the city for the needs of innovation and sustainability. Among these are the establishment of innovation centres (and provision of supportive funding for them) as they are seen as important motivators for urban development.

7.5 Comparison of the Best Practice Cases

The earlier investigated case locations are placed under the microscope to compare them in accordance with statistical resources widely applied in innovation studies in geography and other fields. The statistics concern both sides of innovation (inputs and outputs) as they give an understanding of spatial conditions of education and location-based activity of R&D—these elements are traditionally associated as prerequisites for innovation and they represent input-levels. These indicators are also strongly linked with general income and value-adding variables such as gross regional product, gross-value adding, and enterprise turnovers.

On the other side, then there are the direct innovation and patents counts. Patents are separated to applications and granted patents. In general, innovation activity in a location is better understood through patenting efforts (applications) than granted (reviewed) outcomes of these applications. This is why majority of the patent data applied in this book concerns applications as they have greater numbers and broader coverage. Innovation counts are more difficult to obtain as they require direct queries (surveys) from the companies. Another approach would be to look at the new product launches from different industrial fields but the collection of this type of data is an extensive burden.

Based on these considerations it is time to move to present general statistics concerning macroeconomic conditions of the cases. This, the selected four cities are compared with national and in some cases also the EU reference statistics. The first contextualising indicator is the economic significance (size) of each case. The easiest way to do this is to look at the GDP indicator and corresponding values from the cities and city regions. Table 7.1 presents GDP levels in metropolitan areas. The spatial categories follow the Eurostat classifications.

To summarise, all cases are relatively wealthy, particularly the three Nordic state capitals in terms of their value-adding. Tallinn is currently behind in absolute figures of value-adding per capita and cross-domestic product but its growth rates have been significantly higher compared to those of the other case cities. Table 7.1 is relevant as it gives an understanding of the diversity of the case locations in terms of their overall economic activity. The results are straightforward to interpret. Stockholm is the largest economic urban concentration in the northern Europe with some 25% margin to Copenhagen. Finland has experienced economic recession during the observation years and this is clearly visible in the numbers from Helsinki. Particularly if Helsinki is compared with Copenhagen the growth has been significantly lower (5-year growth has been only 5.5% in Helsinki contrasted with approx. 17% of Copenhagen).

The low growth figures in Helsinki are related to Euro-zone. Finland and Estonia have Euro as their currency whereas Denmark and Sweden have their national currencies that are coupled with the Euro. However, national currencies allow broader monetary freedom to make national adjustments compared to Euro-zone. The used 5-year time period concerns economic slowdown period that has been very severe in Finland even in the context of other Euro-zone countries. The foreign debts of some of the Euro-zone countries (e.g., Greece, Portugal, Italy) are still exceedingly high and have caused negative interest rates on the Euro-zone for the discussed time period.

The other Euro-zone country Tallinn is considerably smaller economy than the other cases, corresponding less than 10% of the figures of Stockholm. Due to the low starting levels Estonia has experienced in the greatest relative growth 2011 till 2015 but in absolute terms the growth has been less than half in comparison to Helsinki that has experienced the slowest growth. Estonia also adopted Euro as the currency relatively late (2011) in comparison to Finland (2002). The Estonian Euro adoption took place during the severe recession indicating also the political goals behind the monetary union. These geopolitical factors are also recognised in the Tallinn's innovation strategy discussed earlier. The presented overall economic activity is a feasible starting point to understand regional wealth and economic realities before innovative and knowledge-intensities of these locations are empirically considered.

Education plays the starting point in terms of verified schooling of population that may be considered as the foundation for innovative or knowledge intensive economies. One of the most widely used indicators of knowledge-intensity is the relative number of persons with tertiary education. Education has also been verified to causally explain innovation levels in spatial contexts and therefore, the search for commonalities or disparities between the cases is reasonable to start with the regional statistics indicating educational attainment on the highest education. Table 7.2 presents EU and national references and regional figures for cases. The data covers years from 2012 to 2016 on the regional scale (NUTS 2).

There are some general characteristics among the cases compared to the EU-28 and Euro-zone averages as all the cases locations and countries have higher levels of tertiary education. The highest education level is a regional figure from the Helsinki region (51.6% in 2016) exceeding the Finnish national average by almost 9%. If the five-year change in education levels is considered, the growth rates have been

7.5 Comparison of the Best Practice Cases

Table 7.2 Tertiary education (% of population) in NUTS 2 regions (Eurostat 2018)

	2012	2013	2014	2015	2016	Average	Growth 2012–2016 (%)
EU-28	27.7	28.6	29.3	30.1	30.7	29.3	1.6
Euro-zone	27.2	28.0	28.5	29.2	29.8	28.5	1.3
Denmark	34.8	35.4	36.1	37.1	38.0	36.3	1.5
Copenhagen	46.2	47.0	48.1	48.5	49.8	47.9	1.7
Estonia (Tallinn)	37.6	37.4	37.6	38.1	38.9	37.9	0.3
Finland	39.7	40.5	41.8	42.7	43.1	41.6	1.9
Helsinki-Uusimaa	48.9	49.3	50.2	51.3	51.6	50.3	1.4
Sweden	35.7	37.0	38.7	39.8	41.1	38.5	2.8
Stockholm	44.4	45.7	47.6	48.7	50.4	47.4	3.0

in accordance with the EU averages even though the starting levels are significantly higher. There are some variations as Sweden and particularly Stockholm region have increased the education levels almost a double compared to others. On the other hand, Estonian figures have experienced a limited growth (0.3% units).

The results clearly indicate that the human capital (referring here to education of population) accumulates in cities. The studied capital cities are the forerunners and their figures should be contrasted with their respective nations. Therefore, it is essential to look at the difference between the national average and capital regions. In the cases of Copenhagen, Helsinki and Stockholm regions the difference to their respective nations is approximately 10%. This indicates the concentration tendency of highly educated people towards larger cities. High education level figures in Nordic countries are also not surprising as their education policies have highlighted the importance of education constantly. For example, in Finland the goal has been that each age group should have tertiary education levels about 70%. Nordic education systems are also part of social policy of the Nordic welfare states. This social policy regime is often referred as 'universal', in which the whole education strata from elementary school till to Ph.D. degrees are cost-free for students and costs are covered by the public sector.

Patents are widely used innovation output indicators and in the following they are presented according to availability from the Eurostat database. It is acknowledged that there are limitations concerning the use of patents (as innovation proxies). However, they provide widely used and applied data resource that is strongly correlated with other innovation measures including R&D investments, personnel and annual years of research. Patents numbers still remain as one of the widely used indicators and there are no better alternatives available (e.g., Chari et al. 2012). Table 7.3 presents important relative indicators of all patent applications with a 12-year time-span. The figures are presented on the national level from the cases. US is added as a reference together with the EU average (all 28 countries and 19 Euro-zone countries).

The most relevant points in Table 7.3 are that Denmark, Finland and Sweden are by far on their own category in terms of applied patents per population. The fourth case location (Tallinn and Estonia) is lacking behind considerably. This is an interesting contradiction to the highly developed citizen services created under the e-Estonia brand and fast adoption phase of moving towards e-citizenship modes (such as e-resident service). The fast development track on one field of e-society therefore does not automatically correlate with innovation statistics as the slow development phase in Estonia. Another interesting issue in the Estonian case is that the country has very low application rate figures compared to the EU average, and if compared to Nordic countries Estonia sends roughly 10–20 times less patent applications per population (in million). However, the growth rate has been the highest between the data period (total difference between 2003 (8.26 applications) and 2014 (18.42) is 122%.

In the case of Denmark, the main observable issue is that it has produced patent applications with significantly lower rate compared to Sweden and Finland. The 11-year growth has been around 18% whereas the other two Nordic countries have experienced growth rates of 37% (Finland) and 52% (Sweden). This is interesting as in 2003 all three countries had almost identical application numbers. The 11-year observation period clearly shows that Finland and Sweden have continued their innovation drive more strongly than Denmark. In addition, Sweden top-ups the chart with more than 350 applications per million persons that is about three times more than the EU average.

The comparison between the US and EU is also relevant as they illustrate a broader spatial scale on which the case countries can be framed on. Two main trends are identifiable: first the patent application intensity has been relatively steady in both cases and in the US the actual number of application has decreased, even though the drop is very small (about 2% negative growth). In addition, if the Euro-zone is compared to EU-28 the application growth has been limited. One explanative factor

Table 7.3 Patents per million people (Eurostat 2018)

	Average 2003–2014	Difference 2003–2014	Growth 2003–2014 (%)	±EU average
EU-28	113.47	4.00	3.70	0.00
Euro-zone	140.27	0.23	0.17	26.79
Denmark	230.16	37.79	18.23	116.68
Estonia	18.63	10.13	122.20	−94.84
Finland	269.84	92.79	37.28	156.36
Sweden	296.65	120.4	52.35	183.18
United States	111.22	−2.44	−2.06	−2.25

7.5 Comparison of the Best Practice Cases

is that countries with relatively high innovation activity (e.g., Denmark and Sweden) belong to the EU-28 group but not in the Euro-zone. The stagnant economic growth of EU in the late 2008 till 2014 is also visible in the annual variations.

To conclude, Tables 7.1, 7.2 and 7.3 are summarising the differences between the case locations. Even though the statistics are available only on national level they may be considered to represent the national context that on the other hand is represented by the most important cities of these countries. Considering the small total populations of the cases (in the global context), these Northern European locations are relatively similar in respect to their regional economies, even though there are exceptions such as Northern regions in Sweden and North-Eastern regions in Finland that experience higher unemployment rates and lower value-adding figures per population (Inkinen 2005; Inkinen and Kaakinen 2016). These background issues have also an impact on the geography of patenting activity.

Next it is necessary to look at patents from sectors point-of-view. Table 7.4 presented relative growth figures of patent applications from all sectors. Eurostat has also produced more detailed statistics on patents concerning particularly 'high-technology' applications on regional (NUTS 2) level. The time series is shorter and the latest data year is 2012. The figures in Table 7.4 are per million inhabitants. Additional good measurement is achieved when the total amount of patent applications is contrasted with the high-technology sector (Table 7.4).

Table 7.5 presents high-technology patent application trends. Helsinki region has experienced to most significant downfall during the observation period as the crude difference between 2008 and 2012 is almost 60 applications less per million inhabitants. The drop is the largest among the cases and also the standard deviation is the highest in the case of Helsinki. A significant explanative factor behind these figures is the decline of Nokia and its dominance in mobile phone manufacturing that started during the observation period. The outcome of that decline was that the whole mobile phone manufacturing segment was sold to Microsoft in 2014 and the production was totally ended in the late 2016. The closure of the production impacted extensively on subcontracting network and smaller companies producing solutions

Table 7.4 High-tech patent applications (per million) to EPO (Eurostat 2018)

	2008	2009	2010	2011	2012	Average	Growth 2008–2012 (%)	Std. Dev.
Denmark	43.96	38.14	40.34	47.16	31.73	40.26	−12.24	5.89
Copenhagen	97.33	84.50	92.13	104.69	66.75	89.08	−30.59	14.50
Estonia	12.66	11.02	9.85	10.06	3.52	9.42	−9.14	3.48
Finland	83.26	73.80	78.59	70.27	66.43	74.47	−16.83	6.65
Helsinki-Uusimaa	168.11	144.62	150.43	123.94	110.82	139.58	−57.28	22.52
Sweden	82.43	72.89	75.00	76.24	71.65	75.64	−10.78	4.19
Stockholm	185.07	150.64	161.48	154.01	142.02	158.64	−43.05	16.34

for Nokia/Microsoft and majority of the patenting was done in Helsinki region where the company headquarters were located.

Unfortunately, the time series ends in 2012 and more recent figures are not available. However, there are clear fluctuations and differences in the application amounts and interestingly year 2012 was the most modest year in terms of application numbers. One explanative reason for declining figures may be found in the overall economic condition of the EU and the financial crises that started from the US real estate market in 2008. It is reasonable to expect that the economic turbulences have had an impact on the demand and production of innovative products during this era. Still, the significant drop in all cases locations from 2011 to 2012 is difficult to be explained. Provided standard deviation figures show that the changes have been most drastic in Helsinki region in Finland. Generally, large variations concern particularly regional scale (capitals) as national statistics experience considerably more stable development.

Tables 7.4 and 7.5 provide a starting point for comparing the share of high-technology patents in relation to all applications. Table 7.5 presents explicitly these shares and the highest relative amount of Estonian applications comes from the high-tech sector exceeding 50%. However, the relative amounts of applications per population are about one to 15 in Estonia compared to Nordic countries. The relative figures show that the Estonian capability of producing patent applications is very limited among the cases but the small amount of applications on the other hand is more focused on the high-technology sector. This is quite understandable, because Estonia lacks large international companies present in the Nordic countries and in several cases high-technology patenting does not require extensive production facilities or structures needed in some other industries, e.g., in construction. This is the case in software designs and programming. Overall, the differences are significant. Finland and Sweden experience rather similar ratios of high-tech applications ending up in an approximately a quarter of all patents. Denmark has the smallest share of high-tech applications (17.49%) that may be considered a relatively low figure. The results give an insight of the diversity of national innovation systems and to the diversity of innovative economy as a whole. Danish innovations are more broadly distributed. The diversity may also have an impact on the regional/national resilience capability as it is widely appreciated that a more diverse and broad industrial profile helps locations to overcome negative shocks in economy.

All case cities highlight environment, sustainability and green growth in their smart city strategies. As presented in the local strategies, clean environment and waste

Table 7.5 Ratio of high-technology patent applications to all patent applications (Eurostat 2018)

	Average (high-tech)	Average (all)	% of high-tech
Denmark	40.26	230.16	17.49
Estonia	9.42	18.63	50.56
Finland	74.47	269.84	27.60
Sweden	75.64	296.65	25.50

management innovations have been extensively present, particularly in the cases of Nordic capitals. Innovations concerning recycling and secondary raw materials in significant topic as environmental innovations have gained a considerable amount of interest and importance particularly in the EU where environmental regulations have become stricter in accordance with global climate agreements. References of the EU and the USA are considerably higher level in environmental patenting than in the case countries. This is also connected to the economies of scale as Nordic economies are small in their absolute sizes. It is logical to consider environmental innovations separately in the strata of innovation activities. Therefore, patent application analysis is concluded here with environmental perspective as Eurostat has produced relevant statistics. They are considered to be adequate proxies to illustrate the actualisation of strategic phrases presented in the smart city strategies. Table 7.6 presents the patent application indicators on national level from 2000 till 2013. These patent applications concern recycling and secondary raw materials. United States is added as a reference together with the EU-28 average.

Considering the strong emphasis on green-tech innovations in city strategies the relative figures in Table 7.6 are surprisingly low. Finland has had the lead position since the first observation data (2010) among the cases. The highest figure from the cases is the Finnish 2013 result (14.42 patents per million inhabitants). However, when compared to patent applications on the field of ICTs these numbers are a total reversal. On average, European Union as a whole has produced ten times more applications than the Nordic leader. On the other side, the numbers of applications are almost non-existing in Estonia during the whole 13-year observation period. In the middle, Denmark and Sweden have relatively similar figures, even though Danish figures have more variation.

This result further gives an indication that environmental patenting requires large economic contexts. Another significant result is that patenting has continued to grow in EU-28 extensively whereas in the US the amounts have remained largely the same or even declined. The results are puzzling as recycling awareness has increased extensively in Nordic countries during the last 30 years and in some cases Nordic countries have been considered to be forerunners also in recycling development. According to

Table 7.6 Patents on recycling and secondary raw materials per million people (Eurostat 2018)

	2000	2005	2010	2011	2012	2013	Average	Growth 2000–2013 (%)	Std. Dev.
EU-28	270.26	254.52	284.93	305.19	346.84	363.78	271.79	93.52	42.85
Denmark	2.75	5.51	0.95	1	3.39	8.17	3.78	5.42	2.21
Estonia	0	0	0.17	0	0.38	0	0.18	0	0.36
Finland	5.27	6.99	10.47	6.87	10.11	14.42	7.91	9.15	2.69
Sweden	3.53	3.3	2.12	6.21	4.32	7.75	3.50	4.22	2.05
United States	184.39	155.81	107.92	92.82	189.88	183.9	132.23	−0.49	36.82

Table 7.7 R&D expenditure (EUR per people) in NUTS 2 regions (Eurostat 2018)

	2011	2012	2013	2014	2015	Average	Std. dev.	Growth 2011–2015 (%)
EU-28	516.7	535.6	542.4	562.3	591.1	549.6	28.35	12.59
Euro-zone	598	615.6	621.6	640.4	660.7	627.3	24.06	9.49
Denmark	1312.7	1360	1371.8	1376.2	1420.1	1368.2	38.46	7.56
Copenhagen	2722	2826.4	2748.3	2790.4	2847	2786.8	52.17	4.39
Estonia (Tallinn)	289.1	287.3	247	217.9	230.3	254.3	32.61	−25.53
Finland	1332.7	1264.9	1231.7	1194.6	1109.5	1226.7	82.90	−20.12
Helsinki-Uusimaa	2024.7	1912.3	1925.5	n/a	1823.8	1921.6	82.25	−11.02
Sweden	1397.4	1464.9	1507.6	1411.3	1504.3	1457.1	51.24	7.11
Stockholm	2139.8	n/a	2465.6	n/a	2465.1	2356.8	187.96	13.20

the Eurostat patenting data this is not the case in terms of innovativeness. It is clear that further studies on environmental and recycling innovations are needed, particularly in urban context where new solutions for recycling and waste management are mostly needed.

Macroeconomic statistics provide foundations for deepening the assessment towards specific economic variables. One of the most widely used innovation input indicators has been R&D expenditure. There are numerous benefits and one of the most important ones is that there is a solid statistical tradition of collecting these numbers starting already in the 1980s in the most advanced countries, including Finland. R&D expenditure data is also available according to private, public, and university sectors. This creates a possibility to evaluate the importance of each sector within economies. In other words, some regional economies are more reliant on public sector R&D whereas some other regions are market driven business R&D concentrations. In addition, R&D expenditure has been applied in numerous regional and urban studies to indicate that growth centres usually are more prone to R&D efforts and they also benefit more of the R&D investments. Table 7.7 highlights the national statistics for case locations and descriptive statistics.

Table 7.7 shows some key-properties of innovative regions in Northern Europe. First, the three Nordic countries and their capital regions are investing considerably more on R&D than the European averages (both Euro-zone and EU-28). Estonia is exception among the cases at it has experienced R&D expenditure levels that are about a half compared to EU averages. The most intensive R&D area is Copenhagen where per population expenditure is approximately 10 times more than the lowest figures observable from Estonia (2015).

Another important difference among the case locations is that all the other nations and regions have had an increasing trend in expenditure figures except Finland, Helsinki and Estonia. The difficult macro-economic current in Finnish (and Euro-

7.5 Comparison of the Best Practice Cases

Table 7.8 Ratio of R&D personnel and researchers per total employment (Eurostat 2018)

	2007	2010	2014	2015	2016	Average	Growth 2007–2016 (%)	Std. Dev.
EU-28	1.094	1.199	1.305	1.336	1.335	1.234	0.241	0.087
Euro-zone	1.185	1.313	1.422	1.456	1.449	1.346	0.264	0.095
Denmark	1.700	2.134	2.211	2.223	2.194	2.118	0.494	0.158
Estonia	0.792	0.963	0.965	0.919	0.943	0.929	0.152	0.072
Finland	2.287	2.319	2.185	2.127	1.993	2.217	−0.294	0.099
Sweden	1.691	1.758	1.816	1.793	1.915	1.783	0.224	0.058
Japan	1.419	1.394	n/a	n/a	n/a	1.385[a]	−0.048[a]	0.020[a]
South Korea	1.150	1.407	n/a	n/a	n/a	1.402[a]	0.452[a]	0.175[a]

[a]Calculated from the latest year figure of 2013

zone) economy has caused cuts particularly in public sector R&D investments. The negative growth figure 2011–2015 indicates that Finland and Estonia have had more difficult road compared to Danish and Swedish examples. Interestingly, the economic down-turn in Europe has not impacted the growth figures of EU-28 that has experienced a growth of 12.6%.

Table 7.8 shows the consistency of R&D expenditure as the standard deviations are relatively small, expect in the cases of Stockholm and Helsinki. These two are, however, opposing as Stockholm has continues positive growth and Helsinki, on the other hand, has a negative trend line. The same applies also to the national figures of Finland that has experienced the second greatest drop (20.1% units) after Estonia (25.5% units). The decreasing figures particularly from Finland are also related to structural changes in ICT industry and the diminishing of Nokia that has been the most intensive R&D investing company together with the subcontracting network that it has supported in Finland.

In addition to R&D expenditure the other common indicator of knowledge-intensity in economy is the employment impact. Table 7.8 presents a 10-year observation period of the relative number of researchers and personnel employed in R&D tasks. The figures concern all sectors (private, public and universities) and they are presented in relation to the total employment figures. This leads to situation where these relative values are very small. The trend figures give indications of the significance of R&D in the respective countries and the growth figures are also connectable to industrial and economic policies implemented in each country.

Table 7.8 has four reference figures (EU-28, Euro-zone, Japan and South Korea). There statistics from the USA are limited and therefore it was left out. The latest statistics from Japan and South-Korea are available till 2013 so their time-series is 8-years long. The highest average employment levels of R&D and research are found in Finland and Denmark that have averages above 2%. The lowest figure, on the other hand, is observed in the case of Estonia (0.9%). Estonian figures are also the most consistent ones and there has been a minimal growth in the significance of employment portion. Estonian growth figures are also smaller than the EU28 average indicating that the R&D fields are not as significant employers there as is the case in other case countries.

The only negative 2007 and 2016 growth figures are found in Finland and Japan. In Finnish case, the final observation (2016) shows a 0.3% unit drop and it is probably connected to government decisions to cut down R&D funding both in research centres and in Universities. In the final figures Sweden has almost caught up the employment levels of Finland. In the case of Japan, the negative growth has been very small and may be connected to stagnant condition of Japanese economy since the 1990s. On the other side, Denmark has the highest growth rate (almost 0.5%) from 2007 till 2016 indicating the increasing significance of research and development in Danish society. The figures in a reference country South-Korea are also indicating clear growth tendency during the observation period. Korean economy is a significant producer of IT technology and the significance of large multinational corporations such as Samsung and LG are probable influencers in this development.

If R&D employment in the case countries is compared to the EU28 and Eurozone references the amounts some 50–70% higher. This indicator also shows that the Nordic countries have had a significant input and expenditure on research and development that is one of the key-doctrines in the national and regional development policies in these countries (e.g., Makkonen and Inkinen 2013). For example, in Finland the set goals and tools to obtain these goals concerning regional and national growth have been strongly associated with the highest education, R&D and innovation. Studies focusing on innovation policies have presented that the rhetoric of innovation, industrial and general economic policies are converging, thus they apply similar concepts and language options. R&D and innovation are at the heart of this policy document rhetoric (also Makkonen and Iniken 2015).

In terms of knowledge-intensive employment it is worthy to look at the employment levels of technology and knowledge-intensive sectors (as defined by Eurostat) and their relative importance in terms of population from 2012 to 2016 (Table 7.9). Population statistics are available from the same period and the relative figures are calculated based on the Eurostat population statistics.

The EU-28 reference shows that an average about little less than 44% employment is currently directed to technology and knowledge-intensive occupations. The overall growth has been slightly increasing. In general, the amount of knowledge intensive employment is approximately 45% of total employment in Europe. The case locations may be considered as forerunners in knowledge-based development and they are approximately on 5–10% higher level than the EU-28. However, there are identifiable differences. The case countries and locations are all above the EU-28 level and the highest portion is observed in Stockholm (53.68% in 2016). The lowest portion among the cases is Finland's national average 2016 that is slightly (less than 1%) above the EU level.

The most significant finding in Table 7.10 is that employment levels in these sectors have been steadily declining in Finland and Helsinki region. The time-period covers the years that have been economically difficult in Finland. Another explanation to this is that Microsoft ended its mobile phone manufacturing during the period and

7.5 Comparison of the Best Practice Cases

Table 7.9 Employment in technology and knowledge-intensive sectors (%) (Eurostat 2018)

	2012 (%)	2013 (%)	2014 (%)	2015 (%)	2016 (%)	Average (%)	Growth 2012–2016 (%)
EU-28	42.82	42.66	42.96	43.32	43.82	43.12	1.00
Denmark	48.18	47.97	48.23	48.62	49.76	48.55	1.58
Copenhagen	50.43	50.64	50.86	51.42	53.22	51.31	2.79
Estonia	46.40	47.06	47.48	48.74	48.98	47.73	2.58
Finland	45.97	45.27	44.89	44.53	44.61	45.06	−1.36
Helsinki-Uusimaa	51.52	50.81	50.22	49.86	49.84	50.45	−1.68
Sweden	49.11	49.23	49.48	49.62	49.84	49.46	0.73
Stockholm	52.88	53.29	53.24	53.43	53.68	53.30	0.80

that impacted to some extent the whole industry field in Finland. Helsinki region itself still has a similar five-year average to other Nordic capitals but it is clear that there will be challenges in efforts to change the direction of declining development. The gap between Helsinki and Copenhagen, which has experienced the most extensive growth (2.79%) during the period, is almost 5% in growth change. This gap is significant and indicates that the maintenance of a steady growth is not automatic if the economy is experiencing negative shocks such is the case in Finland.

An interesting comparison is to look at the results of high-tech employment and tertiary education as the relative figures per population are very similar. This also indicates the joint-need of highly educated people and the number of employment

Table 7.10 Ratio of tertiary educated and employed science and technology workforce to all (Eurostat 2018)

	2012	2014	2016	Average	Growth 2012–2016 (%)	Difference to EU-28 average
EU-28	30.5	31.8	33.2	31.8	2.7	0.0
Euro-zone	30.6	31.7	32.9	31.7	2.3	−0.1
Denmark	36.9	38.0	39.1	38.0	2.2	6.2
Copenhagen	47.1	48.1	49.6	48.2	2.5	16.4
Estonia	37.9	38.9	39.9	38.9	2.0	7.1
Finland	38.7	40.4	41.6	40.2	2.9	8.4
Helsinki-Uusimaa	50.0	50.8	51.9	50.9	1.9	19.1
Sweden	39.3	41.8	44.3	41.7	5.0	9.9
Stockholm	50.0	52.8	55.0	52.6	5.0	20.8
Sweden	48.2	50.7	53.7	50.8	5.5	8.3
Stockholm	59.1	62.0	64.2	61.6	5.1	19.1

positions requiring such skills that may be expected obtainable from university education. A number of studies in economic geography (e.g., Marrocu and Pasi 2012; Makkonen and Inkinen 2013, 2015) have addressed this connection and the correlations between education variables and knowledge-intensity measures are constantly above 90%. These results indicate the intertwining of education and innovation statistics. Employment statistics offer also an interesting opening when looking at the combined figures of tertiary education degree holders and co-existing presence of science and technology employment. For these purposes, Tables 7.10 and 7.11 also provide a comparison between EU-28 reference averages and cases.

Regional knowledge-intensiveness and maturity may perhaps be seen the best in the combined education and employment levels. There are numerous observations to be made from the statistics. Firstly, if total population and active population figures are compared the difference is approximately 10% (higher figures for active population). The most significant indication that the differences between total and active figures provide is about the varying demographic structures of difference countries. However, the differences are small (less than 2% between the 10% level). It is reasonable to focus on the statistics of active population in the following.

Growth rates indicate modest but constant growth within the 5-year period. Denmark and Copenhagen together with Estonia have experienced slower growth than the EU-28. Finland and Helsinki have been around in the similar growth phase with the EU benchmarks. Sweden (5.5%) and Stockholm region (5.1%) have experienced the fastest growth. Difference to EU-28 and Euro-zone averages bring clearly visible the agglomeration of knowledge-intensiveness to capital regions. All Nordic cases show that capital areas are more than 10% higher level than their national averages. Stockholm has almost 20% difference to EU-28 level closely followed by the Helsinki region. The cases of Helsinki and Stockholm also experience the highest percentages in the year 2016 when almost two out of third employees belonged into the knowledge-intensive category either because of education or profession.

The empirical focus moves now from economic and labour market indicators towards the actual use and aspects of 'information society' or 'digital urbanism'. As

Table 7.11 Ratio of tertiary educated and employed workforce to all (Eurostat 2018)

	2012	2014	2016	Average	Growth 2012–2016 (%)	Difference to EU-28 average
EU-28	40.9	42.5	44.1	42.5	3.2	0.0
Euro-zone	41.2	42.5	43.9	42.5	2.7	0.0
Denmark	46.6	47.8	48.1	47.6	1.5	5.1
Copenhagen	57.5	58.4	58.5	58.2	1.0	15.7
Estonia	46.8	47.2	47.4	47.2	0.6	4.7
Finland	48.6	50.5	51.7	50.3	3.1	7.8
Helsinki-Uusimaa	59.8	60.6	62.1	60.8	2.3	18.3

7.5 Comparison of the Best Practice Cases

discussed in literature, e-government is an important field of smart city development and smart government initiatives (e.g., Borins 1998; Reddick 2010; Weerakkody and Reddick 2012; Rodríguez Bolívar 2018). Table 7.12 indicates the development of online (internet) interaction with public authorities in percentages of individuals from 2008 till 2017 (the latest) on the national level and since 2012 also regional statistics of the capital regions. Estonia as a country belongs to one EU-level regional category (NUTS 3) and therefore capital region statistics are not available and the numbers concern the whole country. However, the proportion of Tallinn population of the whole Estonia is more than a quarter and therefore it may be reasonably expected that the figure is suitable for this use as the other regions from Nordic countries also include extensive amount of people living outside the FUAs of capital areas.

The commitment in case countries towards applying e-government tools has been present for a long time. This is an example of the initiators advantage. Table 7.12 presents that all four cases have experienced a steady and, almost, continuous growth every year. The highest figures for the current (2017) government interactions are 91% from Copenhagen and 90% from Helsinki region (Uusimaa). In comparison, the lowest is the national statistic of Estonia of 78%.

The statistics also indicate that the maturing time for e-government took place in Northern Europe between 2008 and 2013. The starting figures were generally below 60% of population and at they had increased to approximately 80 until 85% at the end of the period. After 2013 the growth has been slow as the technology adoption rates become saturated when the 85–90% penetrations are reached. This phenomenon is independent from the technologies as the rule applies to televisions, mobile phones and all extensively used home technologies. Estonia and Tallinn experienced the most significant increase between 2013 and 2014 when the adoption rates increased from 51 to 81%. After the gigantic leap the relative use of e-government has decreased some percent units. Based on the available statistics it seems that e-government use penetrations have reached their potentials in the study countries and regions. In terms of national and regional statistics it is observable that all cases (except Estonia) that the capital regions are running ahead of the national level with a marginal of 2–10%.

Table 7.12 Ratio of internet interaction of individuals with public authorities (Eurostat 2018)

	2008	2009	2010	2011	2012	2013	2014	2015	2016	2017
Denmark	49	73	78	81	83	85	84	88	88	89
Copenhagen	n/a	n/a	n/a	85	87	88	89	88	90	91
Estonia (Tallinn)	37	46	50	53	54	48	51	81	77	78
Finland	62	64	68	68	70	69	80	79	82	83
Helsinki-Uusimaa	n/a	n/a	n/a	n/a	77	80	86	86	88	90
Sweden	59	65	68	74	78	78	81	73	78	84
Stockholm	n/a	n/a	n/a	80	77	81	84	78	86	88

This result also indicates that technological adoption is commonly faster in capital cities than in other parts of the countries.

One of the widely used changes associated with digitalisation and information society development in accordance with the growth of e-government is e-commerce. It is an interesting study topic as it has deep implications for international trade and logistics. In the following some key-indicators concerning online purchasing and shopping are presented (Table 7.13). These figures are supported by the business statistics showing the significance of online trade in the case countries and regions.

The EU average is calculated as an average from all regions listed in Eurostat database and it provides again a suitable reference to compare cases. Table 7.13 clearly illustrates that the greatest variations in online purchasing are found in Estonia that has significantly higher standard deviation figure compared to the others. Moreover, the relative growth has been significant (35% units) considering the short 5-year time-span. However, Estonian 2017 online purchasing activity is matching the EU average and is some 30% units below other cases. EU average has also grown steadily (11%) with the second largest positive variation. The highest online purchasing activity of households is in Copenhagen region reaching up to 84%. Comparison to the lowest figure (Estonia in 2013) of 23% indicates the rapid changes in shopping and ordering practices. Nordic countries have had a high percentage already in the 2013 and the growth levels have been understandably small but continuous.

The results show that the saturation point of e-commerce activity lies somewhere close to 80% level but it may be estimated that figures may grow close to 90% level as other technology adoption indicators have shown (e.g., television, mobile phones and computers in households). An interesting comparison is to look at significance of value-adding of e-commerce sales for companies. Eurostat does not provide regional statistics on this but national averages give an indication applicable also to capital regions. In general, other indicators have shown that all Nordic capitals are some 5–10% units above national averages and it is reasonable to expect this to be the case also with the figures of Table 7.14.

Table 7.13 People (%) having done online purchase during the last 12 months (Eurostat 2018)

	2013	2014	2015	2016	2017	Average	Growth 2013–2017 (%)	Stn. Dev.
EU average	42	47	48	50	53	48	11	4.06
Denmark	77	78	79	82	80	79	3	1.92
Copenhagen	83	84	82	84	84	83	1	0.89
Estonia	23	49	59	56	58	49	35	1.05
Finland	65	68	69	67	71	68	6	2.24
Helsinki-Uusimaa	73	73	74	75	78	75	5	2.07
Sweden	73	75	71	76	81	75	8	3.77
Stockholm	77	79	76	80	80	78	3	1.82

7.5 Comparison of the Best Practice Cases

Table 7.14 indicates that the growth rates have been quite modest in the 7-year period. Considering the fast development phase of applications and e-commerce platforms the value of sales has not increased extensively. There are some positive growth trends observable but in general the annual changes have been small. The highest growth has been in Denmark (6% units) and it's closely followed by Estonia (5% units). However, these two countries also represent the highest portion and the lowest one in total sales. The seven-year average indicates that all cases except Estonia are on approximately 20% level. Estonia has however gained stronger growth following the EU level and is some 5% units behind other case locations.

An interesting observation is that there are negative annual fluctuations. In the case of Sweden, the annual drop from 2013 to 2014 is 6% units that did not recover until 2016 but again there is a drop in 2017. These fluctuations may be understood through the global competition of e-commerce. In general, households buy more and more products online but they do not necessarily impact national companies as the e-commerce is strongly influenced by the presence of transnational corporations such as Amazon (the US) and Alibaba (China). E-commerce have significantly altered the customer possibilities to purchase products abroad and it has impacted also customs and the amount of parcel transport. The increasing amount of parcel transports from international e-commerce companies has already caused policy changes in Nordic countries concerning taxation and customs. For example, Finnish government is planning to establish product taxation rules for small product orders from other than EU countries. This is similar development that took place earlier in Sweden. The main supplier of these small product orders is China.

Technological integration into the business processes is one good way of looking at digitalisation in corporations. Eurostat provides only country level data on this topic but again it may be used to illustrate national developments. Similarly, it is expected that companies locating within capital regions have higher proportions compared to these national statistics. One interesting and relatively unapplied indicator is the adoption of ERP (enterprise resource planning) and CRM (customer relations management) tools such as SAP in information sharing within companies (Table 7.15).

The provided time period of Table 7.15 proves constant positive trends in all case countries and EU references (EU-28 and Euro-zone) except in the case of Sweden.

Table 7.14 Ratio of value of e-commerce sales to all sales (Eurostat 2018)

	2010	2011	2012	2013	2014	2015	2016	2017	Average	Growth 2010–2017 (%)
EU-28	14	14	15	14	15	17	16	18	15	4
Denmark	17	n/a	n/a	n/a	17	20	23	23	20	6
Estonia	11	10	13	12	14	13	15	16	13	5
Finland	18	20	18	16	19	21	n/a	21	19	3
Sweden	18	19	20	14	18	19	21	19	19	1

Approximately above a third of the Nordic companies apply ERP tools and information sharing in their operations whereas the EU average is about a quarter. An interesting observation in the data is 2010 when Sweden was significantly ahead of others with a share of 35%. There might have been data acquisition problems in Sweden since 2015 the information is not available and the latest 2017 figure (31%) drops 14% units compared to 2013 figure (45%). The highest percentage is again in Denmark (47%) in the year 2015. Generally, all the countries except Estonia experience drop of 3–6% units between 2015 and 2017 measurements. This is probably related to the data collection variations. However, this is not discussed or referred in the Eurostat meta-data information.

Standard deviation figures, signalling about the variance of trends, are relatively close in all cases. The most significant growth is observed in Estonia, which had a very low (only 7%) figure of company ERP tool adoption in 2010. In this respect the fastest growth (21% units) is not surprising, but there is still a 10%-unit gap to the Nordic countries. Estonia is also below EU references and it is likely that the ERP and CRM adoption will continue in Estonia in the near future. The ERP system indicator receives high portions also in the Euro-zone (39%) and only Denmark is above the reference figure in 2017 (40%). This indicates that companies in other Euro countries are well equipped in terms of digital corporate management tools.

Table 7.15 results are important also, because 'analytical CRM' has been one of the fast-developing fields within marketing. This refers to the use of Big Data and data mining in customer segmentation. The return of marketing investment (ROMI) modelling is dependent on corporate information collection and cataloguing capabilities of customer databases. It is likely that in the future marketing campaigns will become continuously more individual driven and personal. Consumer identification for the marketing needs has already attracted significant amount of discussion concerning privacy and data security issues.

ERP systems are the first stepping stone in corporate management and planning. They are establishing the road for other branches to take advantage of digital imprints that may be used in operational development such as marketing. However, as

Table 7.15 Ratio of enterprises who have ERP software package to share information (Eurostat 2018)

	2010	2012	2013	2014	2015	2017	Average	Growth 2010–2017 (%)	Std. Dev.
EU-28	21	22	26	31	36	34	28	13	6.28
Euro-zone	25	26	31	36	43	39	33	14	7.23
Denmark	29	33	33	42	47	40	37	11	6.77
Estonia	7	10	15	17	22	28	17	21	7.71
Finland	28	33	37	39	37	39	36	11	4.28
Sweden	35	38	45	43	n/a	31	38	−4	5.73

Table 10.1 results imply, the proportions of companies making benefit of data management are still minority. This implies that only a small fraction of companies may take benefit of the potentials provided by data sourcing in the near future. The second implication is that there are still significant potentials for data technology adoption. It is likely that the adoption figures will rise. However, a major problem has been for a long time that ERP systems and CRM tools are expensive and therefore they are mainly used by large companies.

7.6 Summary

The case locations reveal that similar efforts are being conducted in each location but there are significant variations in the legislation and costs related to entrepreneurship and businesses. Besides, the levels of basic security provided by the governments are significantly different. The cases of Denmark, Finland and Sweden are roughly similar but Estonia provides an interesting case with a different historical background in the 1900s. After regaining independence since the collapse of the Soviet Union in the 1990 Estonia has been able to redesign its entrepreneurial policies and development goals. Estonian salaries are lower together with more flexible regulation of firms including, for example, firm taxation and law-required costs.

Presented statistics are interesting as in majority of cases the growth has been significant and constant during the varying observation periods. However, there are also some indicators that have experienced very slow growth figures or fluctuating condition. These cases may be understood through difficult economic conditions of the periods as majority of the statistics concern years between 2008 and 2015 that may be characterised as slow economic years in the EU. There are several indicators that still have relatively small proportions (%) and perhaps the clearest example is the value of e-commerce sales that averages around 20% in the case countries indicating potentials for growth.

All the cases highlight extensively collaboration and openness in their development. The most widely used collaboration model is triple-helix or its amended version quadruple-helix that includes also citizen participation. Based on the examination it is clear that societal and public-sector roles in the enablement of innovation systems are results of long development paths and path-dependencies. Statistical evidences show similarities among the cases. The first is that studied capital cities and regions are performing on approximately 10% higher levels in innovation activities if compared to their host nations. The second is that the overall economic development is reflected to the innovation indicators particularly in the case of Helsinki and Finland as in several occasions the overall growth in innovation indicators has been negative. All the other cases have a small but observable growth trends. Third, Estonia differs from the Nordic cases in terms of the absolute activity volumes. However, the low starting levels cause that the relative growth is, in majority of indicators, the highest in Estonia.

The presented statistical information on country and region level confirms the high adoption of digitalisation and knowledge-intensiveness in the case locations. The Nordic capitals Copenhagen, Helsinki and Stockholm have very similar statistics regardless of the single indicators applied. They obtain figures that are approximately some 20–30% units higher than the EU references (either EU-28 or Euro-zone). Estonia is more often close to the EU averages and commonly has lower technology adoption rates, particularly in the data concerning private sector enterprises.

One of the most significant topics in the consideration of social media and innovations is the way how local knowledge and citizen information may be integrated into processes of urban planning. The technological development also enables tools to highlight and strengthen cities own historical trajectories and bring forth each location's advantages. This will help locations to preserve and enhance their vitality and attractiveness in the pursuit of investments and skilled knowledge-based labour force. At the same time, strong national growth centres have to think and consider that how they are able to adapt themselves, with as little damage as possible, to the sharp economic fluctuations. This is a question of resilience and economies capability to cover from external shocks.

On the firm level the question of innovation and smart growth boils down to the employees and employer capability of making the best of the educated and innovative workforce. Innovation and creativity is something that is not obtained through education fundamentally but relevant highest education supports innovation activity strongly. This is an interesting notion as commonly innovation is assessed in social sciences according to statistical categories and classifications. However, if a business is classified as a biotechnology or an ICT company in NACE (or SIC) classification, it tells little of the innovativeness of the company automatically. This applies to industrial categories that are usually associated with innovation, knowledge-intensity, and creativity (Bassett-Jones 2005; Stolarick and Florida 2005). Particularly the use of total employees as an innovation indicator is misleading as there is no great difference is logistics person of an ICT company transfers mobile phones around compared to e.g., similar task in forestry and transporting wood. However, on larger spatial scales these distinctions become essential in the understanding of an innovative location.

Ideally, the innovation studies separate the work content from the work process. Both, of course, may be enhanced by innovations (either radical or incremental). This means that work tasks (contents) that require repetition, and that are also the most easily automated, does not involve innovative thinking as such, but if the worker creates a new way of processing the repetition that may be considered as incremental improvement (innovation) in the process itself. This distinction is connected to widely used categorisation of product and process innovations and to the separation of what is being done and how it's being done. These almost always incremental actions improving efficiency and work content are very difficult to be captured into data form. This is also the reason why qualitative methods continue to have important role in the assessment of successful and innovative companies.

The employees work in different types of corporate cultures and operative environments. There are great differences in corporate mind-sets towards innovation and development. Particularly traditional industry fields, such as metal, maritime,

7.6 Summary

and engineering industries, have been in many cases very conservative towards new practices. Digitalisation, however, are impacting also these fields, and particular example may be found in maritime industries that has experienced a rapid change in its internal attitude towards environmental innovations (e.g., scrubber technologies in air exhaust cleaning) and potentials provided by digitalisation in navigation and manoeuvring. These changes have been so extensive that they may be called as radical innovations causing disruptions to the industry as a whole.

Considering practical aspects of an innovative (or smart or knowledge) city it is evident that they manifest themselves through office-spaces and architectures designed for businesses. Business parks, knowledge milieus, science parks, knowledge precincts and innovative districts are common terms that are used to describe these planned entities within urban space. The distinction that whether these innovation locations are mere real-estate project or whether they pose or provide other supportive elements making them hosts of successful businesses is a difficult task.

It is observed that internal structuring of innovation parks and its significance to the economic results is case dependent. In some cases, diverse industry-mix within the location support synergies and economies of scope, and in some cases, there are examples of highly successful locations focusing only on specific industries. The industrial mix has been a great interest in the studies of economic geography, particularly concerning debates on regional resilience and regional capability to recover from sudden economic turbulences. The presented cases are capital cities. That automatically creates employment profile diversification due to an extensive presence of public sector offices and bureaus employing a significant number of government and administrative officials. This is also reflected to the amount of services and presence of supportive institutions (e.g., Borins 2000, 2001).

Innovation is relatively easy to conceptualise according to statics, and particularly concerning innovation inputs (e.g., applied patents, granted patents, R&D investments, R&D personnel, portion of tertiary education, and so on). Innovation outputs are more difficult to define and obtain data. This usually requires the use of survey methods and direct asking from the companies themselves. Another ever existing question is that how companies could benefit from innovative ideas of their employees. Thus, the innovative city is highly dependent on the perspective through which it is looked at. The easiest way is to consider technology aspect and new technological solutions or services for different target groups. The most well-known examples are services for tourists, visitors or residents. The use of LBSs and map-services (such as maps.google.com) are widely applied together with social-media platforms (e.g., TripAdvisor).

Thinking about the network concepts and collaboration taking place among different sets of organisations (as actors within the network), the significance of socially constructed confidence (or trust) networks is essential. These networks require a relatively long time to form and they are quite enduring and robust. The significance of human interaction and relationships are difficult study subjects because it intertwines personal relationships to business relationships. Particularly in smaller countries and fields it is common that key-actors and organisations are familiar and acknowledgeable of each other's presence. They key-personnel on these companies are commonly also aware of the other players.

Arguably, innovative, intelligent and smart cities have become topical concepts in the urban development. It remains important to sharpen and further develop these concepts, particularly, when they are applied as policy guidelines. The following main lessons may be drawn from the studied cases:

- Smart solutions in service and IoT provision will have a significant impact particularly on the fields of urban movement. Telematics and sensor technology integration into the infrastructure together with the development of 'mobility-as-service' (MaaS) continues to be one of the key-development areas in cities.
- Innovations related to clean-tech and environmental sustainability will have continues and increasing importance in the urban development in the near future, particularly in the EU.
- Smart governance that applies digital platforms and applications will continue to be a focal point in smart city development. There will be an increasing potential of clashes between individually targeted data collection and privacy. This concerns also the future development of the social media and copyrights.

These three points are exceedingly present in the studied smart city strategies and they also imply that several of the challenges are not technological but legislative, educational, and environmental. It seems clear that cross-sectional governance development and broad substance-based projects are preferable in comparison to silo-type approach. Even though the need for collaborative action has been recognized for quite some time in innovative urban development the practical implementation of these tasks requires still considerable efforts and coordination.

References

Azadegan, A., & Pai, D. (2008). Industrial awards as manifests of business performance: An empirical assessment. *Journal of Purchasing and Supply Management, 14,* 149–159.
Bassett-Jones, N. (2005). The paradox of diversity management, creativity and innovation. *Creativity and Innovation Management, 14,* 169–175.
Bernier, L., & Hafsi, T. (2007). The changing nature of public entrepreneurship. *Public Administration Review, 67,* 488–503.
Borins, S. (1998). *Innovating with integrity: How local heroes are transforming American government.* Washington, DC: Georgetown University Press.
Borins, S. (2000). What border? Public management innovation in the United States and Canada. *Journal of Policy Analysis and Management, 19,* 46–74.
Borins, S. (2001). Innovation success and failure in public management research: Some methodological reflections. *Public Management Review, 3,* 3–17.
Chari, V. V., Golosov, M., & Tsyvinski, A. (2012). Prizes and patents: Using market signals to provide incentives for innovations. *Journal of Economic Theory, 147,* 781–801.
City of Copenhagen. (2017). *The city of Copenhagen's business and growth policy 2015–2020.* Available online https://international.kk.dk/sites/international.kk.dk/files/uploadedfiles/Business_and_%20Growth_%20Policy.pdf.

References

City of Helsinki. (2017). *The most functional city in the world. Helsinki city strategy 2017–2021*. Available online https://www.hel.fi/static/helsinki/kaupunkistrategia/strategia-en-2017-2021.pdf.

City of Stockholm. (2017). *Strategi för Stockholm som smart och uppkopplad stad*. Available online http://www.stockholm.se/PageFiles/130302/Strategi-for-en-smart-och-uppkoppladstad-Stockholms-stad.pdf.

City of Tallinn. (2013). *Tallinn enterprise and innovation strategy 2014–2018*. Available online http://www.tallinn.ee/eng/investor/Tallinn-Enterprise-and-Innovation-Strategy-2014-2018.

Dodge, M., & Kitchin, R. (2001). *Mapping cyberspace*. London: Routledge.

Eurostat. (2018). Database. Available at https://ec.europa.eu/eurostat/data/database.

Georghiou, L., Smith, K., Toivainen, O., & Ylä-Anttila, P. (2003). *Evaluation of the Finnish innovation support system*. Publications 5. Helsinki: Ministry of Trade and Industry Finland.

Government of Finland. (2006). *A renewing, human-centric and competitive Finland. The national knowledge society strategy 2007–2015*. Helsinki: Prime Minister's Office.

Hisrich, R., Peters, M., & Shepherd, D. (2005). *Entrepreneurship* (6th ed.). New York: McGraw-Hill.

Inkinen, T. (2005). European coherence and regional policy? A Finnish perspective on the observed and reported territorial impacts of EU research and development policies. *European Planning Studies, 13*(7), 1113–1122.

Inkinen, T. (2015). Reflections on the innovative city: Examining three innovative locations in a knowledge bases framework. *Journal of Open Innovation, 1*(8), 1–23.

Inkinen, T., & Kaakinen, I. (2016). Economic geography of knowledge intensive technology clusters: Lessons from the Helsinki metropolitan area. *Journal of Urban Technology, 23*(1), 95–114.

Inkinen, T., & Vaattovaara, M. (2007). *Technology and knowledge-based development. Helsinki metropolitan area as a creative region*. ACRE report 2.5. Amsterdam: University of Amsterdam.

Inkinen, T., & Vaattovaara, M. (2010). Creative urban region in the Nordic country. Combining tradition with development in Helsinki. In K. Metaxiotis, F. J. Carrillo & T. Yigitcanlar (Eds.), *Knowledge-based development of cities and societies: Integrated multi-level approaches* (pp. 196–210). Hershey: IGI Global.

Kiuru, J., & Inkinen, T. (2017). Predicting innovative growth and demand with proximate human capital: A case study of the Helsinki metropolitan area. *Cities, 64*, 9–17.

Kusiak, A. (2007). Innovation: The living laboratory perspective. *Computer-Aided Design and Applications, 4*(6), 863–876.

Laforet, S. (2008). Size, strategic and market orientation effects on innovation. *Journal of Business Research, 61*, 753–764.

Laforet, S. (2009). Effects of size, market and strategic orientation on innovation in non-high-tech manufacturing SMEs. *European Journal of Marketing, 43*, 188–212.

Makkonen, T., & Inkinen, T. (2013). Innovative capacity, educational attainment and economic development in the European Union: Causal relations and geographical variations. *European Planning Studies, 21*(12), 1958–1976.

Makkonen, T., & Inkinen, T. (2014). Innovation quality in knowledge cities: Empirical evidence of innovation award competitions in Finland. *Expert Systems with Applications, 41*(12), 5597–5604.

Makkonen, T., & Inkinen, T. (2015). Geographical and temporal variation of regional development and innovation in Finland. Fennia. *International Journal of Geography, 193*(1), 134–147.

Marrocu, E., & Paci, R. (2012). Education or creativity: What matters most for economic performance? *Economic Geography, 88*(4), 369–401.

OECD. (2018). *Programme for international student assessment*. Available online www.oecd.org/pisa/.

Ramstadt, E. (2009). Expanding innovation system and policy: An organizational perspective. *Policy Studies, 30*, 533–553.

Reddick, C. (Ed.). (2010). Comparative e-government: An examination of e-government adoption across countries. *Integrated Series in Information Systems, 25*. New York: Springer.

Rodríguez Bolívar, M. P. (Ed.). (2018). *Smart technologies for smart governments. Transparency, efficiency and organizational issues*. New York: Springer.

Sotarauta, M., & Kautonen, M. (2007). Co-evolution of the Finnish national and local innovation and science arenas: Towards a dynamic understanding of multi-level governance. *Regional Studies, 41*(8), 1085–1098.

Stolarick, K., & Florida, R. (2005). Creativity, connection and innovation: A study of the linkages in the Montréal region. *Environment and Planning A, 38,* 1799–1817.

Wessner, C. (2009). Government programs to encourage innovation by start-ups & SMEs: The role of innovation awards. In S. Nagaoka, M. Kondo, K. Flamm, & C. Wessner (Eds.), *21st century innovation systems for Japan and the United States: Lessons from a decade of change* (pp. 77–95). Washington, DC: The National Academies Press.

Weerakkody, V., & Reddick, C. G. (Eds.). (2012). Public sector transformation through e-government: Experiences from Europe and North America. London: Routledge.

Chapter 8
Conclusion to Part II

Abstract The studied cases in the previous chapter generate wealth of knowledge and empirical evidence. These case studies also indicate the importance of tradition, strategic management, and mindset towards innovative development. All the cases are strongly committed to innovation and digitalisation. The cases perform differently according the statistical evidence giving indications for potential areas of improvement. Cases also share the significance of the Baltic Sea as one of their main international trade routes. This chapter provides a conclusion to this part of the book that looks into empirical evidence in better understanding the characteristics of geographies of innovation and disruption.

Keywords Innovation activity drivers · Innovation policy · Interpretations Planning · Future prospects · Reflections

There are certain commonalities between the presented smart cities and their innovation activity drivers. The first notable issue is that all locations are highly important hubs in their nations and national contexts. The size issue is also relevant. There are variations in the absolute and relative sizes of the cities in terms of their population but most commonly the most innovative locations are not giant cities but they do have enough critical mass to attract other industries and services to the same vicinities. The smallest case location in terms of absolute population is the Tallinn with an approximate population of 426,000. The largest on the other hand is Stockholm with around 950,000 million people. If FUAs are considered the sizes of each location roughly doubles in each case.

Planning is an important dimension in the urban development and clustering. As indicated in the cases of Copenhagen, Helsinki, and Stockholm planning and location decisions done by companies have changed and are still changing under the change of economic structure of post-industrial cities. Particularly companies producing digital services and products are able to locate themselves relatively freely. The clustering of software and social-media companies close to city centres and mixed environments with a significant number of residents living in the same locations are examples of the changing needs of immaterial industries.

Spatial clustering increases the economic efficiency. There are solid evidences that close proximities help and create synergies among related and supportive industries. However, close proximities do not necessarily contribute to the innovativeness of companies as such. As an example, the initial idea resulting into an innovation may be conceived location free, but the actualisation of this initial idea is done more efficiently in a cluster location where other knowledge resources are more likely to be available. This problematic concern also subcontracting and network clusters.

The future development paths of innovative and smart cities come intertwined with the broad societal development taking place in ICT and media-technologies. One major issue will be the role of data security and privacy linked to the Big Data. Identification, monitoring and segmenting tools have already developed to the level where individuals are easily identifiable if so wanted. The main challenge lies in the small and medium sized data collection houses that larger information appliers use as subcontractors. These companies may disappear from the market and other companies will in some point acquire their data storages. The possibility for continues ownership changes causes problems. The second main issue connects environmental development and urban growth tightly together. It is likely that environmental innovations become more and more important in the future as there is continuous urbanisation in progress.

To conclude and summarise the Part II the following points are put forward:

- All case cities are strongly putting efforts in order to develop their smart city services and innovative growth agenda;
- The cases are constantly above the reference averages making them suitable targets for further studies in smart city and innovation development;
- In economic terms, all the Nordic capitals may be considered to be quite similar. Tallinn, on the other hand, is behind them particularly in indicators depicting general economic activity and innovation;
- All the cases indicate that cities should find their own strength areas and focus on their endogenous growth. This, however, requires broad enough economic profile in general that profiling can be done in the first place;
- Environmental innovations and production of cleaner environment is currently strongly present in the development agendas of the cases. They will continue to do so in the foreseeable future.

Part III
Place Making for Knowledge Cities

Chapter 9
Introduction to Part III

Abstract In the age of global knowledge economy cities and region are in a tough competition for fostering, attracting and retaining investment and talent. Developing strategies to help cities to initiate and complete their knowledge economy transformation has become a hot agenda in urban circles. In this context space and place making for such transformation is deemed critical. In consequence knowledge-based urban development and place making have become fundamental approaches to form prosperous knowledge cities. This chapter provides an introduction to the part of the book that focuses on place making for knowledge cities as critical approaches for the knowledge-based formation of our cities and their regions.

Keywords Innovation · Place making · Knowledge cities · Knowledge economy · Knowledge society · Knowledge precinct

Many cities around the world, today, are facing the prospect of a major transformation as the world moves towards a global information order. In the 21st century, dynamic processes of economic and spatial restructuring are radically altering urban economies, and knowledge has become a precious local and regional resource for post-industrial territories, which were severely affected by the economic and social implications of structural change. In order to compensate for the loss of industrial monopolistic power, and to remain competitive in a world economy, the new global division of labour forces cities to concentrate on developing knowledge as a critical local and regional resource. Subsequently, preparing for a knowledge economy and society has become a new political challenge. The result of this process is the creation of 'knowledge cities' (Yigitcanlar 2010).

Knowledge cities are defined as cities that search for the creation of value in all of its areas and develop high standards of life, cultural vibrancy, and economic development, among other aspects including a higher level of income, education, training and research (Carrillo 2006). At the same time, knowledge cities are regional knowledge economy driven cities with high value-added exports created through research, technology, and brainpower and purposefully designed to encourage the nurturing of knowledge. Achieving the transformation of a city into a knowledge city requires leveraging urban competitiveness in many areas, such as the economy, society, built and natural environments, and governance, through holistic and integrated policy and planning mechanisms (Yigitcanlar and Bulu 2015).

In recent years, as part of the knowledge city transformation initiatives, construction of 'knowledge precincts' has become a highly popular trend in many parts of the world for fostering, attracting or retaining innovation and knowledge generation activities, industries and workers. Knowledge precincts are defined as integrated centres of knowledge generation, learning, commercialisation and lifestyle. In other words, they are high-growth knowledge industry and worker clusters, and distinguish the functional activity in an area, where an agglomeration of knowledge and technological activities has positive externalities for the rest of the city and the region as well as companies located there. Besides being spaces of innovation and knowledge generation, these precincts have been considered as a solution to complex political and economic issues in societies—e.g., regional industry problems, under-commercialisation of publicly financed research, shortage of new product development, productivity issues, and unemployment. Urban administrators often view these precincts as a significant policy instrument for promoting or supporting urban development and economic growth. Many of these administrators have pursued the idea of creating knowledge precincts that are able to attract global flows of ideas, knowledge, people, and capital, and become trendsetting knowledge and innovation hubs/clusters (Yigitcanlar and Dur 2013).

Even though knowledge city and knowledge precinct concepts are highly popular amongst the many cities of the world that are seeking a knowledge-based urban development in the era of the global knowledge economy and rivalry, only a number of cities to date, however, has managed to become a prosperous knowledge city. For instance, Amsterdam, Austin, Barcelona, Boston, Eindhoven, Helsinki, Melbourne, San Francisco, and Singapore are among the best practice knowledge city examples that performed a strong commitment to developing effective knowledge-based urban development strategies to achieve their desired outcomes—concerning economic, societal, spatial and institutional progress. It is not a coincidence that these cities also home for some of the world's most successful knowledge precincts/corridors—to name a few: 22@Barcelona, Arabianranta, Boston Innovation District, One-North, Silicon Hills, Silicon Valley, and Strijp-S (Pancholi et al. 2015).

This Part III of the book concerns of knowledge cities and their place making efforts. The part particularly focuses on generating insights on the potential pathways for cities—for them to adopt an effective knowledge-based urban development approach and to become a prosperous knowledge city. As working on a specific case city naturally makes generation of sound lessons and insights possible and easier, the part of the book concentrates its attention on a vibrant case study from a developed country context. Namely, Brisbane, capital city of the Queensland state of Australia—an emerging knowledge city—is selected as the case city for this knowledge city investigation. The selection of the case study city is justified as follows.

In the context of Brisbane, the foundations of knowledge city aspiration were seeded with a number of government policies, including Smart State and City Strategies and most recently Advance Queensland innovation initiative, dating back as early as to the late 90s. As part of these policies, some investment has been made to achieve knowledge-based urban development through various urban and regional planning schemes in and around Brisbane. These investments include the development of a number of knowledge precincts—such as Brisbane Technology Park, Dutton Park Knowledge Precinct (also known as Diamantina Knowledge Precinct), and Herston Health Precinct. However, the end of the Smart State Strategy along with recent economic recession in the country (due to the global financial crisis, and end of the mining boom) has slowed down the concentrated efforts—in spite of the State Government's 'Advance Queensland' initiative, the Commonwealth Government's 'National Science and Innovation Agenda', and Brisbane's new city brand 'Australia's New World City'. At present, there seems to be a lack of a strong point of reference and a long-term vision for the orchestration of Brisbane's transformation into a knowledge city, along with better planning, designing, and fostering Brisbane's existing and emerging knowledge precincts through a city-wide holistic and integrated perspective. In other words, as this study reveals, Brisbane might be becoming vulnerable and disadvantaged in the global rivalry for investment and talent, as the city has not yet clearly figured out a robust knowledge-based development strategy. Learnings from the Brisbane case could be of interest to many emerging knowledge cities of the world to better strategize and support their development and progress.

9.1 Rationale and Scope

Against the aforementioned backdrop, this part of the book aims to develop a thorough understanding of the underlining conceptual and practical issues on the transformation of cities into knowledge cities, and planning, designing, and fostering existing and emerging knowledge precincts. The study, reported in this book, intends to place Brisbane under the microscope scrutinising its potential and progress from the lens of knowledge-based urban development—focusing on economic, societal, spatial and institutional development domains—targeting the city and its region. In terms of the geographical boundary of the investigation area, a flexible approach is adopted as some of the issues relate to Brisbane at the precinct (neighbourhood), some city (local government area), some greater city (Greater Brisbane), and some metropolitan region (South East Queensland) scales. This part of the book particularly focuses on the knowledge city development of Brisbane and attempts to address the following research questions.

- Why is pursuing a knowledge city goal important for cities of the global knowledge economy era?
- What are the key conditions and policy drivers to transform cities into prosperous knowledge cities?
- How can lessons learned from global knowledge city best practice be evaluated for their potential in Brisbane?

- How does Brisbane stand as an emerging knowledge city internationally and nationally?
- How can a holistic long-term knowledge city vision suitable for Brisbane be developed?

9.2 Methodology

The methodology of the study on knowledge cities investigation contains the following systematic approaches:

- Review of academic and grey literature to determine the key conceptual and practical issues concerning knowledge cities concept.
- Global best practice analyses to determine lessons from the best practice knowledge cities—i.e., Austin, Barcelona, Melbourne, and Singapore—by focusing on the success and failure factors and analysing their suitability for Brisbane.
- Empirical analysis to compare Brisbane with major cities from Australia and overseas—i.e., Birmingham, Boston, Helsinki, Istanbul, Manchester, Melbourne, San Francisco, Sydney, Toronto, and Vancouver—by focusing on the key economic, social, spatial and governance performance indicators relevant to knowledge cities.
- Recommendations and strategic directions to provide insights, suggestions and potential strategic directions in the light of the findings from conceptual and practical analyses targeting both hard and soft factors and measures.

References

Carrillo, F. J. (Ed.). (2006). *Knowledge cities: Approaches, experiences and perspectives*. New York: Routledge.

Pancholi, S., Yigitcanlar, T., & Guaralda, M. (2015). Place making facilitators of knowledge and innovation spaces: Insights from European best practices. *International Journal of Knowledge-Based Development, 6*(3), 215–240.

Yigitcanlar, T. (2010). Making space and place for the knowledge economy: Knowledge-based development of Australian cities. *European Planning Studies, 18*(11), 1769–1786.

Yigitcanlar, T., & Bulu, M. (2015). Dubaization of Istanbul: Insights from the knowledge-based urban development journey of an emerging local economy. *Environment and Planning A, 47*(1), 89–107.

Yigitcanlar, T., & Dur, F. (2013). Making space and place for knowledge communities: Lessons for Australian practice. *Australasian Journal of Regional Studies, 19*(1), 36–63.

Chapter 10
Theory and Practice of Knowledge Cities and Knowledge-Based Urban Development

Abstract The practice in the age of global knowledge economy has created a hostile competition between urban locations to secure a sustained economic and societal development. The fast running vehicle of being successful in this competition is widely agreed as generation of creativity, knowledge and innovation and then translating them into high value added commercial products or services. In this context, besides economy and society of knowledge, spatial and governance domains are deemed fundamental. Knowledge-based urban development has come to the scene to deliver a knowledge city that brings all these strengths under the same roof. This chapter elaborates and provides insights on the theory and practice of knowledge cities and knowledge-based urban development.

Keywords Knowledge-based urban development ·
Sustainable urban development · Knowledge cities · Smart cities
Knowledge and innovation spaces

Many cities around the world, today, are facing the prospect of a major transformation as the world moves towards a global information order. In the 21st century, dynamic processes of economic and spatial restructuring are radically altering urban economies and increasing the importance of local economic development in nations' progress (Portney 2013). Even though cities have always played a vital role in the national economy, during the past couple of decades their importance has significantly increased. At present, cities are the major drivers of national economic growth, and national progress is closely tied to the performance of their large metropolitan regions and their robust city centres (Sassen 2011).

In this context, knowledge has become a precious local and regional resource for post-industrial territories, which were severely affected by the economic and social implications of structural change. In order to compensate for the loss of industrial monopolistic power, and to remain competitive in a world economy, the new global division of labour forces cities to concentrate on developing knowledge as a critical resource. This makes cities to possess an economy driven by high value-added exports created through research, technology, and brainpower (Anttiroko 2015).

Knowledge, although, in itself is a necessity, it is not a sufficient condition for further progress in societies. Innovation—not only as a way of doing things faster, better and cheaper, but also as a creative mode of thinking and acting—is based on new insights and applications, which help industry and science to gain a strong competitive position (Van Geenhuizen and Nijkamp 2012). From this perspective, preparing for a creative/innovative knowledge economy and society is a necessity and it has become a new political challenge (Cooke and Leydesdorff 2006; Huggins and Strakova 2012; Yigitcanlar et al. 2012). The result of this process is the creation of knowledge-intensive and innovation-oriented localities—namely knowledge cities (Metaxiotis et al. 2010; Gabe et al. 2012; Madanipour 2013). However, the pathway of achieving a knowledge city transformation is full of obstacles and challenges.

Against this background, this chapter of the book aims to address the following two critical questions by determining the key conceptual and practical issues based on the review of the academic and grey literature.

- Why is pursuing a knowledge city goal important for cities of the global knowledge economy era?
- What are the key conditions and policy drivers to transform cities into prosperous knowledge cities?

Please note that in this part of the book (and the following Part IV) rather than 'smart cities' the concept of 'knowledge cities' is placed under the microscope. There are several reasons for this conscious choice.

The first one is that there are many urban brands have been used interchangeably—with minor differences—in urban studies throughout the last few decades. For example, these include, but not limited to: sustainable city, digital city, intelligent city, ubiquitous city, techno-centric city, creative city, knowledge city, and smart city (Lara et al. 2016; Yigitcanlar and Kamruzzaman 2018; Yigitcanlar et al. 2018). In contrast to the common understanding on the smart city notion—mostly a techno-centric view—the knowledge city notion brings a more comprehensive urban development conceptualisation to the table (Trindade et al. 2017; Yigitcanlar et al. 2017). Secondly, knowledge cities form an overarching umbrella concept that places knowledge and innovation spaces located in an urban locality under its wings for nurturing and fostering them (Esmaeilpoorarabi et al. 2018a). In other words, knowledge and innovation precincts/districts/hubs are the drivers of the global knowledge economy and knowledge city concept offers more to these locations than of smart cities notion (Chang et al. 2018; Esmaeilpoorarabi et al. 2018b). Lastly, there is no fully-fledge smart cities in the world, so far, but on the contrary, there are a number of globally recognised fully-fledged knowledge cities. Investigating these concrete knowledge city best practices contributes to our understanding and policy development domain way more than those hypothetical or in small-scale smart city examples (for extensive smart city analyses, see Yigitcanlar 2016).

10.1 Evolution of Knowledge Cities

The last few decades witnessed the rise of the knowledge economy as the main driver of global and local economic development. Particularly during the first decade of the 21st century, strong evidence has been presented substantiating an argument for an 'urban turnaround' taking place (Asheim 2012). While the discourse on industrial districts, clusters, and new economic spaces was dominant in the 1980s and 1990s (Piore and Sabel 1984; Porter 1990; Storper 1997), in the 2000s the focus has been directed towards cities as the node of new economic activities, new forms of collaboration and organisation as well as the new preferred location of the creative class of knowledge workers (Florida 2002; Storper and Venables 2004; Asheim et al. 2007; Yigitcanlar et al. 2007; Asheim 2012).

During this period, the notion of 'knowledge city' has evolved from similar concepts such as 'knowledge clusters' (Arbonies and Moso 2002; Huggins 2008), 'ideopolis' (Garcia 2004), 'technopolis' (Smilor et al. 1988; Scott 1993), 'science city' (Anttiroiko 2004), 'learning city' (Larsen 1999), 'intelligent city' (Komninos 2002), and 'sustainable city' (Camagni et al. 1998). The earliest reference to knowledge cities dates back to 1990 to a report prepared by the Institute for Spatial Organisation of the Netherlands Organisation for Applied Scientific Research (TNO-INRO 1990). With over two decades of research and practice, knowledge city is still a relatively new concept with no clear definition that is fully agreed on by scholars. The followings are the most commonly accepted views on how knowledge city is perceived—presented in chronological order:

- According to Ihlanfeldt (1995, p. 129), knowledge cities are "hypothesized to expedite the transfer of knowledge and the creation of knowledge that affects the growth factors identified by the neoclassical model, in particular, the quality of labour and technical change … [these] dense urban environments can lead to unexpected combinations of seemingly unrelated ideas that may provide important leaps forward in knowledge".
- Dvir (2006, p. 245) frames the knowledge city from an individual's point of view as "a milieu, which triggers and enables an intensive, ongoing, rich, diverse, and complex flow of knowledge moments … [where] a knowledge moment is a spontaneous or planned human experience in which knowledge is discovered, created, nourished, exchanged, and transformed into a new form".
- As for Edvinsson (2006, p. 6), a knowledge city is "purposely designed for encouraging and nourishing the collective knowledge, i.e., intellectual capital, as capabilities to shape efficient and sustainable actions of welfare over time".
- Ergazakis et al. (2006, p. 4) underline the concept of a knowledge city as a broad one referring to all aspects of social, economic, and cultural life of a city, and state "a knowledge city aims at a knowledge-based [urban] development by encouraging the continuous creation, sharing, evaluation, renewal, and update of knowledge. This can [only] be achieved through the continuous interaction" with all stakeholders, including citizens.

- Musterd and Deurloo (2006, p. 92) view knowledge cities as "cities that create the right conditions that act as a magnet for research institutes and highly-educated knowledge workers, not only by offering an attractive working environment, but also by creating a favourable living environment, one that attracts and manages to retain creative talent. Places where talented employees are found—pools of creative talent—are becoming increasingly decisive in determining where businesses choose to locate".
- Yigitcanlar (2009, p. 239) sees knowledge cities as localities that are "tailored to the needs of a knowledge economy where ideas rule and there are infinite recipes for innovation and wealth creation. Their growth is based on the generation of value using common assets with the purpose of achieving sustainability".
- Carrillo (2015, p. 1) perceives knowledge cities as the engines of innovation and growth, and states, "the association of the terms 'knowledge' and 'city' conveys the conglomeration of technological, academic, cultural, scientific, and innovation capabilities in cities and regions operating as engines of economic growth".
- As for Yigitcanlar (2015, p. 7484), a knowledge city is a "city that searches for the creation of value in all its areas and develops high standards of life, cultural support and economic development, among other aspects including higher level of income, education, training and research, at the same time it is a regional knowledge economy driven locality with high value-added exports created through research, technology, and brainpower and purposefully designed to encourage the nurturing of knowledge".

Following the implementation of the Institute for Spatial Organisation's knowledge cities study (TNO-INRO 1990) in Delft (The Netherlands), the city has become the first one to adopt knowledge city as a city brand—Delft, the knowledge city—in 1996. The knowledge city strategy of the city mainly focused on emphasising physical planning, forming a supportive economic policy, advancing the position of the city as an international knowledge nod, upgrading the cultural level of the city, creating a broad carrying capacity for the city, and improving city marketing efforts significantly (Van Geenhuizen et al. 1997). Delft's success encouraged a number of cities to follow the same pathway. For example, in 1999 Barcelona used the 'city of knowledge' as the brand for the city for its transition into a knowledge economy and society (Romein et al. 2011). Another Dutch city Eindhoven set the development lines of the city's future as a knowledge city in 1999 (Ergazakis et al. 2004). Amsterdam has been using the label of 'creative knowledge city' since 2000 (Musterd 2004). In 2002, Munich has started marketing itself as a 'city of knowledge' (Van Winden 2010). Around the same time, Melbourne has initiated its knowledge city transformation efforts (Yigitcanlar et al. 2008a). Many cities followed the footsteps of these pioneering cities since then (Yigitcanlar and Sarimin 2011).

Today knowledge city is a widely adopted policy across many developed and developing country cities. Besides, many cities without using this deliberate brand are also working towards achieving a knowledge-based urban development. From Europe a number of cities, including Birmingham and Manchester from the UK, Dublin from Ireland, Valencia and Madrid from Spain, and Vienna from Austria, have been characterising their post-industrial image as a knowledge city (Carrillo et al. 2014). Additionally, from the European context, Istanbul-Turkey is also aiming a knowledge city vision by heavily investing in major infrastructure upgrades and mega-project developments (Yigitcanlar and Bulu 2015). In the North American context, Austin, Boston and San Francisco from the US, Montreal, Ottawa, Toronto and Vancouver from Canada, and Monterrey from Mexico are among the exemplary cities with high knowledge cities ambitions. In the South American context, examples include Curitiba and Rio de Janeiro from Brazil. In the Middle Eastern context, Doha from Qatar, Dubai and Abu Dhabi from the UAE are preparing their economies and cities to the post-oil era by emerging as knowledge cities. In the Asian context, Bangalore and Hyderabad from India, Kuala Lumpur from Malaysia, Seoul from Korea, Shenzhen from China, and Singapore are in a large pool of cities with different levels of invested interests in knowledge city transformation. Melbourne, Sydney, and Brisbane from Australia and Auckland from New Zealand are among the cities that have either established themselves as a knowledge city or emerging as one from the Oceania context.

In order to celebrate the achievements of knowledge cities from all across the globe, an international think tank (The World Capital Institute—www.worldcapitalinstitute.org) has been conducting the annual Most Admired Knowledge City Awards (MAKCi) since 2007. The awards are given in two categories, i.e., Knowledge City-Regions and Knowledge Metropolises, to the cities around the world that are successfully engaging in formal and systematic knowledge-based urban development processes under the flagship of knowledge cities. The MAKCi framework includes eight knowledge capital dimensions to stand as indicators as the visible drivers of collective capital creation in knowledge-based development of cities. These capital dimensions are: (a) Identity capital; (b) Intelligence capital; (c) Financial capital; (d) Relational capital; (e) Human individual capital; (f) Human collective capital; (g) Instrumental-material capital, and; (h) Instrumental-knowledge capital (for further info see Garcia 2010).

Brisbane was nominated for this award in 2008, 2009, 2012, 2013, 2015 and 2016. These nominations were done by the first author of this book and his colleagues. Further information on the nomination process is available at the MAKCi platform. During the latest award round in 2016, Dublin and Melbourne have received the MAKCi awards (Table 10.1) in the aforementioned categories respectively (see www.worldcapitalinstitute.org/makciplatform).

10.2 Essential Assets of Prosperous Knowledge Cities

The popularity of the knowledge cities concept, particularly during the first decade of the 21st century, led to investigations into determining the key conditions and drivers to transform cities into prosperous knowledge cities. Stated by Johnston and Huggins (2016), the knowledge city concept takes a broad view of knowledge as a key construct for understanding the means and modes of urban development, with

Table 10.1 Recipients of the most admired knowledge city (MAKCi) awards

	Knowledge city category[a] (2007–2008)	Knowledge city-region category[b] (2009–present)	Knowledge metropolis category[c] (2009–present)	Emergent knowledge city category[d] (2017–present)
2007	Singapore	n/a	n/a	n/a
2008	Singapore	n/a	n/a	n/a
2009	n/a	Manchester	Barcelona	n/a
2010	n/a	Ottawa	Melbourne	n/a
2011	n/a	Holon	Singapore	n/a
2012	n/a	Austin	Singapore	n/a
2013	n/a	Copenhagen	Melbourne	n/a
2014	n/a	Ottawa	Seoul	n/a
2015	n/a	Vienna	Montreal	n/a
2016	n/a	Dublin	Melbourne	n/a
2017	n/a	Austin	Boston, Barcelona	Arequipa

[a]Knowledge city category discontinued after the 2008 round and replaced with two new categories of knowledge city-regions and knowledge metropolis
[b]Knowledge city-region is considered medium size ($\geq 0.5M \leq 3M$) cities/towns acting as a regional centre for political, economic or cultural activities
[c]Knowledge metropolis is considered largest to medium (over 3M) urban conglomerates often acting as international hubs for political, economic or cultural activities
[d]Emerging knowledge city category considers the best performing emergent knowledge cities regardless of the population size

an explicit focus on identifying the 'knowledge assets' of urban environments. In other words, forming prosperous knowledge cities requires understanding the diverse socioeconomic and socio-spatial forms of creative urban regions and their valuable knowledge assets (Edvardsson et al. 2016).

Urban knowledge assets are unique to a locality and can be defined as the common knowledge capabilities of stakeholders to implement strategies that aim at long-term sustainable development (Arbolino et al. 2017, 2018; Kozak 2011). Furthermore, adapting the originally national-level definition by Bontis (2004) to cities, urban knowledge assets can be characterised as the hidden values of individuals, enterprises, institutions and communities that are the current and future sources of wealth creation. In their recent researches, Lonnqvist et al. (2014) and Yigitcanlar et al. (2014) find that the management of urban knowledge assets is critical as it deals with both development of a knowledge city's internal knowledge assets and gaining access to regional, national and international assets. Although knowledge assets are critical for knowledge cities, they heavily depend on other complementary assets. Table 10.2 lists a selection of essential assets of knowledge cities.

'Local and regional contexts', directly linked with the aforementioned assets, are critical determinants of the success of knowledge city formation. A number of schol-

10.2 Essential Assets of Prosperous Knowledge Cities

Table 10.2 Essential assets of a knowledge city (derived from Yigitcanlar and Dur 2013)

Assets	Capitals	Descriptions	Examples
Cultural assets	Cultural capital	Capability to take advantage of unique local culture and value systems to generate social mobility and cohesion	Historical and archaeological sites, handicrafts, cuisine, ethnography, cultural identity, diversity, openness, and tolerance
Financial assets	Financial capital	Capability to generate and sustain a healthy monetary base for development	Financial support, institutions and resources available to people, firms and cluster formation
Human assets	Human capital	Capability to create conditions for the full biological and psychological development of residents and employees	People's capacity and skills to work, education and training centres, the thickness of the labour market
Instrumental assets	Instrumental capital	Capability to take advantage of location and to build and renew a world-class physical infrastructure	Natural and constructed amenities, flora and fauna, technical and hard infrastructure
Knowledge assets	Knowledge and intellectual capitals	Capability to generate and transfer knowledge and foster creativity and innovation in all major areas of city life	Intellectual property rights, research and development centres, universities, project partners, patents, trademarks, copyrights, and other results of human innovation and thought
Relational assets	Relational capital	Capability to develop quality interactions and collaborations with all significant internal and external agents	Management, governance, institutions, networks, interactions, collaboration, orchestration of the development
Social assets	Social capital	Capability to enhance the goal achievement potential of its constituent communities	Civic initiatives, community centres, communities, social amenities and infrastructures
Symbolic assets	Identity capital	Capability to create and evolve a distinctive and well-positioned urban positioned urban personality/locality	City brands, geographic trademarks, landmark buildings, endemic plants, city reputation, unique characteristics and pull factors

Table 10.3 Success factors of knowledge cities (derived from Ergazakis et al. 2006; Yigitcanlar 2009)

Success factors
→ Political and societal will and good governance—leadership and stewardship
→ Strategic vision and long-term development planning—dynamic planning
→ Financial support, incentives, partnership and strategic investments—support mechanisms
→ Agencies to orchestrate and promote knowledge-based urban development—management
→ The international and multicultural character of the city—openness and tolerance
→ Metropolitan web-portal and apps—e-services and participatory decision-making
→ Value creation to citizens—skill development, employment, social outcomes
→ Urban innovation engines—places to trigger, speed up, host and facilitate innovation
→ Knowledge society rights of citizens—accessibility, information, education, participation
→ Public access to communication networks—low-cost fast broadband, free public Wi-Fi
→ Research excellence—leading world class universities and R&D institutions
→ Public library networks—public free access to information in hard and soft forms
→ The quality of life and place, and affordable housing and urban services/amenities—quality measures

ars acknowledge the role of contextual features—such as economic, societal and natural assets and conditions, and policy, planning and institutional support mechanisms—in the success of knowledge city transformation (Carrillo 2006; Yigitcanlar et al. 2008b; Carrillo et al. 2014). Researchers in the field underline a number of key success factors in building prosperous knowledge cities. Ergazakis et al. (2006) and Yigitcanlar (2009) compile these success factors as listed in Table 10.3. These factors could be seen as the key ingredients for success. However, as the best practice analyses reveal (see the next section of the book) 'good will, transparency, and social consensus' are also needed to work as a glue that sticks all these factors together (Sarimin and Yigitcanlar 2012).

As knowledge cities are complex entities, and they attempt to transform cities into knowledge cities would likely result in failure unless they are carefully planned, guided and governed. Such planning, guidance or governance practice requires making or incorporating effective and efficient knowledge-based policies. According to Van Geenhuizen and Nijkamp (2012), these policies, for example, should target the following desired outcomes listed in Table 10.4.

Van Winden et al. (2007) conceptualise a knowledge city with seven critical foundations: a knowledge base; an economic base; quality of life; accessibility; urban diversity; urban scale; and, social equity. In addition to these, they also mention the organisational capacity to effectively attract knowledge workers, create knowledge, apply knowledge, and develop growth clusters (see Fig. 10.1). Moreover, Romein et al. (2011) highlight the key elements of a knowledge city as: (a) social climate; (b) buzzing atmosphere; (c) employment opportunities; (d) built environment; (e) urban amenities; (f) knowledge clusters and incubation spaces, and; (g) policy and

10.2 Essential Assets of Prosperous Knowledge Cities

Table 10.4 Desired policy outcomes (derived from Van Geenhuizen and Nijkamp 2012)

Desired policy outcomes
→ Fostering, attracting and retaining knowledge workers and industries
→ Regarding higher education institutes as primary sources of growth and prosperity
→ Encouraging personal ties as a source of knowledge flow and social capital development
→ Creating and fostering knowledge/innovation/creative quarters as an integrated part of the urban fabric
→ Producing and preserving the unique identity or image of the city
→ Developing and applying effective and transparent governance and management practices
→ Empowering citizens as knowledge creators and innovators

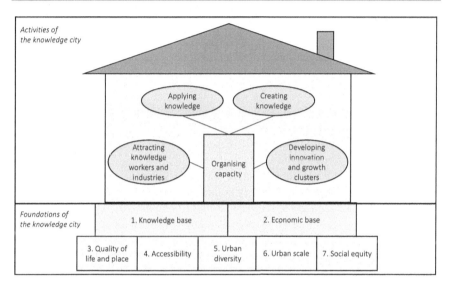

Fig. 10.1 Foundations and activities of knowledge cities (derived from Van Den Berg et al. 2005)

governance. Furthermore, a review of knowledge city practice shows that the most successful knowledge cities specialise in a few sectors only—but never a single one—based on their endogenous strengths but set ambitious goals for each. They also develop their supporting knowledge-based policies carefully to form a creative environment to achieve these goals (see Yigitcanlar 2009). The next section of the book presents some of the most common knowledge-based development policies.

10.3 A New Planning and Development Paradigm: Knowledge-Based Urban Development

The last two decades have presented strong evidence that an urban turnaround is taking place (Asheim 2012). Cities with ambitious development goals have started to interlink their knowledge-based activities with urban planning practices (Van Geenhuizen and Nijkamp 2012). Planning and development of urban knowledge precincts or innovation districts are the most obvious example of this linkage. A much less common, but highly important, practice is the coordinated actions through connecting city innovation systems and urban planning practices—such as Singapore. A number of cities, today, are being recognised globally as knowledge cities even though they have not purposely placed strategic planning efforts targeting specifically the knowledge city formation—such as London, New York, and Tokyo. Some cities basically benefit from their primate positions in their countries and qualify as a knowledge city. Some others have already gained the recognition as they have the global or world city reputation. However, not all cities have the advantages of those global, world or primate cities. However, for many cities, a careful design of the planning process for knowledge city transformation is required. Indicated by Knight (1995, p. 227), "city development must become more intentional, less accidental and less determined by the external forces [and] shaped by endogamous processes". KBUD provides an opportunity to achieve this.

The idea of KBUD has emerged as a means of systematically examining the role of knowledge and networks as key components of urban economic evolution (Knight 1995; Kunzmann 2009; Yigitcanlar and Lonnqvist 2013). In essence, this approach takes the view that knowledge is a key construct for understanding the means and modes of urban development (Yigitcanlar 2009), with an explicit focus on identifying the knowledge assets of urban environments (Lonnqvist et al. 2014). KBUD has become a fashionable term in terms of urban policymaking. This is not only the case in post-industrial cities of developed countries, but also in cities of developing countries, which are still in the process of industrialisation. In responding to this, urban development policies have been redirected in recent years to enhance and to build on the strengths of cities' knowledge resources regardless their state of development (Wang 2009; Zhao 2010).

Like many other theories, KBUD was also born imperfect, but has a constantly evolving nature. The early attempts of KBUD initiatives can be traced back to the 1980s when the success of Silicon Valley in the US and Cambridge Science Park in the UK in the 1960s and 1970s triggered international enthusiasm of development in this kind. Such as the Japanese Technopolis Program, Hsinchu Science and Industrial Park in Taiwan, Taedok Science Town in South Korea, Singapore Science Park, French Technopoles Program and so on are seen as typical examples of efforts, which Castells and Hall (1994) would assign a general term—technopole—to represent this phenomenon. Theoretically, all these initiatives were economy centred and based on an assumption that there was a linear relation between technology advancement and economic growth (Wang 2009).

10.3 A New Planning and Development Paradigm: Knowledge-Based Urban Development

According to Wang (2009), despite the enthusiastic, studies into the actual performance of technopoles show mixed results. Massey et al. (1992) argue that technopoles are a form of prestigious urban development with few productive synergies generated and these are no significant evidence supporting that geographical proximity between universities and a science park accounting for the promotion of technology transfer. Storey and Tether's (1998) study shows that most science parks in Europe have made only a modest direct contribution to employment and it was difficult to assess their contribution to technology transfer. Westhead and Storey (1995) find that the actual links between university and firms, which are located in the science park, were less significant than that anticipated. Similarly, Vedovello (1997) shows that, based on a comparative analysis, geographical proximity between partners is not an important influence on the existence or strength of formal links between university and industry.

The abovementioned adequacies of the science city and technopoles along with the rapid advancement of information technology in the 1990s and the global trend of transforming societies into knowledge society brought a new form of KBUD—i.e., knowledge city (Wang 2009). The notion of knowledge cities quickly replaced the mostly technology and economy focused technopole concept. In the late 1990s, the academic, political and societal discourse about urban and regional development changed radically. After the shift from Fordist to post-Fordist economy and society, a new paradigm shift is witnessed around the turn of the century. According to Bontje et al. (2011a, p. 1), "[t]his upcoming [development] paradigm suggests that the economic future of cities and city-regions increasingly depends on the capacity to attract, generate, retain and foster creativity, knowledge, and innovation". This paradigm, namely KBUD, has started to become highly popular during the last years of the 20th century considering the impacts of the global knowledge economy on urban localities and societies (Yigitcanlar 2011).

The followings are among the commonly accepted views on how KBUD is perceived—presented in chronological order:

- In 1995, Richard Knight published his illuminating article, 'Knowledge-based development: policy and planning implications for cities', arguing the need and emergence of a new approach to city development focusing on KBUD (Knight 1995). He defined: "knowledge-based [urban] development [as] the transformation of knowledge resources into local development [which] could provide a basis for sustainable development" (Knight 1995, pp. 225–226).
- In 2000, KBUD is seen as a crucial set of strategies for achieving the quality of life. According to AEUB (2000, p. 1), the aim of KBUD is "to develop urban settlements that are gradually evolved to [become] more in line with sustainability objectives and improve [their] quality of life [by accommodating] knowledge-based urban development strategies as opposed to [exclusively] physical resource-based strategies".
- Although not directly referred exactly as KBUD, since the beginning of the 21st century, OECD (2001) has been adopting knowledge management frameworks in its strategic directions regarding glocal (global and local) development, and this

strategy strongly indicates that a link to be urgently established between knowledge management and urban development. The end result of this linkage is KBUD.
- Later on, in 2004, KBUD was emphasised as a fundamental medium for the development of knowledge cities. As for Entovation (2004, p. 2), KBUD is "the perfect new medium in which to grow more liveable, stimulating, cleaner, intelligent, enlightened, tolerant and meaningful communities worldwide … [and] the knowledge city is the first new urban formation tailored to the needs of a knowledge economy where ideas rule and there are infinite recipes for innovation and new wealth creation."
- The mid-2000s was the period that KBUD was coined as an emerging urban and regional development phenomenon and started to be widely seen as a development strategy tool for enhancing the competitiveness of cities within the context of expanding knowledge-based economy and society and forming prosperous knowledge cities. Yigitcanlar (2005, p. 3) stated that "[t]he significant increase of the knowledge-based development strategies for the pursuit of metropolitan competitiveness of regions is encouraging city administrations to adopt these strategies for moving towards and establishing knowledge cities".
- Along with the increasing popularity of knowledge cities, from the mid-2000s onwards the term KBUD has started to receive larger attention and gained wider recognition. In late 2000s KBUD has, for the first time, started to be seen as a development process rather than solely a development strategy and defined as not only "a powerful strategy for economic growth and the post-industrial development of cities and to participate in the knowledge economy, [but also] a strategic management approach, applicable to creative urban regions" (Yigitcanlar et al. 2008c, p. 10).
- The most recent view on KBUD is that it is a process with a set of policies targeting of building a place to form perfect 'climates' for 'business, people, space/place and governance', and emphasize on the balance and integration of these climates (Yigitcanlar 2014a, b). In other words, it is the new development paradigm of the global knowledge economy era that aims to bring economic prosperity, environmental sustainability, a just socio-spatial order, and good governance to cities, and produces a prosperous knowledge city.

In addition to the abovementioned views Carrillo (2015) elucidates the evolution of KBUD—that is defined earlier as: "the collective identification and enhancement of the value set whose dynamic balance furthers the viability and transcendence of a given community" (Carrillo 2014, p. 416). Carrillo (2014) introduces three generations of knowledge-based development. However, this clarification does not directly focus on the urban or spatial dimension of development. Therefore, in a new classification attempt, this Chapter brings together the works of Carrillo (2014, 2015), Carrillo et al. (2014) and Yigitcanlar et al. (2016) and reconceptualises the evolution of the KBUD concept as below and in Table 10.5.

- The first generation or 'product-centred' KBUD approach focuses on object attributes of the development (e.g., manufactured knowledge/product), and places lesser attention on the talent-base (i.e., knowledge worker) involved in the produc-

10.3 A New Planning and Development Paradigm: Knowledge-Based Urban Development

tion. This type of KBUD approach forms isolated and gated exurban or suburban knowledge and innovation spaces, and mostly concentrates on tangible outcomes (e.g., financial gain, infrastructure development) or hard factors. Science and technology parks are examples of knowledge spaces generated as a result of the first generation KBUD approach. For instance, Longhua Science and Technology Park in Foxconn City, Shenzhen, China (iPads for Apple are being manufactured in this park, and it was subjected to media attention for the poor work conditions that led to numerous worker suicides) is an example of this type of development.

- The second generation or 'talent-centred' KBUD approach particularly concentrates on the agent attributes of the development such as knowledge producing structures (e.g., roles, hierarchy, networks) as well as relevant agent or knowledge worker competencies (e.g., knowhow, adaptability, creativity). This type of KBUD approach forms more connected and accessible exurban, suburban or urban knowledge and innovation spaces. It, along with hard factors, also concentrates on intangible outcomes (e.g., human capital) or soft factors. While since the late 1990s the clustering of companies and technologies outside metropolitan areas was conceived as one of the most effective strategies to create innovation and economic value in a region, with the dawn of the 21st century, the pendulum began to change direction, and innovation ecosystems are now emerging in the heart of cities. Inner city knowledge precincts are examples of knowledge spaces gener-

Table 10.5 Evolution of the knowledge-based urban development concept (derived from Carrillo 2015)

Concepts	Generations		
	1st generation	2nd generation	3rd generation
Period	1960s–1990s	1990s–2000s	2000s–2010s
Focus	Product-centred	Talent-centred	Community-centred
Knowledge perspective	Information content	Flow capacity	Value alignment
Socioeconomic perspective	Accumulate and retain knowledge stock/investment	Facilitate and increase knowledge circulation and investment	Dynamically adjust for sociocultural, spatial and economic viability
Clustering perspective	Exurban/suburban science, technology, and business parks	Exurban/suburban/urban knowledge and innovation precincts/districts	Suburban/urban knowledge and innovation community precincts/districts
Development leveraging instrument	Infrastructures to increase the stock of knowledge and investment	Platforms for facilitating the flow of knowledge and encouraging/marketing investment	Strategies for achieving the dynamic balance of all urban capitals for a sustainable development

ated as a result of the second generation KBUD approach. For instance, Brisbane's Dutton Park Knowledge Precinct (where no community building activities and no onsite accommodation options are available for the workers and their families) is an example of this type of development.

- The third generation and 'community-centred' KBUD approach focuses on the context, meaning and value, and thus brings to the forefront the received distinction between tangible or traditional capitals (e.g., physical and monetary) and intangible or contemporary capitals (e.g., social, human, intellectual capitals and knowledge assets). This type of KBUD approach effectively utilises both hard and soft factors in forming highly connected and accessible community knowledge and innovation spaces that are integrated with rest of the city and their surroundings. This KBUD approach not only builds a space, but also a place for knowledge and innovation generation in a cohesive urban community setting. Inner city knowledge community precincts are fine examples of knowledge spaces generated as a result of the third generation KBUD approach. For instance, Singapore's One-North (where the precinct has a reputation for making place for knowledge communities) is a rare example of this type of development.

During the last five decades or so the KBUD concept has evolved and used for shaping technopoles, science cities, knowledge cities and the like. In time moving away from mostly infrastructure-focused view (the first generation KBUD), it has become a more sustainable and balanced development type (the third generation KBUD) concerning many developmental aspects beyond economy and technology—i.e., sociocultural, spatial and institutional development (Carrillo et al. 2014). Even though there have been some developments in the theorisation of KBUD and advocating its prospects for cities and societies, the practice of KBUD to form flourishing knowledge cities is a complicated process with numerous challenges. KBUD is conceptualised in the framework illustrated in Fig. 10.2 and some of the common policies are listed in Table 10.6 (Yigitcanlar 2014a, b).

Along with KBUD, place making has become a popular approach particularly to create space and place attractive to knowledge industry and workers. In the context of KBUD place making considers place not only as location, but also as: (a) physical characteristics of its design that help develop a sense of place; (b) wider context that influences this shaping up; (c) relationally constituted multidimensional processes that are embedded in broader set of social, economic, and organisational relations shaping it, and; (d) image that is developed by virtue of this context, processes, physical characteristics and meanings associated by people (Pancholi et al. 2015). Placemaking is a critical mechanism of delivering KBUD, and to provide quality measures particularly in knowledge and innovation spaces of cities, where these spaces are the primary KBUD nexus of cities.

10.3 A New Planning and Development Paradigm: Knowledge-Based Urban Development 123

Table 10.6 Common knowledge-based urban development policies (derived from Yigitcanlar 2014a, b)

Policy categories	Policy sets	Policies	Descriptions
Economic development	Macro-economic foundations	Gross domestic product	Gross domestic product (GDP) per capita in USD purchasing power parities
		Major international companies	Number of global top 500 companies located
		Foreign direct investment	Ratio of international share in foreign direct investments
		Urban competitiveness	Global urban competitiveness index ranking
	Knowledge economy foundations	Innovation economy	International city ranking in innovation economy
		Research and development	Ratio of research and development expenditure in GDP
		Patent applications	Patent Cooperation Treaty patent applications per million inhabitants
		Knowledge worker pool	Ratio between professionals and managers and all workers
Socio-cultural development	Human and social capitals	Education investment	Ration between public spending on education and GDP
		Professional skill base	Ratio of residents over 18 years with tertiary degree

(continued)

Table 10.6 (continued)

Policy categories	Policy sets	Policies	Descriptions
		University reputation	World university rankings
		Broadband access	Ratio of access to fixed broadband subscribers per capita
	Diversity and independency	Cultural diversity	Ratio of people born abroad
		Social tolerance	International country tolerance ranking
		Socio-economic dependency	Ratio between the elderly population and the working age
		Unemployment level	Ratio of unemployment
Enviro-urban development	Sustainable urban development	Eco-city formation	International city ranking in eco-city
		Sustainable transport use	Ratio of sustainable transport mode use for commuting
		Environmental impact	CO_2 emissions in metric tons per capita
		Urban form and density	Population density in persons per sqkm
	Quality of life and place	Quality of life	International city ranking in quality of life
		Cost of living	International city ranking in cost of living

(continued)

10.3 A New Planning and Development Paradigm: Knowledge-Based Urban Development

Table 10.6 (continued)

Policy categories	Policy sets	Policies	Descriptions
Institutional development	Governance and planning	Housing affordability	Ratio between GPD per capita and median dwelling price
		Personal safety	International city ranking in personal safety
		Government effectiveness	Level of government effectiveness
		Electronic governance	International city ranking in e-government
		Strategic planning	Level of KBUD strategies in strategic regional and local development plans
		City branding	International city ranking in city branding
	Leadership and support	Effective leadership	Level of institutional and managerial leadership in overseeing KBUD
		Strategic partnership and networking	Level of triple-helix and PPPs and global networking-global city ranking
		Community engagement	Level of institutional mechanisms for community building and public participation
		Social cohesion and equality	Level of income inequality in gini coefficient

10.4 Prospects and Constraints in the Formation of Knowledge Cities

Today, a number of planned knowledge and innovation clusters across the globe sits virtually empty or in varying stages of completion. According to Zhuravlyova (2017), the main reason is that cities of these clusters are not welcoming places to live and work and lacking the offerings of the quality of life and place, beauty and diversity. In other words, these cities that aiming to house thriving knowledge precincts that lack of a knowledge city potential are destined to failure (Yigitcanlar and Sarimin 2015).

The premise of knowledge cities is not limited to economic prosperity and quality of living. Ideally, they are the localities of intellectual advancement, environmental sensitivity, social inclusiveness and cohesiveness, and participatory and transparent governance (Goldberg et al. 2006; Bontje et al. 2011b; Yigitcanlar et al. 2012). The global financial and European migrant crises along with the impacts of the global climate change have proven that establishing and sustaining a knowledge city status is a highly challenging business. In other words, despite its prospects initiatives and efforts towards the formation of knowledge cities are constantly being challenged by a number of constraints. Yigitcanlar et al. (2014) list some of the prospects and

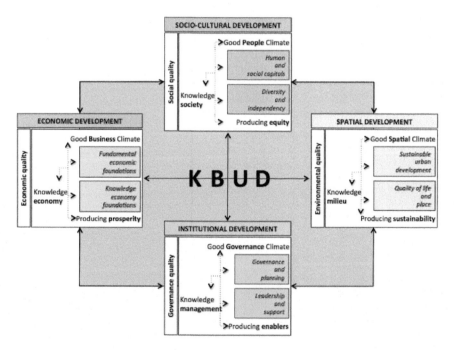

Fig. 10.2 Conceptual framework of knowledge-based urban development (derived from Yigitcanlar 2014a, b)

10.4 Prospects and Constraints in the Formation of Knowledge Cities

Table 10.7 Prospects and constraints in the formation of knowledge cities

Prospect examples	Constraint examples
• Ability to network and cooperate with other cities	• Absence of incentive programs for innovation and R&D
• Active citizens in the development of the city, participating in decision-making	• Cost of world-class infrastructure and amenities
• Agility and capability to react immediately to the key challenges	• Declining industry-based development and new business creation not at the desired level
• Bringing universities, industries, and government in knowledge precincts	• Differing level of inequality in the subregions, disparity between the city and others
• Building new knowledge-intensive brands	• Inadequate investment in skill development
• Capability to further building on existing strengths	• Increasing import dependency particularly in value-added products
• City branding and marketing initiatives	• Increasing long-term unemployment
• Collaboration between municipalities in the region and joining forces for development	• Insufficient innovation and support mechanism to become an international player
• Community spirit, lots of volunteer work and team spirit	• Inward looking culture and demotivation for collaboration
• Encouraging the mix of people and ideas	• Lack of a vibrant urban social scene
• Focused institutional support and focusing on the right business sectors	• Lack of business and employment attractiveness particularly to SMEs
• High-level accessibility and safety	• Lack of entrepreneur spirit
• High-level cutting-edge technology adoption and use	• Lack or limited collaboration between three-tier of government
• High-quality education and growing young population	• Lack or limited industry-government-academia partnership
• Increasing economic diversity of the city by investing in more than one sector	• Limited influence of national level decision-making and poor lobbying
• Increasing research base and relevant knowledge base	• Limited institutional perspective and governance structures and flexibility
• International collaboration in knowledge intensive areas	• Limited natural environmental quality and impacts on health
• Knowledge worker quality and young professional skill-base of the city	• Limited networks for innovation collaboration and establishing joint ventures

(continued)

constraints in the formation of knowledge cities (see Table 10.7). The items in the list can also be regarded as measures that provide an indication whether a city is heading in the right direction.

Table 10.7 (continued)

Prospect examples	Constraint examples
• Low threshold between institutions that is encouraging partnership	• Limited planning, coordination and leadership for the long run
• Increasing investment efforts on the internationalisation	• Low foreign direct investment and attraction to global knowledge industries and workers
• New open-source innovation platforms to boost the competitive edge	• Low level of internationalisation especially in the science and education areas
• Open culture in social domains and to new ideas to turn them into marketable products	• Low-level service sector employment due to lower salaries
• Partnership between numerous organisations	• No target setting in R&D and the global focus
• Regional actors' collaboration and joining forces for development	• Not so attractive region for highly talented immigrants
• Rich natural environment, attractive to residents and visitors	• Poor quality or unaffordable housing and services
• Strong academy and industry collaboration and partnership	• Raising environmental challenges (e.g., impacts of global climate change)
• Strong city with sociocultural activities, and high level of infrastructure and quality of life	• Social and economic disparities, and low quality of life for some residents
• Strong network for collaboration for urban development	• Staying in the shadow of the primate city and no strong brand of the region
• Targeting the development based on endogenous assets and knowledge	• Traditional SMEs lacking the science and technology knowledge

10.5 Summary

This chapter focused on providing an understanding of the knowledge cities and knowledge-based urban development concepts and their evolution—with the following two central questions in mind.

The first question, why pursuing a knowledge city goal is important for cities of the global knowledge economy era, is addressed through explaining the changes and critical challenges of the 21st century—including dynamic processes of economic and spatial restructuring while the world is moving towards a global information order. This process is not only pushing cities and nations to become more innovative, but also exacerbating the global rivalry for gaining an economic competitive edge. Highlighted by Esmailpoorarabi et al. (2016, p. 14), "globalisation and the digital revolution have entirely changed the nature of the economy, which [now] heavily relies on the knowledge-based activities. This economy type is engaged with knowledge, learning, and innovation processes mainly in urban spaces. Transition to a knowledge economy has shaped a new form of urban competitiveness in which every level of urban spaces—i.e., regions, cities, clusters—are in harmony for enhancing

10.5 Summary

their mostly intangible creative and innovative values". The result is the pursuit of a knowledge city for many urban locations (Sarimin and Yigitcanlar 2011).

The second one, what the key conditions and policy drivers are to transform cities into prosperous knowledge cities, has been a subject of extended discussion in the literature. In recent years, the role of 'contextual features'—such as economic and societal assets and conditions, and policy, planning and institutional support mechanisms—in the success of knowledge cities have been widely acknowledged (Asakawa and Lehrer 2003). Besides, 'place-making'—a people-centred and multi-faceted approach for planning, designing and managing urban spaces—has become a prominent strategy to attract footloose capital and talent into knowledge cities. The most prominent of these conditions (such as local and regional context and essential assets) and drivers (such as knowledge-based policies) are introduced and elaborated along with the primary prospects and constraints of knowledge cities in this chapter of the book.

References

AEUB (Agència d'Ecologia Urbana de Barcelona). (2000). Research description. Accessed on September 6, 2011 from http://www.bcnecologia.net/en.
Anttiroiko, A. V. (2004). Science cities: Their characteristics and future challenges. *International Journal of Technology Management, 28*(3–6), 395–418.
Anttiroiko, A. V. (2015). City branding as a response to global intercity competition. *Growth and Change, 46*(2), 233–252.
Arbolino, R., Carlucci, F., Cira, A., Ioppolo, G., & Yigitcanlar, T. (2017). Efficiency of the EU regulation on greenhouse gas emissions in Italy: the hierarchical cluster analysis approach. *Ecological Indicators, 81*(1), 115–123.
Arbolino, R., Carlucci, F., Simone, L., Yigitcanlar, T., & Ioppolo, G. (2018). The policy diffusion of environmental performance in the European countries. *Ecological Indicators, 89*(1), 130–138.
Arbonies, A., & Moso, M. (2002). Basque country: The knowledge cluster. *Journal of Knowledge Management, 6*(4), 347–355.
Asakawa, K., & Lehrer, M. (2003). Managing local knowledge assets globally: The role of regional innovation relays. *Journal of World Business, 38*(1), 31–42.
Asheim, B. T. (2012). Innovating: Creativity, innovation and the role of cities in the globalising knowledge economy. In T. Yigitcanlar, K. Metaxiotis, & F. J. Carrillo (Eds.), *Building prosperous knowledge cities: Policies, plans and metrics* (pp. 3–23). Cheltenham: Edward Elgar.
Asheim, B., Coenen, L., Moodysson, J., & Vang, J. (2007). Constructing knowledge-based regional advantage: Implications for regional innovation policy. *International Journal of Entrepreneurship and Innovation Management, 7*(2–5), 140–155.
Bontis, N. (2004). National intellectual capital index. *Journal of Intellectual Capital, 5*(1), 13–39.
Bontje, M., Musterd, S., & Pelzer, P. (2011a). *Inventive city-regions: Path dependence and creative knowledge strategies*. UK: Ashgate.
Bontje, M., Musterd, S., Kovacks, Z., & Murie, A. (2011b). Pathways toward European creative-knowledge city-regions. *Urban Geography, 32*(1), 80–104.
Camagni, R., Capello, R., & Nijkamp, P. (1998). Towards sustainable city policy: An economy-environment technology nexus. *Ecological Economics, 24*(1), 103–118.
Carrillo, F. J. (Ed.). (2006). *Knowledge cities: Approaches, experiences and perspectives*. New York: Routledge.

Carrillo, F. J. (2014). What 'knowledge-based' stands for? A position paper. *International Journal of Knowledge-Based Development, 5*(4), 402–421.

Carrillo, F. J. (2015). Knowledge-based development as a new economic culture. *Journal of Open Innovation: Technology, Market, and Complexity, 1*(15), 1–17.

Carrillo, F. J., Yigitcanlar, T., García, B., & Lönnqvist, A. (2014). *Knowledge and the city: Concepts, applications and trends of knowledge-based urban development*. New York: Routledge.

Castells, M., & Hall, P. (1994). *Technopoles of the world: The making of the 21st century industrial complexes*. New York: Routledge.

Chang, D., Sabatini-Marques, J., Costa, E., Selig, P., & Yigitcanlar, T. (2018). Knowledge-based, smart and sustainable cities: a provocation for a conceptual framework. *Journal of Open Innovation: Technology, Market, and Complexity, 4*(1), 5.

Cooke, P., & Leydesdorff, L. (2006). Regional development in the knowledge-based economy: The construction of advantage. *The Journal of Technology Transfer, 31*(1), 5–15.

Dvir, R. (2006). Knowledge city, seen as a collage of human knowledge moments. In F. J. Carrillo (Ed.), *Knowledge cities: approaches, experiences and perspectives* (pp. 245–259). New York: Routledge.

Edvardsson, I. R., Yigitcanlar, T., & Pancholi, S. (2016). Knowledge city research and practice under the microscope: A review of empirical findings. *Knowledge Management Research & Practice, 14*(4), 537–564.

Edvinsson, L. (2006). Aspects on the city as a knowledge tool. *Journal of Knowledge Management, 10*(5), 6–13.

Entovation. (2004). Our knowledge city manifesto. Accessed on September 6, 2011 from http://www.entovation.com.

Ergazakis, K., Metaxiotis, K., & Psarras, J. (2004). Towards knowledge cities: Conceptual analysis and success stories. *Journal of Knowledge Management, 8*(5), 5–15.

Ergazakis, K., Metaxiotis, K., & Psarras, J. (2006). An emerging pattern of successful knowledge cities' main features. In F. J. Carrillo (Ed.), *Knowledge cities: Approaches, experiences and perspectives* (pp. 3–16). New York: Routledge.

Esmaeilpoorarabi, N., Yigitcanlar, T., & Guaralda, M. (2016). Place quality and urban competitiveness symbiosis? A position paper. *International Journal of Knowledge-Based Development, 7*(1), 4–21.

Esmaeilpoorarabi, N., Yigitcanlar, T., & Guaralda, M. (2018a). Place quality in innovation clusters: An empirical analysis of global best practices from Singapore, Helsinki, New York, and Sydney. *Cities, 74*(1), 156–168.

Esmaeilpoorarabi, N., Yigitcanlar, T., Guaralda, M., & Kamruzzaman, M. (2018b). Evaluating place quality in innovation districts: A Delphic hierarchy process approach. *Land Use Policy, 76*(1), 471–486.

Florida, R. (2002). *The rise of the creative class: And how it's transforming work, leisure, community and everyday life*. New York: Basic Books.

Gabe, T., Abel, J., Ross, A., & Stolarick, K. (2012). Knowledge in cities. *Urban Studies, 49*(1), 1179–1200.

Garcia, B. (2004). Developing futures: A knowledge-based capital for Manchester. *Journal of Knowledge Management, 8*(5), 47–60.

Garcia, B. (2010). Making MAKCi: An emerging knowledge-generative network of practice in the Web 2.0. *VINE, 40*(1), 39–61.

Goldberg, M., Maya, E., & Levin-Sagi, M. (2006). Citizen participation in decision-making processes: Knowledge sharing in knowledge cities. *Journal of Knowledge Management, 10*(5), 92–98.

Huggins, R. (2008). The evolution of knowledge clusters: Progress and policy. *Economic Development Quarterly, 22*(4), 277–289.

Huggins, R., & Strakova, L. (2012). Knowledge-based economic development in emerging regions: Policy issues and implications in the Balkan Peninsula. *Regional Studies, 46*(7), 961–975.

References

Ihlanfeldt, K. R. (1995). The importance of the central city to the regional and national economy: A review of the arguments and empirical evidence. *Cityscape, 1*(2), 125–150.

Johnston, A., & Huggins, R. (2016). The spatio-relational nature of urban innovation systems: Universities, knowledge intensive business service firms, and collaborative networks. *Journal of Urban Technology, 23*(1), 29–52.

Knight, R. V. (1995). Knowledge-based development: Policy and planning implications for cities. *Urban studies, 32*(2), 225–260.

Komninos, N. (2002). *Intelligent cities: Innovation, knowledge systems, and digital spaces.* New York: Taylor & Francis.

Kozak, M. (2011). Strategic approach to intellectual capital development in regions. *International Journal of Learning and Intellectual Capital, 8*(1), 76–93.

Kunzmann, K. (2009). The strategic dimensions of knowledge industries in urban development. *disP-The Planning Review, 45*(2), 40–47.

Lara, A., Costa, E., Furlani, T., & Yigitcanlar, T. (2016). Smartness that matters: comprehensive and human-centred characterisation of smart cities. *Journal of Open Innovation: Technology, Market, and Complexity, 2*(1), 8.

Larsen, K. (1999). Learning cities: The new recipe in regional development. *Organisation for Economic Cooperation and Development. The OECD Observer,* (217/218), 73–76.

Lönnqvist, A., Käpylä, J., Salonius, H., & Yigitcanlar, T. (2014). Knowledge that matters: Identifying regional knowledge assets of the Tampere region. *European Planning Studies, 22*(10), 2011–2029.

Madanipour, A. (2013). *Knowledge economy and the city: Spaces of knowledge.* New York: Routledge.

Massey, D. B., Quintas, P., & Wield, D. (1992). *High-tech fantasies: Science parks in society, science and space.* London: Routledge.

Metaxiotis, K., Carrillo, J., & Yigitcanlar, T. (Eds.). (2010). *Knowledge-based development for cities and societies: An integrated multi-level approach.* Hersey: IGI Global.

Musterd, S. (2004). Amsterdam as a creative cultural knowledge city: Some conditions. *Built Environment, 30*(3), 225–234.

Musterd, S., & Deurloo, R. (2006). Amsterdam and the preconditions for a creative knowledge city. *Tijdschrift Voor Economische En Sociale Geografie, 97*(1), 80–94.

OECD. (2001). *The new economy: Beyond the hype.* Paris: OECD Press.

Pancholi, S., Yigitcanlar, T., & Guaralda, M. (2015). Place making facilitators of knowledge and innovation spaces: Insights from European best practices. *International Journal of Knowledge-Based Development, 6*(3), 215–240.

Piore, M., & Sabel, C. (1984). *The second industrial divide: Possibilities for prosperity.* New York: Basic Books.

Porter, M. E. (1990). *The competitive advantage of nations.* London: Macmillan.

Portney, K. E. (2013). *Taking sustainable cities seriously: Economic development, the environment, and quality of life in American cities.* Boston: MIT Press.

Romein, A., Fernández-Maldonado, A. M., & Trip, J. J. (2011). Delft blues: The long road from University Town to knowledge city. *International Journal of Knowledge-Based Development, 2*(2), 148–165.

Sarimin, M., & Yigitcanlar, T. (2011). Knowledge-based urban development of Multimedia Super Corridor, Malaysia: an overview. *International Journal of Knowledge-Based Development, 2*(1), 34–48.

Sarimin, M., & Yigitcanlar, T. (2012). Towards a comprehensive and integrated knowledge-based urban development model: status quo and directions. *International Journal of Knowledge-Based Development, 3*(2), 175–192.

Sassen, S. (2011). *Cities in a world economy.* New York: Sage Publications.

Scott, A. J. (1993). *Technopolis: High-technology industry and regional development in Southern California.* San Francisco: University of California Press.

Smilor, R., Gibson, D., & Kozmetsky, G. (1988). Creating the technopolis: High-technology development in Austin. *Journal of Business Venturing, 4*(1), 49–67.

Storey, D. J., & Tether, B. S. (1998). Public policy measures to support new technology-based firms in the European Union. *Research Policy, 26*(9), 1037–1057.

Storper, M. (1997). *The regional world: Territorial development in a global economy*. New York: The Guilford Press.

Storper, M., & Venables, A. J. (2004). Buzz: Face-to-face contact and the urban economy. *Journal of Economic Geography, 4*(1), 351–370.

TNO-INRO (Netherlands Organisation for Applied Scientific Research, Institute for Spatial Organisation). (1990). *Delft knowledge city concept*. Technical Report, Delft.

Trindade, E., Hinnig, M., Costa, E., Sabatini-Marques, J., Bastos, R., & Yigitcanlar, T. (2017). Sustainable development of smart cities: a systematic review of the literature. *Journal of Open Innovation: Technology, Market, and Complexity, 3*(1), 11.

Van Den Berg, L., Van Winden, W., & Pol, P. (2005). *European cities in the knowledge economy: The cases of Amsterdam, Dortmund, Eindhoven, Helsinki, Manchester, Munich, Münster, Rotterdam and Zaragoza* (Vol. 1013). Gower Publishing, Ltd.

Van Geenhuizen, M., & Nijkamp, P. (Eds.). (2012). *Creative knowledge cities: Myths, visions and realities*. Cheltenham: Edward Elgar.

Van Geenhuizen, M., Rijckenberg, H., & Nijkamp, P. (1997). Universities and knowledge-based economic growth: The case of Delft (NL). *GeoJournal, 41*(4), 369–377.

Van Winden, W. (2010). Knowledge and the European city. *Tijdschrift Voor Economische En Sociale Geografie, 101*(1), 100–106.

Van Winden, W., Van Den Berg, L., & Pol, P. (2007). European cities in the knowledge economy: Towards a typology. *Urban Studies, 44*(3), 525–549.

Vedovello, C. (1997). Science parks and university-industry interaction: Geographical proximity between the agents as a driving force. *Technovation, 17*(9), 491–502.

Wang, X. (2009). *Knowledge-based urban development in China* (Unpublished Ph.D. thesis). Newcastle University, UK.

Westhead, P., & Storey, D. J. (1995). Links between higher education institutions and high technology firms. *Omega, 23*(4), 345–360.

Yigitcanlar, T. (2005). The making of knowledge cities: Lessons learned from Melbourne. In *International Symposium on Knowledge Cities*, November 28–30, pp. 1–18, The World Bank and Arab Urban Development Institute, Medina, Saudi Arabia.

Yigitcanlar, T. (2009). Planning for knowledge-based urban development: Global perspectives. *Journal of Knowledge Management, 13*(5), 228–242.

Yigitcanlar, T. (2011). Position paper: Redefining knowledge-based urban development. *International Journal of Knowledge-Based Development, 2*(4), 340–356.

Yigitcanlar, T. (2014a). Position paper: Benchmarking the performance of global and emerging knowledge cities. *Expert Systems with Applications, 41*(12), 5549–5559.

Yigitcanlar, T. (2014b). Valleys, alleys and roundabouts: Innovating beyond a precinct. *The Conversation*. Accessed on March 17, 2017 from http://theconversation.com/valleys-alleys-androundabouts-innovating-beyond-a-precinct-24290.

Yigitcanlar, T. (2015). Knowledge based urban development. In M. Khosrow-Pour (Ed.), *Encyclopedia of information science and technology* (3rd ed., pp. 7475–7485). Hersey, PA: IGI Global.

Yigitcanlar, T. (2016). *Technology and the city: Systems, applications and implications*. New York: Routledge.

Yigitcanlar, T., Baum, S., & Horton, S. (2007). Attracting and retaining knowledge workers in knowledge cities. *Journal of Knowledge Management, 11*(5), 6–17.

Yigitcanlar, T., & Bulu, M. (2015). Dubaization of Istanbul: Insights from the knowledge-based urban development journey of an emerging local economy. *Environment and Planning A, 47*(1), 89–107.

Yigitcanlar, T., & Dur, F. (2013). Making space and place for knowledge communities: Lessons for Australian practice. *Australasian Journal of Regional Studies, 19*(1), 36–63.

References

Yigitcanlar, T., Edvardsson, I. R., Johannesson, H., Kamruzzaman, M., Ioppolo, G., & Pancholi, S. (2017). Knowledge-based development dynamics in less favoured regions: Insights from Australian and Icelandic University Towns. *European Planning Studies, 25*(12), 2272–2292.

Yigitcanlar, T., Guaralda, M., Taboada, M., & Pancholi, S. (2016). Place making for knowledge generation and innovation: Planning and branding Brisbane's knowledge community precincts. *Journal of Urban Technology, 23*(1), 115–146.

Yigitcanlar, T., & Kamruzzaman, M. (2018). Does smart city policy lead to sustainability of cities? *Land Use Policy, 73*(1), 49–58.

Yigitcanlar, T., Kamruzzaman, M., Buys, L., Ioppolo, G., Sabatini-Marques, J., Costa, E., et al. (2018). Understanding 'smart cities': Intertwining development drivers with desired outcomes in a multidimensional framework. *Cities, 81*(1),145–160.

Yigitcanlar, T., & Lönnqvist, A. (2013). Benchmarking knowledge-based urban development performance: Results from the international comparison of Helsinki. *Cities, 31*(1), 357–369.

Yigitcanlar, T., Lonnqvist, A., & Salonius, H. (2014). Analysis of a city-region from the knowledge perspective: Tampere, Finland. *VINE, 44*(3), 445–466.

Yigitcanlar, T., Metaxiotis, K., & Carrillo, J. (Eds.). (2012). *Building prosperous knowledge cities: Policies, plans and metrics*. Cheltenham: Edward Elgar.

Yigitcanlar, T., O'Connor, K., & Westerman, C. (2008a). The making of knowledge cities: Melbourne's knowledge-based urban development experience. *Cities, 25*(2), 63–72.

Yigitcanlar, T., Velibeyoglu, K., & Baum, S. (Eds.). (2008b). *Knowledge-based urban development: Planning and applications in the information era—Planning and applications in the information era*. Hersey: IGI Global.

Yigitcanlar, T., Velibeyoglu, K., & Martinez-Fernandez, C. (2008c). Rising knowledge cities: The role of knowledge precincts. *Journal of Knowledge Management, 12*(5), 8–20.

Yigitcanlar, T., & Sarimin, M. (2011). The role of universities in building prosperous knowledge cities: the Malaysian experience. *Built Environment, 37*(3), 260–280.

Yigitcanlar, T., & Sarimin, M. (2015). Multimedia Super Corridor, Malaysia: knowledge-based urban development lessons from an emerging economy. *VINE: The Journal of Information and Knowledge Management, 45*(1), 126–147.

Zhao, P. (2010). Building knowledge city in transformation era: Knowledge-based urban development in Beijing in the context of globalisation and decentralisation. *Asia Pacific Viewpoint, 51*(1), 73–90.

Zhuravlyova, S. (2017). Breath of fresh air: Cote d'Azur. *The Forecast, 5*(1), 90–93.

Chapter 11
Global Knowledge City Best Practice

Abstract There are a number of cities so far has achieved a globally recognised prosperous knowledge-based urban development. This chapter presents some of the leading examples of global knowledge city best practice. These examples include Barcelona, Austin, Singapore and Melbourne. The chapter provides a detailed analysis on these prominent knowledge cities and generates useful insights for other cities targeting such a successful knowledge-based urban development practice to turn their cities into knowledge-based ones. Brisbane city from Australia is used as an example of one of these aspiring cities. The chapter concludes with a commentary on how the lessons captured from global best practices could be useful for this particular city.

Keywords Knowledge city · Smart city · Sustainable urban development
Best practice · Knowledge-based urban development · Barcelona · Austin
Singapore · Melbourne

In order to develop effective and efficient policies for transforming cities into prosperous knowledge cities, it is essential to learn from the other high-performing knowledge cities' experiences. A best practice analysis is a commonly used method or technique to determine success factors and produce lessons from the outstanding knowledge city practices that set the standards. Findings of such best practice analysis are useful to form ideas, policies, and guidelines that represent the most efficient or prudent course of action. However, careful translation of the best practice lessons to the local context and meticulous tailoring of the potential actions that have worked in the investigated cases are of the utmost importance.

Against this background, this chapter of the book aims to address the following critical question through drawing lessons from the best practice knowledge cities by focusing on the success and failure factors and analysing their suitability for Brisbane.

- How can lessons learned from global knowledge city best practice be evaluated for their potential in Brisbane?
- How these finding could be used by other cities and regions with similar characteristics, contexts and ambitions?

11.1 A Best Practice Knowledge City from Europe: Barcelona

Europe is the continent where the knowledge cities concept was coined in the 1990s. Since then many cities adopted the concept in their vision and brands—such as Amsterdam, Barcelona, Delft, Eindhoven, Helsinki, and Munich. From this context with numerous rich knowledge cities, this section investigates Barcelona, Spain (winner of the 2009 Most Admired Knowledge City Award—MAKCi) as a potential role model city for Brisbane.

At the turn of the new millennium, Barcelona finds itself facing an outbreak of a new scientific and technical revolution entailing great political, economic and social challenges. The appearance of a new wave of technological innovations and acceleration in the economy's internationalisation process make traditional forms of organisation in the city obsolete. Faced with the globally competitive environment of the knowledge economy, Barcelona has undertaken a profound technological and cultural regeneration in order to position itself among the major metropolises of the global knowledge society (Yigitcanlar 2009).

After the 1992 Olympic Games, which provided the city with a better infrastructure, in 1999, Barcelona City Council developed a strategic plan for KBUD of the city with an aim of transforming Barcelona into a 'City of Knowledge'. This plan emphasised the necessity of the cultural and creative sectors to become the motor of a new transformation of the metropolis. A large number of residents and around 200 public institutions volunteered for the development and implementation of the knowledge city strategy. Private sector's initiatives and actions, mainly in the development of infrastructures and knowledge businesses, played an important role in the success of the whole KBUD process (Yigitcanlar 2009).

According to Bakici et al. (2013, p. 139), the development perspective of the city "implies a high-tech intensive and an advanced city that connects people, information and city elements using new technologies in order to create a sustainable, greener city, competitive and innovative commerce and a recuperating life quality with a straightforward administration and a good maintenance system. [Moreover, it] is a collaborative movement among its corporations, academic institutions, government authorities and the residents of Barcelona, aimed at becoming a reference program for economic engines and urban development".

Barcelona received the City Climate Leadership award, European Capital of Innovation prize, and Bloomberg Mayor's Challenge award in 2014 in recognition of its knowledge city progress (Capdevila and Zarlenga 2015). The knowledge city strategy of Barcelona aims to develop productive districts at human speed inside a hyper-connected and zero-emission city. Most importantly the city's commitment to avoid social exclusion on the basis of race and income while undertaking major urban regeneration projects are commendable (Griffith 2015). The key objectives of the strategy are (Barcelona City Hall 2012):

- Establishment of a new organisational unit to deliver knowledge city principles and objectives;

11.1 A Best Practice Knowledge City from Europe: Barcelona

- Legal framework for public-private-partnership to flourish;
- Urban area for public-private-partnership to settle and grow;
- Feed the urban innovation ecosystem;
- Living lab, development of a community of citizen developers;
- Facilities for start-ups and small- and medium-sized enterprise (SME) experimentation;
- Opening new opportunities for citizens to be more active and participative, and;
- Strong international linkages.

Today Barcelona is one of the most successful knowledge cities and marked its name as 'the Culture Capital' of Europe. Among the various initiatives shaping this vision for a culture of excellence in Barcelona, the development of innovation districts is a notable one (Yigitcanlar 2016). Particularly, the 22@Barcelona KBUD project stands out on account of its ambitious scope and innovative conceptualisation (Charnock and Ribera-Fumaz 2011). This innovation district project transformed 200 ha of old industrial land in the city centre into a privileged urban environment for creation, transfer, and attraction of knowledge, and establishes a creative urban core for Barcelona (Yigitcanlar 2009).

The success of Barcelona's innovation district development model for knowledge city formation comes from: (a) Establishing sound foundations for dialogue on urban space, economy, and society, and; (b) Creating a unique development model that puts universities, companies, and the city administration at the heart of economic and technological transformations (Yigitcanlar 2016). Particularly the accomplishments of the Barcelona model have made it a benchmark of urban, economic and social transformation for many other cities. This model is studied and followed by numerous knowledge cities and knowledge and innovation districts around the world (Pareja-Eastaway and Piqué 2011).

Besides Barcelona's strong creative sectors, the knowledge sectors of the city are specialised in media, energy, medical technology, ICT, and design, and contributes to the economic development significantly. The Torre Agbar—a 38-story skyscraper/tower located in the 22@Barcelona innovation district and designed by famous French architect Jean Nouvel—is associated with Barcelona's economic development ambition, and direction and symbolises the KBUD of the city (Fig. 11.1).

KBUD features that have characterised the new development perspective of the city constitute the Barcelona model. The model combines strategic policymaking (urban policy), urban planning (urban transformation) and urban discourse (the need for a City of Knowledge). The Barcelona model provides confidence and certainty on the issues of: (a) Good will; (b) Transparency, and; (c) Social consensus (Clua and Albet 2008). These three strengths along with the followings can be seen as the core ingredients of success of the Barcelona model (see Garcia-Ramon and Albet 2000):

- Establishing a competent municipal leadership in the design and management of the development, even though the investment is mainly of private origin;

Fig. 11.1 Symbol of Barcelona's knowledge-based urban development

- Receiving unconditional support, as a result of good governance and mayor's charisma, from public administrations, financial institutions, and socioeconomic entities;

- Keeping a global vision for the city in spite of mega international projects and events (e.g., Expo, Olympic Games);
- Building the development on the basis of existing and unique tangible and intangible assets of the city (e.g., architecture, history, culture);
- Diligently working on strategic urban marketing and city branding including networking and lobbying;
- Strictly following existing urban development plans and planning regulations so as to maintain coherence, credibility, and legitimacy of the development;
- Understanding the role of public spaces (e.g., streets, squares, facilities) as characteristic elements of generating identity and establishing social and cultural integration;
- Avoiding gentrification and introducing mixed land-use to avoid marginal social zones and to maintain social coherence;
- Creating a powerful coalition of professionals, technicians, developers, neighbourhood associations, and local councillors, and;
- Involving citizens in the planning and development processes with the role of urban volunteers.

During the last decade, a new development movement has appeared in Barcelona, like many other knowledge cities, to build a 'smart city' on the foundations of the existing knowledge city—smart city is an urban environment which, supported by pervasive information and communication technology systems, is able to offer advanced and innovative services to citizens in order to improve the overall quality of their life (Piro et al. 2014). Therefore, in recent last years, a plethora of successful projects have emerged from the 22@Barcelona district and the rest of Barcelona to support both knowledge city and smart city ambitions of the city (see Yigitcanlar 2016).

11.2 A Best Practice Knowledge City from North America: Austin

North America is a perfect region for knowledge city best practice analysis as it hosts many prominent knowledge cities such as Austin, Boston, San Francisco, and Toronto. Particularly San Francisco metropolitan area with its famous Silicon Valley stands out as a major role model and benchmark for many cities to pursue and realise similar achievements. As San Francisco and Silicon Valley will be examined as a knowledge precinct/corridor best practice in Part III of this book, this section of the book focuses on Austin, Texas, USA (winner of the 2012 Most Admired Knowledge City Award—MAKCi).

Over the past three decades, Austin has emerged as a prosperous knowledge city, with an explosive growth of technology companies large and small, established firms and start-ups, and a vibrant social atmosphere (Fig. 11.2). There have been many contributors to this evolution of a passive government-dominated state capital to a

Fig. 11.2 Snapshot from a vibrant social scene of Austin

thriving innovation metropolis, spawning companies and products of national and international significance.

Austin was one of the first US cities to recognise both the emerging economic importance of knowledge work and the possibilities of attracting footloose industry (Long 2010). The city, led by business organisations, having decided to lure these new clean industries to the area, developed a plan to attract large corporations by touting the relatively low cost of living and the quality of university graduates. The university of Texas at Austin has long been considered as an elite public university, comparable in quality to private Ivy League institutions such as Yale and Harvard (Yigitcanlar 2009).

Austin's dual promise of cost savings and high-quality mental labour proved effective and by the late 1950s, the city boasted a fledgling electronics production sector with instrument maker Tracor—a major North American defence electronics contractor. The following decade saw the beginnings of the production of personal computing commodities. Austin was an early beneficiary of this new market attracting Texas Instruments to its growing knowledge industry base (Chapple et al. 2004). Despite building success in the 1970s by attracting Unisys, Westinghouse and Compaq established headquarters in Austin; the city did not rest on its laurels. In 1980 the city's local government, with important contributions from the Austin Chamber of Commerce and representatives from wider the business community, drew up the 'Austin 2010 Plan'. The plan marked an evolution in KBUD policy (Yigitcanlar 2009).

The traditional focus on the production process continued. At the same time, the Austin 2010 Plan committed to the systematic development of research capabilities in technology and computer science at the University of Texas (Tretter 2013). The plan also launched a tradition of local government working hand-in-hand with leading figures in the regional private sector to create an exemplary entrepreneurial climate and an open, flexible interface between government and business. Communication, between different sectors of society and around questions of the production innovation, was a central feature of Austin's creative environment. It was given effect via a local government-sanctioned network of companies and individuals (Yigitcanlar 2009).

The plan, in marked innovation for 1980, committed the city to redevelop its urban and cultural life in ways that would attract and retain knowledge workers. Economic development policy, in summary, widened its purview and started to explicitly consider the living conditions of workers. In one of its most prominent initiatives, the city started to self-consciously promote its legacy of music. In 1991, Austin capped this aspect of its lifestyle promotion with the adoption of an official city slogan, 'Live Music Capital of the World'. The slogan found ready acceptance in the US thanks largely to the television series of Austin City Limits. In its early recognition of the importance for the creative class of lifestyle choices, Austin was again in the vanguard of knowledge work/worker development (Long 2016). This dual-focus urban development policy helped launch high-skill, value-adding urban development in Austin on a path of sustained growth (Yigitcanlar 2009).

By 2002, the city began to look more seriously at how to diversify economic development. In the wake of a significant economic downturn—the dot-com bust stalled Austin's tech economy—there was an urgent need to change the existing economic development path. Recognising the ability of Austin's reputation to attract talent, capital and industry, in 2003 mayor Will Wynn formalised one of the largest financial investments of the city under the shelter of the Austin Chamber of Commerce and several businesses and community leaders. The plan aimed at fostering job-creating investment in Central Texas. The plan was first launched in 2004—after a thorough assessment of Austin's competitive position—as 'Opportunity Austin 1.0' with subsequent development of two five-year strategies called Opportunity Austin 2.0 (2009–2013) and Opportunity Austin 3.0 (2014–2018). All of them have registered solid success, surpassing expected goals. Top priorities of the current version of the Opportunity Austin initiative include 'boosting economic diversification to proactively strengthen the economy, deepening the talent pool through development and attraction, and keeping Austin attractive to entrepreneurs, business leaders and site selectors through expanded advocacy on issues such as a comprehensive regional transportation system and regional collaboration' (Austin Chamber 2016).

The uniqueness of Opportunity Austin 3.0 comes from presenting a 'checklist' of high impact strategies that provide the Austin Chamber with a core set of activities to continue fuelling the ascent of Austin—rather than including dozens of actions focused on multiple goal areas and objectives. The checklist (Austin Chamber 2012a) focuses on 'what' Austin must do to be most competitive for future jobs, investment, and talent. The implementation plan guidelines (Austin Chamber 2012b) focuses on 'how' they will be accomplished, including prioritisation of strategies and action steps based on their regional impact and viability, identification of lead and support entities, cost projections, and identification of whether the proposed action is ongoing, enhanced, or new.

The principal goal areas of Opportunity Austin 3.0 are divided into three key competitiveness areas that are critical to Austin's future success: These areas along with relevant strategies are as follows (Austin Chamber 2012a):

- Economy—creating high-value jobs for workers at all skill levels in sectors that will further diversify Austin's economy;

  - Strategy 1: Support high-value growth in the Austin's existing businesses.
  - Strategy 2: Showcase Austin in targeted domestic and international markets.
  - Strategy 3: Capture opportunities to build clusters in emerging target business sectors.
  - Strategy 4: Foster a world-class support system for technology start-up and commercialisation activities.
  - Strategy 5: Become a leading voice for policies and programs that enhance greater Austin's competitive position.

- Talent—preparing students effectively for university and careers while providing regional employers with the talent they need to grow and succeed;

  - Strategy 6: Effectively prepare Austin students for university and careers.
  - Strategy 7: Optimise Austin's workforce capacity.

- Place—addressing challenges related to Austin's strong and sustained population growth that is critical to ensuring that the quality of life continues to attract top talent and the infrastructure supports economic development goals.

  - Strategy 8: Continue to aggressively pursue multi-modal transportation development to mitigate traffic congestion in Austin.
  - Strategy 9: Foster regional dialogues on the benefits and positive impacts of Austin's development.

The key initiatives highlighted in the Opportunity Austin 3.0 are as follows (Austin Chamber 2012b):

- Leverage Austin's future medical school and teaching hospital as catalysts for the development of a regional life sciences and biomedical cluster.
- Create a non-partisan Central Texas Transportation Alliance as a business/citizen partnership to advance transportation improvements in the region.
- Develop tools to formalise career-preparation pipelines in Austin.
- Reorient Portfolio Austin based on Opportunity Austin 2013 target sectors.
- Continue the evolution of the Austin Technology Partnership.
- Focus targeted attraction efforts on Austin's highest-value domestic markets.
- Continue efforts to maximise the benefit of technology commercialisation at the University of Texas-Austin and its partners.

- Ensure that Austin's business costs are supportive of continued regional growth and investment.
- Design a branded campaign to focus public attention on Austin's development dynamics and their positive effect on economic and quality of life issues.

In 2008, the global economic downturn hampered Austin's high-tech sector. Big corporations tightened their belts, tech start-ups started failing and computer sales substantially decreased. By the end of 2009, the number of jobs in this sector slumped (IBM 2012) the lowest registered since the dot-com bust. As a response, tech firms began to diversify their businesses and become less reliant on hardware and computer sales. Instead, they started to develop a diversified portfolio in hardware, but also in software and tech services. Tech giants such Dell and IBM drove this move from manufacturing to services. Today, Austin is top in chip design, with companies such as Intel, Apple, Samsung, ARM, Freescale and Qualcomm, all of which have design operations in the city. The city is also strong in mobile and cloud technology. According to data compiled by Economic Modelling Specialists International, in 2014 there were almost 100,000 jobs in tech industries, slightly higher than before the recession. As of today, the Metro Monitor elaborated by the Brookings Institute lists this city as one of the top ten recovery economies in the US.

Today, the city also enjoys of a healthy start-up cluster ecosystem, consisting of approximately 2200 active tech start-ups. In this respect, Austin has a solid background in promoting new business creation. In 2016, Austin is ranked first in the start-up index of the Kauffman Foundation in the US. According to the Global Start-up Ecosystem Ranking, Austin's consistent performance in the tech sector has led to a tech-savvy workforce of over 100,000 people, yet Austin's supply can still not keep up with demand. Recent success stories include a number of global companies founded in the city such as HomeAway and RetailMeNot, which have together created over 5000 new jobs. In terms of venture capital, Austin has been called the 'new destination of choice' because of its organic and sustainable ecosystem for technological innovation and in popular rankings of most innovative cities in 2014.

In tandem with an attractive physical environment, Austin also promotes human activity. On the bedrock of a pleasant and dependable climate, the city offers: a wide range of outdoor recreation opportunities; a diverse range of ethnic and cultural settings; a vibrant nightlife; and a thriving live music scene. Insofar as Austin stands out as a city of music, alive to the possibilities of creativity and innovation, its policies in the area bear closer inspection. Again, the city has not rested content with past achievements but has set in place structures to support: Novel performance genres with strong new-technology content; and strong ties between artists, the business community and government to encourage and reward musical innovation and avant-garde technique. Besides the quality of life and place, Austin has also developed a culture of social tolerance that is attractive to creative people with ideas and skills that diverge from the norm. The drive for diversity and tolerance is, however, not an ethical issue. Rather, it is argued, the gathering together of difference, the concentration of diversity, is an efficient mechanism for hot-housing innovation (Yigitcanlar 2009). An example of the initiatives for building a culture of social tolerance in Austin is the 'Peace Day Austin'—celebrating International Peace Day 21th September throughout Austin by people participating, sharing and inspiring one another to make peace during the International Peace Week.

Austin is a clear example of a city with a locally driven creativity and innovation policy, the result of collaboration among an active citizenry, private enterprise, government, and university. With the convergence of music, film, and entertainment into a digital media sector, the urban core has seen an unprecedented residential development, and streetscape improvement becoming one of the most vibrant in the country. Austin's recreational amenities and music scene, a low cost of living, and a skilled workforce are core characteristics that make the city a stunning place to live and work. The city has received numerous awards and recognitions for constantly embarking on a variety of initiatives to strengthen its creative collaboration between business interest, city governance, and higher education. Some recent endeavours include the Imagine Austin Comprehensive Plan 2012, and the innovation district anchored by the Dell Medical School called Capital City Innovation Inc. 2016.

Austin also hosts the South by Southwest Conferences and Festivals, which offer the unique convergence of original music, independent films, and emerging technologies (Grodach 2012). Begun in 1987, this festival has continued to grow in scope and size every year, being comparable to some of the major forums of the 21st century—TED talks, COMDEX, Burningman or Sundance. The emergence of such communities has been accelerated by the internet. However, virtual communities are no longer enough. People want to physically gather together. In these terms, the festival can be conceived as a 'pop-up community', that is, the physical representation of technocrats, who are seeking the new ideas and creative technologies that will shape the future.

11.3 A Best Practice Knowledge City from South East Asia: Singapore

South East Asian is another colourful context with a number of cities targeting knowledge-based development, where some of them have already started to set trends for the rest of the world's cities. Particularly some of the region's countries have managed to establish and maintain high-growth economies since the 1960s, fuelled by exports and rapid technology-driven industrialisation. Hong Kong and Singapore are among the biggest financial centres worldwide, while South Korea and Taiwan are important hubs of global manufacturing in automobile and electronic components as well as information technology, respectively. These four Asian Tigers and their major cities heavily invested in KBUD. From this context, this section focuses on Singapore (winner of 2007, 2008, 2011, 2012 Most Admired Knowledge City Award—MAKCi).

Singapore city-state became independent in 1965. Since then, the country's economy has rapidly grown from a traditional entrepot (transshipment) model to one of the most modern manufacturing and financial centres in Asia. According to Wong (1986), Singapore's economic development path can be classified into four stages. The first stage was the industrial take-off phase that occurred during 1965 until the mid-1970s. This period was characterised by high dependence on technology transfer from multinational corporations (MNCs). The second stage is also known as local technological deepening (the mid-1970s to late-1980s), in which Singapore's economic growth was focused on developing local technology capabilities enabled by the MNCs. The third stage was the stage of applied research and development (R&D) expansion that occurred during the late-1980s to late-1990s. It was characterised by the rapid growth of applied R&D activities by MNCs and the establishment of local public R&D institutions. Finally, the 1990s onward mark the fourth stage of Singapore's economic development, i.e., a shift towards high technology entrepreneurship and basic R&D. This stage was characterised by a strong emphasis on indigenous technological capabilities, the formation of local high-tech start-ups, and a shift towards science-based industries. As for Koh and Wong (2005), these four phases summarise Singapore's strategy to shift from a user of technology to a creator of technology.

As a newly industrialised country that has among the world's highest per capita income, Singapore has been highly successful in its quest for rapid economic growth. Singapore's success in developing national technological capability is evidenced by the 32.9% increase in the country's gross domestic product (GDP) per capita, the expenditure on R&D as a percentage of GDP, the number of R&D personnel per 10,000 labour force, the received foreign direct investment, and the number of patents applied (Maitra 2016). Singapore offers capital opportunities for innovative companies, which seek investment to scale-up their production and distribution networks. It is attempting to build capabilities in all stages of the innovation process, from exploration stage to exploitation stage, to create an ecosystem for high-technological innovations. Such ambitious attempts and interactions with firms abroad have led to the systematic expansion of patenting portfolios of Singaporean companies (Cheah and Yu 2016).

Singapore pursues a highly organised pathway where firms perform on common research agendas with public research institutes performing as bridging institutions to ensure science and technology are aligned for systemic industrial development. Singapore is committed to both upstream (e.g., research, prototyping, integration, pilot scale) and downstream (e.g., scaling of production) production (Yeung 2006). These different pathways in city development may be attributed to government interventions, such as the establishment of specific techno-entrepreneurial capabilities, the creation of science parks, and knowledge precincts/districts (Wong and Bunnell 2006).

Besides the multicultural nature of Singapore's workforce and ability to attract global talent, the city is paving the way to become a global knowledge capital that will drive its leadership position in key knowledge-intensive industries. The world's top labour force sustains Singapore's leadership position in key knowledge-intensive industries. Singapore has a large base of engineering talent, with the number of engineers in universities and polytechnics expanding steadily each year, in addition to the sizeable pool of skilled technicians. Besides an open immigration policy has served to enhance Singapore's talent pool, which gives companies the opportunity to source for the best personnel from anywhere in the world (Yigitcanlar 2009).

One of the unique features of Singapore's KBUD is space and place making and planning tasks have been handled through a single authority—JTC Cooperation. JTC is the lead agency in Singapore to spearhead the planning, promotion, and development of a dynamic industrial landscape. Since its inception in 1968, JTC has played a major role in Singapore's economic development journey by developing land and space to support the transformation of industries and create quality jobs. Over the decades, JTC pioneered cutting-edge industrial infrastructure solutions to meet the evolving needs of companies with each phase of industrialisation. As the knowledge industry infrastructure specialist, JTC has developed over 7000 ha of industrial land and 4 million m^2 of ready-built facilities. Apart from developing Singapore's first industrial township, Jurong Industrial Estate, from swampland, other key iconic projects that JTC has developed include a chemical hub in Jurong Island; business and industrial parks such as Airport Logistics Park of Singapore, International and Changi Business Parks, Seletar Aerospace Park, CleanTech Park, Tuas Biomedical Park, and One-North, a cluster for knowledge-based industries with key developments like Biopolis, Fusionopolis and Mediapolis. A snapshot from a knowledge-intensive commerce and innovation space of Singapore is shown in Fig. 11.3—photo taken from Changi International Business Park indicating incorporation of natural environment in the precinct design.

The Singapore Economic Development Board was founded over 50 years ago, right after Singapore became an independent nation. The aim of the board was creating new jobs, attracting multinational companies and stimulating the creation of export-oriented industries to help Singapore get off its feet. Today, Singapore has extensive experience in industrial land development or in other words in developing knowledge and innovation spaces (e.g., science and technology parks, knowledge precincts). Limited land availability in the city-state has made Singapore transfer its know-how abroad and develop new knowledge precincts in China, such as China-Singapore Suzhou Industrial Park, and Sino-Singapore Tianjin Eco-City (see Yigitcanlar 2016).

One of the important success factors behind Singapore's achievements is the policy and political continuity over a very long time. The People's Action Party has dominated Singaporean politics since the first general election held in 1959. The same party has been in government since then without interruption. According to Hwee (2002, p. 213) reasons for the prolonged electoral dominance of the ruling party include:

- Under the governance of the ruling party, there has been impressive economic growth, rising employment and the provision of basic social services such as education, housing, and health.
- The ruling party's pragmatism and ability to mediate the diverse interests of a multiracial nation.

11.3 A Best Practice Knowledge City from South East Asia: Singapore

Fig. 11.3 Snapshot from a knowledge and innovation space of Singapore

- The exemplary conduct of the ruling party leadership in running an incorrupt government and civil service.
- The structures of the ruling party authoritarian government have been implemented with moderation.

The continuity of government KBUD policies over the half-century in Singapore that prioritise economic development, cultural diversity, and show no tolerance to corruption resulted in the rapid advancement of Singapore to become one of the world's most innovative nations. Huff (1995) highlights four critical elements in the success of the Singapore's governance model that helped the city-state to become a leading knowledge city: (a) Development commitment and exclusion of interest groups; (b) Sequencing, stability and economic progress; (c) Calibre of leadership, and; (d) Planning for development. Today, home ownership in Singapore is over 90%, and the unemployment rate is under 2%. Furthermore, Singapore comes first in the latest 2015 OECD PISA global education survey in all categories (i.e., science, mathematics, and reading)—see http://www.oecd.org/pisa.

Singapore has also benefited mightily from its geographic location close to the rising giant, China. With China now poised to be the world's number one economy, Singapore has positioned itself as the place to go for Western companies who want exposure to China, but not necessarily the costs and complexity of setting up shop on the Chinese mainland. As a result, you have Procter and Gamble making Singapore a hub for consumer and beauty products or Nielsen making Singapore the location of an innovation hub. Large US multinationals such as DuPont, General Motors and ADM continue to make Singapore an important part of their global strategy, attracted by Singapore's proximity to China and the ability to tap into Asia's future growth (Basulto 2015).

Singapore has achieved its earlier vision to become an international technology investment and innovation hub for foreign companies. Singapore's next challenge is to create a brand of its own other than Singapore Airlines—also ST Engineering that is listed in the Top 100 most innovative companies in the world by Forbes. Today, a vast new entrepreneurial ecosystem is being created in Singapore, which aspires to be the Silicon Valley of Southeast Asia. Now it claims to be filled with over 42,000 start-ups, and according to estimates provided by the Singapore Economic Development Board, nearly one in 10 working-age people in Singapore is trying to start a company or has already established one (Basulto 2015).

Singapore has become one of the world's most important suppliers of innovation services as the city-state has been investing in the right ingredients—e.g., a strong commitment to science and education, financial incentives, world-class laboratories, prestigious universities, entrepreneur environment, talent- and business-friendly ambiance, education institutions and much more. Most importantly, Singapore has a sound understanding of the theory of national or city innovation systems and urban planning that the knowledge city-state is eagerly putting them into practice (Frenkel and Maital 2014).

11.4 A Best Practice Knowledge City from Oceania: Melbourne

Oceania is the smallest continent in the world, yet still remains a recognised innovator in many areas of information science and technology, leading the world in industries such as solar power development, sustainable resources, and space science research. With a population of approximately 38 million, its number of universities is around 70—a few of them are among the top-50 universities in the world. Current areas of trending research include the fields of health, business, digital media, and environmental development. Despite its small population, Oceania has been the source of many innovations, scientific discoveries, and inventions. Some of the greatest achievements in science and technology include the invention of Wi-Fi, black box flight recorder, spray-on regenerative skin technology, the electronic pacemaker, disposable syringes, and Google maps. From this context, this section focuses on Melbourne, Australia (winner of 2010, 2013, 2016 Most Admired Knowledge City Award—MAKCi).

The development of Melbourne as a knowledge city has long been a priority for the City of Melbourne, which is a recognised as an early adopter of KBUD strategy. During the last several decades, urban administrators of Melbourne have

11.4 A Best Practice Knowledge City from Oceania: Melbourne

recognised the value of knowledge generation and innovation for establishing a competitive local economy and society. They have, particularly, aimed to shape Melbourne's urban development processes by considering the 21st century economic and social trends—i.e., knowledge economy and society. In order to succeed in this urban turnaround effort, city administrators have effectively used the strategic development planning approach. The city of Melbourne already has an extensive experience in utilising planning as an effective tool to increase the competitiveness of the city, where the first strategic plan of the city dates back to 1929. Building on this planning tradition, city administrations of Melbourne have been applying KBUD as an effective strategy through planning and policymaking to manage the spatial urban change processes of the city since the late 1990s (Yigitcanlar et al. 2008a).

KBUD policies that aim to design Melbourne as a knowledge city date back to early 1990. For instance, Social Justice Coalition's (1991) report on Melbourne's Docklands revealed that Melbourne had a vision of forming knowledge precincts, and the development of these precincts was seen to provide an effective solution to economic problems. Similarly, the Department of Planning and Development (1994) saw the prosperity increasingly depending on the ability of Melbourne to compete in the global knowledge economy. Melbourne metropolitan strategy acknowledged that the performance of Victoria is depending to a large extent on Melbourne's global economic competitiveness along with its ability to operate efficiently as an urban system focused on knowledge generation and innovation (Yigitcanlar 2009).

Melbourne, through KBUD policies, generated State and Local government support for the communities in keeping up with the knowledge economy, and shaping their future is among the key aspects of Melbourne success. In this regard, the Department of Victorian Communities was established in 2002 to commit to working with and across all levels of government, community and business to provide the support and resources communities need to shape their own future and support the knowledge-based development of Melbourne's creative urban region (Department of Victorian Communities 2004).

In terms of planning, in 2001 'Melbourne City Plan 2010' was put into action to particularly position the city as a global city with a knowledge economy, focused not only on traditional strengths, but also on emerging knowledge-intensive sectors, such as biotechnology; ICTs; specialist finance, legal, and administration services; educational services; and environmental technologies. The plan aimed to shape the future of the city as a prosperous, innovative, culturally vital, attractive, people focused, and sustainable city (Shaw 2003). The objectives of the plan revealed hints about how the city's future is planned as a knowledge city. These objectives were (City of Melbourne 2001):

- Developing a gateway for biotechnology in the Asia-Pacific region;
- Redressing skill shortage and building reputation as the ICT capital of Australia;
- Attracting strategic knowledge-based industry businesses to support and facilitate innovative start-up businesses;
- Promoting growth in the tertiary education services;

- Developing and promoting as a place that understands, respects and operates successfully with other business cultures;
- Developing and promoting diverse and highly skilled workforce to attract global projects, and;
- Enhancing and promoting liveability and lifestyle options, including affordable, high-quality housing and educational centres, and rich and diverse culture.

In 2002, a metropolitan strategy plan for Melbourne was introduced—'Melbourne 2030'. The plan builds on a vision by focusing on nine key directions: a more compact city, better management of metropolitan growth, networks with the regional cities, a more prosperous city, a great place to be, a fairer city, a greener city, better transport links, and better planning decisions and careful management (Victorian Government 2002). Melbourne 2030—and its 2008 update 'Melbourne @ 5 million'—provides directions for a strong and innovative economy, based on the view that all sectors of the economy are critical to economic prosperity, where knowledge clusters play a critical role in the success of KBUD of Melbourne (Victorian Government 2008).

Melbourne 2030 offers opportunities for internationally competitive industry clusters, and it expands logistics and communications infrastructure, including broadband telecommunications services, to underpin the development of the innovation economy. Melbourne Central Activities District and Docklands are planned to remain a key location for high-order commercial and knowledge-intensive development, and entertainment core of the metropolitan area. Ongoing housing development in central Melbourne takes advantage of this area's unmatched accessibility to jobs, facilities, recreational and cultural opportunities, and adding to the after-hours vibrancy of the inner areas (Victorian Government 2002).

In 2013, the knowledge city concept remained a central component of the City of Melbourne's vision of the city. It was included in the Council Plan 2013–17 as one of eight strategic goals, explicitly detailing the Council's aspiration. Goal 4 of the plan sees knowledge city as a long-term community goal and states, "Melbourne will be a place where ideas and talent thrive, based on a well-resourced education and research sector and knowledge networks … It will have a dynamic online culture to amplify our knowledge capability" (City of Melbourne 2013, p. 24).

In 2014, Melbourne released its knowledge city strategy as a single policy document—'A Knowledge City Strategy 2014–2018'. The strategy outlines the City of Melbourne's contribution to the ongoing development of the municipality's knowledge capacity, culture, and reputation. Through this strategy, the City of Melbourne has assumed four principal roles that seek to create a future in which the knowledge sector is better branded, understood, valued and experienced. The roles are (City of Melbourne 2014):

- Promoter: To position Melbourne as the region's knowledge capital and raise public and sectoral awareness of its strengths and opportunities;
- Leader: To lead by example in innovation, knowledge transfer and related knowledge sector activities;
- Partner: To use existing networks to involve the knowledge sector in the implementation of initiative, and;

11.4 A Best Practice Knowledge City from Oceania: Melbourne

- Enabler: To assist the growth of emerging businesses by providing infrastructure, information, financial support and connections to networks.

In 2014, a new metropolitan strategy plan for Melbourne was introduced—'Plan Melbourne'. The plan aims to improve access to a diversity of employment opportunities, including knowledge jobs in six designated precincts in metropolitan Melbourne (Fig. 11.4). This is to say the plan has a strong focus on infrastructure development and creation of knowledge clusters with national significance. In addition to the existing prominent knowledge and innovation clusters for Melbourne—i.e., Parkville, Monash and Dandenong South Employment Clusters—the plan encourages the further development of three emerging knowledge-intensive clusters—i.e., La Trobe, East Werribee, and Sunshine Employment Clusters. These National Employment Clusters are designated geographic concentrations of interconnected businesses and institutions that make a major contribution to the national economy and Melbourne's position as a global knowledge city (Victorian Government 2014). Monash Employment Cluster is the second largest knowledge precinct in Australia after Sydney's Macquarie Knowledge Park (Pancholi et al. 2017a).

In order to boost sustainable business and trade in Melbourne Federal, State, and Local governments have some business development and support funds and incentive programs available for SMEs. Although rather limited, this financial support still helps Melbourne to become one of the most notable concentrations of advanced

Fig. 11.4 Snapshot from one of Melbourne's knowledge precincts—Parkville

industrial and scientific research in the Asia-Pacific region. The depth of research available is evolving into clusters of cutting-edge expertise not only in academia, but in sectors as diverse as nanotechnology, biotechnology, automotive, aeronautics, financial services, and design. There are eight universities operating in Melbourne. They deliver highly relevant and accessible higher education courses and also conduct collaborative research with multinational companies such as Toyota, NEC, Ford, Glaxo Smith Klein, GE Holden, IBM, Boeing, and Hawker de Havilland (Yigitcanlar 2009).

Melbourne's success is not only limited to bringing all business, education, research and development clusters together, other clusters (i.e., tourism, sports, art, and culture) have also great contributions to its transition into a knowledge city. Cultural and international sportive activities are among the major factors of Melbourne's tourism attraction. While having a large and vibrant sports life, Melbourne is perhaps best known as 'the art and culture capital' of the Asia-Pacific region as it is the home of a large number of international art and cultural activities, which contributes significantly to Melbourne's transformation into a knowledge city (Yigitcanlar et al. 2008b).

The urban administrators are aware of the importance of frequently measuring and benchmarking its progress in critical KBUD areas. For that purpose, the 'Eye on Melbourne' has been developed to provide a visual representation of the approach to monitoring Melbourne's progress. The Eye on Melbourne takes its frame from the vision that was set out for Melbourne in Plan Melbourne. At the core of the Eye on Melbourne are the six key outcomes. Proposed measures are included for the outcomes that Plan Melbourne is seeking to contribute to or influence. All proposed measures link to more than one outcome and these linkages are shown by the coloured boxes next to each measure, which matches the colour of the relevant outcome segments (Victorian Government 2014).

Melbourne hosted the 3rd Knowledge Cities World Summit in 2010, and the city received their first Most Admired Knowledge City Award (MAKCi) during this international event. During this summit, Melbourne's 1st Knowledge Week was also celebrated with numerous activities open to the public. After its success, the event has become the City of Melbourne's unique annual festival for a smart and innovative city, where the event brings the future city to life, today. Melbourne's Office of Knowledge Capital convened both of these events. The Office of Knowledge Capital was founded based on the 'Manchester: Knowledge Capital' model that aimed to develop and promote the city as the nation's knowledge capital. The members of the Melbourne Vice Chancellors Forum were the founding partners in the Office of Knowledge Capital, which were: Australian Catholic University, City of Melbourne, Committee for Melbourne, Deakin University, La Trobe University, Monash University, RMIT University, Swinburne University of Technology, University of Melbourne, and Victoria University.

The strategic objectives of the Office of Knowledge Capital were to: (a) Develop and promote Melbourne locally, nationally and globally to achieve recognition as 'Australia's Knowledge Capital' and a 'Global University City; (b) Facilitate collaboration across and between the university sector with governments, industries and communities, and; (c) Form links with similar knowledge cities to enable mutually beneficial interactions. The office also handled the tasks of promotion of Melbourne's knowledge capabilities, brokering of collaborative knowledge initiatives, and management of Melbourne's knowledge information. In a way, the Office of Knowledge Capital served as the organisation to orchestrate KBUD in Melbourne. In 2012, it was merged with Enterprise Melbourne—the new business development initiative of Melbourne.

11.5 Lessons from the Global Knowledge City Best Practice

Previous sections of this chapter placed four prosperous knowledge cities from different country and continent contexts—namely, Barcelona, Austin, Singapore, and Melbourne—under the microscope. While there are some commonalities between them—e.g., strong governance, long-term planning, knowledge economy and society vision, support mechanisms—all four of these cities have built their achievements on their unique strength area bases.

Barcelona is an exemplary case for successful urban and economic regeneration. The city has managed to develop new and upgrade existing urban infrastructure through organising the 1992 Olympics. At the same time, Barcelona formed its knowledge city strategy for an economic and social turn/progress (Walliser 2004). Besides the infrastructural developments in the city another urban transformation has taken place particularly in a brownfield inner urban area—Poblenou district. The district has been transformed into a world-famous innovation district—22@Barcelona. However, this development also led to a property bubble, increasing property prices and rent significantly in and around the 200 ha district (see Charnock et al. 2014). Moreover, Barcelona symbolises its knowledge economy ambition with a landmark building that communicates this passion with the public and visitors on a daily basis.

Austin highly benefited from the innovation system and culture of the country that innovation and entrepreneurship are structured to open the economy to new technologies and industries. According to Abramson et al. (1997) the main strengths of the US innovation system are: (a) its diversity; (b) the decentralised and properly distributed organisation of universities and research institutions across the nation, which are the agents of technological innovation; (c) the high mobility of scientists and researchers, permeability of trained workers between private and public institutions; (d) the workforce flexibility and adaptability to changing circumstances, and; (e) the scale and openness of the national economy that is divided into homogeneous segments of the domestic market. On the downside, popularity and the rapid growth of Austin led to housing unaffordability issue that the city is now trying to find an effective solution.

Singapore, through strong governance, managed to develop strategies to attract the enviable level of foreign direct investment into the country (Baum et al. 2007). Today many of the major knowledge-intensive international companies' regional headquarters are located in Singapore—for example, Google, J.P. Morgan, PricewaterhouseCoopers, KPMG, IBM, Apple, ExxonMobil, Rolls Royce, Citibank, Ernst & Young, Procter & Gamble, General Electric, Hewlett-Packard, Thomson Reuters, and Seagate. This investment has been an important factor in Singapore's economic competitiveness, and its success as a knowledge economy. Moreover, the tasks of planning and space and place formation for knowledge companies are single-handily taken care of JTC Cooperation. JTC was established in 1968 as a semi-independent organisation under the Ministry of Trade and Industry to develop industrial estates across Singapore. Most importantly the investment on education has resulted in National University of Singapore ranking 12th university of the world in 2016 and outperforming the rest of the world in the OECD's 2015 PISA survey, which evaluates the quality, equity, and efficiency of school systems.

Melbourne has been using strategic planning mechanisms effectively to develop and implement KBUD policies to turn the city into a prosperous knowledge city. The exemplary strategies include forming an alliance with the neighbouring city councils for a better impact of the implement policies and investing in the knowledge precinct development in the city and the metropolitan region by forming national employment clusters. Moreover, the vibrant urban and social life Melbourne offers, as one of the World's most liveable cities, is a significant drawcard for global talent and investment that follows talent concentrations. Not resting on its past achievements, the city develops new policies and more importantly performance monitoring tools such as the Eye on Melbourne.

In the light of the global knowledge city best practice analysis, the following generic lessons could be useful for Brisbane to consider:

- Establishing transparent, less bureaucratic, participatory and dynamic governance practice—Brisbane could learn from San Francisco's experiences in establishing participatory governance model and making the government agencies more open, accountable, and responsive to citizens;
- Founding and orchestrating a (semi-)government organisation unit for KBUD—Brisbane could learn from Manchester and Melbourne's experiences in establishing an organisation to orchestrate the knowledge agenda development of the city;
- Preparing simple and highly effective long-term KBUD strategy and implementation plans—Brisbane could learn from Austin's sustained policy success evolved as a result of dynamic planning view and practice (particularly Opportunity Austin strategy checklist and implementation guidelines) to always keep the city as a trend-setter one;
- Creating a political culture for action-oriented bipartisan politics for KBUD—Brisbane could learn from Melbourne's experiences in managing the city and state authorities working together effectively for creating a bright future for the city;
- Developing a strong vision for knowledge city formation—Brisbane could learn from Austin, Barcelona, Melbourne and Singapore's experiences in developing a unique vision, and then concentrating on its delivery;

11.5 Lessons from the Global Knowledge City Best Practice

- Sustaining a KBUD policy continuity—Brisbane could learn from Austin and Singapore's uninterrupted and constantly evolving strong knowledge-based development policy focus;
- Integrating urban development plans with regional/city innovation systems—Brisbane could learn from Austin and Singapore's achievements in running the knowledge and innovation agenda in conjunction with sustainable urban development practices;
- Getting community support behind the KBUD projects—Brisbane could learn from Barcelona's successful public campaign for the 'city of knowledge' transformation supported by the active participation of majority of citizens;
- Stocktaking and mapping the endogenous strengths and assets of the city, and then building the development based on these strengths and assets—Brisbane could learn from Austin and Singapore's experiences in building on endogenous strengths and assets, and then creating opportunities for new areas of strengths;
- Developing attractive urban spaces for knowledge generation and innovation activities—Brisbane could learn from Austin, Barcelona, Melbourne and Singapore's experiences in making space and place for the knowledge economy;
- Providing high-quality and affordable urban infrastructures, amenities, and services;
- Establishing world-class top universities and promoting university-industry-government (triple helix model) partnership—Brisbane could learn from Austin, Melbourne and Singapore's experiences in accommodating universities in top-100 world's universities and their involvement in commercial R&D partnership;
- Providing incentives for knowledge-based activities, start-ups, and SMEs, and forming a highly entrepreneurial culture—Brisbane could learn from Austin and Singapore's experiences in turning their city into an entrepreneurial heaven though innovation support policies and mechanisms;
- Branding and marketing the city and promoting it internationally—Brisbane could learn from Austin, Barcelona, Melbourne and Singapore's experiences in turning their cities into an international brand and destination, and;
- Attracting foreign investment and mega events and hosting major forums to build and upgrade infrastructure—Brisbane could learn from Austin, Barcelona and Melbourne's experiences in hosting international mega events, and developing a major forum.

Lastly, while closely following the success and failure factors in international best practice is a highly useful approach to form strategies and actions in order to narrow the gap between with these leading exemplar knowledge city cases, it is crucial for any city to preserve and consolidate their own uniqueness (Pancholi et al. 2017b). Brisbane, therefore, needs to form a development pathway through carefully tailored strategies that are suitable to the unique characteristics—such as climate, location, culture, architecture, governance structure—of the city (Yigitcanlar 2011a, b).

11.6 Summary

This chapter of focused on providing lessons for Brisbane drawn from the prosperous knowledge city best practices with a central question in mind—how lessons learned from global knowledge city best practice can be evaluated for their potential in Brisbane. The analysis has revealed some potentially useful generic lessons for Brisbane to consider in its knowledge city journey. However, as mentioned earlier, these lessons need to be carefully considered for adoption to the local context. The potential actions also need to be meticulously tailored for Brisbane to make sure that they would work effective and efficiently to provide desired outcomes. Furthermore, these lessons are also useful for other emerging knowledge cities or cities that are aiming to move towards a knowledge city formation to consider. Likewise, such consideration should also include a careful evaluation of the local characteristics of the city and its society so that a locally customised solution could be developed.

References

Abramson, H.N., Encarnacao, J., Reid, P.P., & Schmoch, U. (Eds.) (1997). *Technology transfer systems in the United States and Germany: Lessons and perspectives.* National Academies Press.

Austin Chamber. (2012a). *Opportunity Austin 3.0: Strategic checklist.* Austin Chamber, Austin, TX.

Austin Chamber. (2012b). *Opportunity Austin 3.0: Implementation guidelines.* Austin Chamber, Austin, TX.

Austin Chamber. (2016). *Opportunity Austin.* Accessed on November 24, 2016 from http://www.austinchamber.com/the-chamber/opportunity-austin.

Bakıcı, T., Almirall, E., & Wareham, J. (2013). A smart city initiative: The case of Barcelona. *Journal of the Knowledge Economy, 4*(2), 135–148.

Barcelona City Hall. (2012). Barcelona smart city. Accessed on August 26, 2015 from https://www.dropbox.com/s/h2vxzfd7g5oa5fs/helsinki201_barcelona.pdf.

Basulto, D. (2015). The secrets to Singapore's track record of innovation excellence. *The Washington Post.* Accessed on December 2, 2016 from https://www.washingtonpost.com/news/innovations/wp/2015/05/26/the-secrets-to-singapores-track-record-of-innovation-excellence/?utm_term=.0d88e6e3234c.

Baum, S., Yigitcanlar, T., Mahizhnan, A., & Andiappan, N. (2007). Singapore government online: a consideration of e-government outcomes. *Journal of E-Government, 3*(4), 65–84.

Capdevila, I., & Zarlenga, M. I. (2015). Smart city or smart citizens? The Barcelona case. *Journal of Strategy and Management, 8*(3), 266–282.

Chapple, K., Markusen, M., Schrock, G., Yamamoto, D., & Yu, P. (2004). Gauging metropolitan "high-tech" and "i-tech" activity. *Economic Development Quarterly, 18*(1), 10–29.

Charnock, G., & Ribera-Fumaz, R. (2011). A new space for knowledge and people? Henri Lefebvre, representations of space, and the production of 22@ Barcelona. *Environment and Planning D, 29*(4), 613–632.

Charnock, G., Purcell, T. F., & Ribera-Fumaz, R. (2014). City of rents: The limits to the Barcelona model of urban competitiveness. *International Journal of Urban and Regional Research, 38*(1), 198–217.

References

Cheah, S., & Yu, C. (2016). Assessing economic impact of research and innovation originating from public research institutions and universities—Case of Singapore PRIs. *Triple Helix, 3*(1), 1–36.
City of Melbourne. (2001). *Melbourne city plan 2010*. Melbourne: City of Melbourne.
City of Melbourne. (2013). *Melbourne City Council Plan 2013–2017*. Melbourne: City of Melbourne.
City of Melbourne. (2014). *Knowledge city strategy: Strengthening Melbourne's knowledge sector through collaboration 2014–2018*. Melbourne: City of Melbourne.
Clua, A., & Albet, A. (2008). 22@barcelona plan. In T. Yigitcanlar, K. Velibeyoglu, & S. Baum (Eds.), *Knowledge-based urban development* (pp. 132–147). Hersey, PA: IGI Global.
Department of Planning and Development. (1994). *Melbourne metropolitan strategy*. Melbourne: Department of Planning and Development, State Government of Victoria.
Department of Victorian Communities. (2004). *Annual report 2003–2004*. Melbourne: Department of Victorian Communities.
Frenkel, A., & Maital, S. (2014). *Mapping national innovation ecosystems: Foundations for policy consensus*. Cheltenham: Edward Elgar.
Garcia-Ramon, M., & Albet, A. (2000). Pre-olympic and post-olympic Barcelona. *Environment and Planning A, 32*(8), 1331–1334.
Griffith, J. C. (2015). Barcelona, Spain as a model for the creation of innovation districts and sustainable social housing without spatial segregation. *Revista De Derecho Urbanístico y Medio Ambiente, 251*(1), 1–37.
Grodach, C. (2012). Before and after the creative city: The politics of urban cultural policy in Austin, Texas. *Journal of Urban Affairs, 34*(1), 81–97.
Huff, W. G. (1995). The developmental state, government, and Singapore's economic development since 1960. *World Development, 23*(8), 1421–1438.
Hwee, Y. L. (2002). Electoral politics in Singapore. In A. Croissant (Ed.), *Electoral politics in Southeast and East Asia* (pp. 203–217). Friedrich-Ebert-Stiftung: Singapore.
IBM (2012). Analysts: Austin tech scene is back from recession. IBM News. Accessed on November 24, 2016 from http://ibmnews.tmcnet.com/news/2014/05/12/7823722.htm.
Koh, W. T., & Wong, P. K. (2005). Competing at the frontier: the changing role of technology policy in Singapore's economic strategy. *Technological Forecasting and Social Change, 72*(1), 255–285.
Long, J. (2010). *Weird city: Sense of place and creative resistance in Austin, Texas*. Austin, TX: University of Texas Press.
Long, J. (2016). Constructing the narrative of the sustainability fix: Sustainability, social justice and representation in Austin, TX. *Urban Studies, 53*(1), 149–172.
Maitra, B. (2016). Investment in human capital and economic growth in Singapore. *Global Business Review, 17*(2), 425–437.
Pancholi, S., Yigitcanlar, T., & Guaralda, M. (2017a). Governance that matters: Identifying place-making challenges of Melbourne's Monash employment cluster. *Journal of Place Management and Development, 10*(1), 73–87.
Pancholi, S., Guaralda, M., & Yigitcanlar, T. (2017b). Context, contribution and characteristics of public spaces for place making in contemporary knowledge and innovation spaces: Observations from Brisbane, Australia. *The Journal of Public Space, 2*(4), 91–102.
Pareja-Eastaway, M., & Piqué, J. M. (2011). Urban regeneration and the creative knowledge economy: The case of 22@ in Barcelona. *Journal of Urban Regeneration & Renewal, 4*(4), 319–327.
Piro, G., Cianci, I., Grieco, L. A., Boggia, G., & Camarda, P. (2014). Information centric services in smart cities. *Journal of Systems and Software, 88*(1), 169–188.
Shaw, K. (2003). Discretion vs. regulation and the sorry case of Melbourne City Plan 2010. *Urban Policy and Research,* 21(4): 441–447.
Social Justice Coalition. (1991). *Picking winners*. Melbourne: Social Justice Coalition.
Tretter, E. (2013). Sustainability and neoliberal urban development: The environment, crime and the remaking of Austin's downtown. *Urban Studies, 50*(11), 2222–2237.

Victorian Government. (2002). *Melbourne 2030: Planning for sustainable growth*. Melbourne: Victorian Government Department of Sustainability and Environment.

Victorian Government. (2008). *Melbourne 2030: A planning update—Melbourne @ 5 million*. Melbourne: Victorian Government Department of Planning and Community Development.

Victorian Government. (2014). *Plan Melbourne*. Melbourne: Victorian Government.

Walliser, A. (2004). A place in the world: Barcelona's quest to become a global knowledge city. *Built Environment, 30*(3), 213–224.

Wong, K. W., & Bunnell, T. (2006). New economy discourse and spaces in Singapore: A case study of one-north. *Environment and Planning A, 38*(1), 69–83.

Wong, P. K. (1986). Saving, capital inflow and capital formation. In C. Y. Lim & P. J. Lloyd (Eds.), *Singapore: Resources and growth*. Oxford: Oxford University Press.

Yeung, H. W. (2006). Innovating for global competition: Singapore's pathway to high-tech development. In B. A. Lundvall, P. Intarakumnerd, & J. Vang (Eds.), *Asia's innovation systems in transition* (pp. 257–292). Cheltenham: Edward Elgar.

Yigitcanlar, T. (2009). Planning for knowledge-based urban development: Global perspectives. *Journal of Knowledge Management, 13*(5), 228–242.

Yigitcanlar, T. (2011a). Knowledge-based development processes of an emerging knowledge city: Brisbane, Australia. *A|Z ITU Journal of the Faculty of Architecture, 8*(1), 53–67.

Yigitcanlar, T. (2011b). Moving towards a knowledge city? Brisbane's experience in knowledge-based urban development. *International Journal of Knowledge Based Organizations, 1*(3), 22–38.

Yigitcanlar, T. (2016). *Technology and the city: Systems, applications and implications*. New York: Routledge.

Yigitcanlar, T., O'Connor, K., & Westerman, C. (2008a). The making of knowledge cities: Melbourne's knowledge-based urban development experience. *Cities, 25*(2), 63–72.

Yigitcanlar, T., Velibeyoglu, K., & Baum, S. (Eds.). (2008b). *Creative urban regions: Harnessing urban technologies to support knowledge city initiatives*. Hersey: IGI Global.

Chapter 12
Benchmarking City Performance

Abstract In the global knowledge economy era, benchmarking urban performance has become a routine practice for many cities. Such exercise is vital for cities to know about their performance and what the competitors are doing and how this performance compares their own. This chapter focuses on the city benchmarking topic and presents an international city comparison study. As the exemplar base city Brisbane, Australia is selected. Brisbane's knowledge-based urban development performance in numerous areas is benchmarked against global knowledge cities. These cities include Birmingham, Boston, Helsinki, Istanbul, Manchester, Melbourne, San Francisco, Sydney, Toronto, and Vancouver. As for the cross-comparison tool Knowledge-Based Urban Development Assessment Model (KBUD-AM) is utilised.

Keywords Benchmarking performance · Knowledge-based urban development
Knowledge city · Smart city · Sustainable urban development
Knowledge-Based Urban Development Assessment Model (KBUD-AM)
Brisbane · Birmingham · Boston · Helsinki · Istanbul · Manchester · Melbourne
San Francisco · Sydney · Toronto · Vancouver

The highly competitive nature of the global knowledge economy era puts cities in almost a desperate need to closely monitor progress and achievements of their competitors. At this point, city benchmarking is a useful method for following others and formulating improvements by making a comparative identification of the key elements, peculiarities, and deficits (Luque-Martinez and Munoz-Leiva 2005). A city benchmarking exercise provides lessons learned from comparisons—in a process whereby cities look beyond their boundaries as a means of learning and stimulating development—and helps in identification of future development and problem-solving strategies (Huggins 2010). This way, benchmarking allows urban policy organisations to become learning organisations through the identification, comprehension, and implementation of successful urban policy and development practices (Greene et al. 2007; Yigitcanlar 2014a, b).

In recent years, city benchmarking exercises have become increasingly widespread within the sphere of urban policy making, with many scholars arguing that careful and meaningful benchmarking is an essential prerequisite for informed and strategic policy making—that may provide a catching up opportunity for newly emerging knowledge cities (Luque-Martinez and Munoz-Leiva 2005; Malecki 2007). A number of international city benchmarking indices and studies have been published to assess the relative performance of cities globally focusing on different urban characteristics. These studies have coincided with a more sophisticated understanding of the elements of competition amongst cities. This has resulted in benchmarking against a competitor or successfully competing cities becoming an essential tool for progressive city planning, management, and development (Stokie 1999; Rondo-Brovetto and Saliter 2007; Yigitcanlar 2014a, b).

City benchmarking is seen as an invaluable method that can inform strategic planning and urban policy making, and thus can improve competitive positions of cities. Basically, benchmarking allows cities to: (a) Take stock of the current situation; (b) Compare itself with cities that are performing better; (c) Identify strategies for improvement; (d) Set targets for future performance; (e) Monitor and review progress; (f) Prioritise infrastructure and service funding; (vii) Build networks amongst cities, and; (g) Provide opportunity for increased collaboration amongst cities (Holloway and Wajzer 2008). Even though, many city benchmarking studies are undertaken, thus far there has been only very limited application of benchmarking studies specifically on knowledge cities investigating their competitive KBUD edges (see Yigitcanlar and Lonnqvist 2013; Yigitcanlar 2014a, b).

Against this background, this chapter of the book aims to address the following critical question through comparing Brisbane with major cities from Australia and overseas by focusing on the key economic, social, spatial and governance performance indicators relevant to knowledge cities.

- How does Brisbane stand as an emerging knowledge city internationally and nationally?

12.1 Knowledge-Based Urban Development Assessment

Evaluating the KBUD performance of a city requires measurement information that is produced by capturing the values of relevant measurement variables. Today, around the world many public, private, academic and not-for-profit agencies have been developing city benchmarking and performance analysis models, tools or indices (Yigitcanlar 2014a, b). For instance, the following list (Table 12.1) is only a small part of these city indices that are developed for conducting such measurements on one or a few relevant aspects of KBUD.

The above-listed indices are used to benchmark and rank cities globally, nationally or regionally. They are highly useful to determine the performance of a city based on a single or a combination of several KBUD characteristics—for example, knowledge economy or quality of life. However, there are a number of limitations of these indices, which are listed below (Yigitcanlar 2014a, b).

12.1 Knowledge-Based Urban Development Assessment

Table 12.1 Popular city performance indices

City performance indices
→ 2thinknow's Innovation City Index
→ A.T. Kearney, Inc's Global Cities Index
→ Arcadis Sustainable Cities Index
→ Between's Smart City Index
→ Boston Consulting Group: Decoding Global Talent
→ Brookings Global Metro Monitor
→ CBRE How Global is the Business of Retail?
→ Charles Landry's Creative Cities Index
→ CITI Foundation Accelerating Pathways
→ City RepTrak Topline Report
→ ECA's Cost of Living and Location Rating Surveys
→ EIU Hotspots, Global Liveability Ranking, and Safe Cities Index
→ Ericsson's Networked Society City Index
→ Euromonitor Top 100 City Destinations Ranking
→ fDi Global Cities of the Future - fDi Asia-Pacific Cities of the Future
→ GaWC's Globalization and World Cities Index
→ GE's Sustainable City Index
→ Global Financial Centres Index
→ Grosvenor Resilient Cities
→ GUCP's Global Urban Competitiveness Index
→ ICCA Country and City Rankings
→ INSEAD's Global Innovation Index
→ Insight Australia's Regional Competitiveness Index
→ ISiM's Knowledge City Index
→ JLL City Momentum Index
→ Knoema's Global City Competitiveness Index
→ KPMG Competitive Alternatives and Competitive Alternatives Focus on Tax
→ MasterCard Global Destination Cities Index
→ MERCER's Personal Safety, Quality of Life, and Eco-City Indices
→ Milken Institute's Best-Performing Cities Index
→ Monocle's Most Liveable Cities Index
→ NEF's Happy Planet Index
→ Numbeo Cost of Living Index
→ OECD's Better Life Index
→ PwC Cities of Opportunity
→ QS's World University Rankings and Best Student Cities Index
→ Richard Florida's Creative Class Index
→ RMIT's Global University City Index
→ Siemens's Green City Index
→ Simon Anholt's City Brands Index
→ Smart City Assessment Model (SCAM)
→ Solidance Most Innovative Cities in Asia Pacific
→ The Economist Group's Global City Competitiveness Index
→ The Mori Memorial Foundation's Global Power City Index
→ TomTom Traffic Index
→ Toronto Board of Trade Scorecard on Prosperity
→ UBS Prices and Earnings
→ ULI Emerging Trends in Real Estate Asia Pacific
→ UN-HABITAT's City Development, City Prosperity, and Urban Governance Indices
→ World Bank's Knowledge Economy Index
→ Xinhua-Down Jones International Financial Centres Index

- Use of aggregate indices and indicators: Most indicators underlying the aggregate indices are based on variables measured in pieces or weights—rather than using aggregate indices as variables, the composition of an index should have individual indicators or variables as a basis, classified in dimensions;
- Subjectivity of weightings: In most cases, a subjective weighting has been applied to build the index—and these weights are not necessarily ground-truthed or a sensitivity analysis is conducted;
- Partial performance evaluation: Individually these indices do not provide the full picture of a city's KBUD performance or achievements—that are indeed multi-facetted rather than single or limited focus;
- Difficulties in application: The application of these indices is not straightforward—many issues, including the choice of factors to measure and data availability, must be considered, and;
- Unreliable benchmarking ability: A key feature in the existing measurement indices is the use of benchmarking to provide a point of reference for the interpretation of the measurement results—nevertheless, as these indices do not provide the complete KBUD performance picture; benchmarked results are questionable when KBUD of cities is concerned.

12.2 Methodology of the Knowledge-Based Urban Development Assessment Model

For the aforementioned reasons, this chapter of the book adopts an assessment model/index that is specifically designed for comprehensive KBUD performance analysis. This model is applied to assess the global standing of Brisbane as an emerging knowledge city—The KBUD Assessment Model (KBUD-AM). This indexing model consists of a composite indicator, four indicator categories, eight indicator sets, and 32 KBUD indicators. Four of the indicator categories correspond to the four development pillars of KBUD—i.e., economic, societal, spatial and institutional (Yigitcanlar 2014a, b). These pillars and the eight indicator sets are derived from the literature and the KBUD conceptualisation earlier shown in Fig. 10.2. The KBUD-AM's methodological approach includes:

- Utilising an indexing framework for KBUD assessment;
- Selecting indicators of the framework;
- Determining the weightings of the indicators;
- Collecting data via primary and secondary data collection techniques;
- Using statistical techniques to scale and normalise data for comparison, and;
- Conducting statistical and descriptive analyses of the findings.

The index has a large and flexible indicator base that is specifically determined for each comparative knowledge city study. Hence, the indicator set provides a valid broad picture view of KBUD despite the potential limitations of individual measures. These indicators are selected from the prominent KBUD literature (see Carrillo et al.

12.2 Methodology of the Knowledge-Based Urban Development …

2014; Yigitcanlar 2014a, b) with involvement of 25 international KBUD experts on the basis of the following key principles: (a) Measurability; (b) Analytical soundness; (c) Comparability; (d) Geographic coverage; (e) Data availability, and; (f) Relevance and suitability (see Carrillo et al. 2014).

KBUD-AM, as default, uses an equal weighting for its indicators. However, to consider the potential benefits of assigning alternative weightings that may potentially improve the accuracy of the index findings; suitable weighting options for the indicators of the index are determined. The weighting options are assigned as a result of a three round Delphi exercise conducted with the abovementioned 25 international KBUD experts (for more information see Carrillo et al. 2014). In this Delphi exercise, the experts suggested the followings be considered: (a) Equal category and indicator weighting systems; (b) Equal category and variable indicator weighting, and; (c) Variable category and indicator weighting systems. Table 12.2 illustrates the KBUD-AM structure and alternative indicator weightings.

Following the selection of indicators, KBUD-AM, firstly, requires standardisation or in other words normalisation of the indicator values. The index utilises the *z-score* normalisation technique to reflect the specific distribution of the indicator values and present a relative scale (i.e., the value of between 0 and 1) according to the best and worst performers. The z-score normalisation of indicator values is calculated in accordance with the following formula (12.1):

$$z = \frac{(x - \mu)}{\sigma} \quad (12.1)$$

In 12.1, z corresponds to the normalised indicator value, x, μ and σ subscripts denote data, mean and standard deviation values, respectively. Once normalised values are entered into the index, all indicators are assigned their weightings to calculate the indicator set scores, as specified by the following Eq. (12.2):

$$I_{MEF} = \sum_{i=1}^{n} \frac{MEF_i}{n} * w_i; I_{KEF} = \sum_{i=1}^{n} \frac{KEF_i}{n} * w_i; I_{HSC} = \sum_{i=1}^{n} \frac{HSC_i}{n} * w_i;$$

$$I_{DI} = \sum_{i=1}^{n} \frac{DI_i}{n} * w_i;$$

$$I_{SUD} = \sum_{i=1}^{n} \frac{SUD_i}{n} * w_i; I_{QLP} = \sum_{i=1}^{n} \frac{QLP_i}{n} * w_i; I_{PL} = \sum_{i=1}^{n} \frac{GP_i}{n} * w_i;$$

$$I_{SP} = \sum_{i=1}^{n} \frac{LS_i}{n} * w_i \quad (12.2)$$

In 12.2, *I* and *w* correspond to the indicator score and the weight, and *MEF, KEF, HSC, DI, SUD, QLP, GP* and *LS* subscripts represent macroeconomic foundations, knowledge economy foundations, human and social capitals, diversity and independency, sustainable urban development, quality of life and place, governance

Table 12.2 Structure of KBUD-AM and alternative indicator weightings (derived from Yigitcanlar 2014a, b)

Composite indicator	Indicator categories	Indicator sets	Indicators	Indicator descriptions	Equal category and indicator weightings	Equal category and variable indicator weightings	Variable category and indicator weightings
Knowledge-based urban development	Economic development	Macro-economic foundations	Gross domestic product	Gross domestic product (GDP) per capita in USD purchasing power parities	0.03125	0.03114	0.0340
			Major international companies	Number of global top 500 companies located	0.03125	0.02793	0.0279
			Foreign direct investment	Ratio of international share in foreign direct investments	0.03125	0.02791	0.0290
			Urban competitiveness	Global urban competitiveness index ranking	0.03125	0.03220	0.0323
		Knowledge economy foundations	Innovation economy	International city ranking in innovation economy	0.03125	0.03258	0.0351

(continued)

Table 12.2 (continued)

Composite indicator	Indicator categories	Indicator sets	Indicators	Indicator descriptions	Equal category and indicator weightings	Equal category and variable indicator weightings	Variable category and indicator weightings
			Research and development	Ratio of research and development expenditure in GDP	0.03125	0.03415	0.0391
			Patent applications	Patent Cooperation Treaty patent applications per million inhabitants	0.03125	0.03202	0.0349
			Knowledge worker pool	Ratio between professionals and managers and all workers	0.03125	0.03208	0.0324
			Category total		0.25000	0.25000	0.2647
	Societal development	Human and social capitals	Education investment	Ration between public spending on education and GDP	0.03125	0.03583	0.0376

(continued)

Table 12.2 (continued)

Composite indicator	Indicator categories	Indicator sets	Indicators	Indicator descriptions	Equal category and indicator weightings	Equal category and variable indicator weightings	Variable category and indicator weightings
			Professional skill base	Ratio of residents over 18 years with tertiary degree (certificate, bachelor, master, Ph.D.)	0.03125	0.03254	0.0324
			University prestige	World university rankings	0.03125	0.03329	0.0314
			Wireless broadband coverage	Ratio of access to fixed broadband subscribers per capita	0.03125	0.02768	0.0277
		Diversity and independency	Cultural diversity	Ratio of people born abroad	0.03125	0.02984	0.0302
			Social tolerance	International country tolerance ranking	0.03125	0.03271	0.0320
			Socio-economic dependency	Ratio between the elderly population and the working age (15–64 years population)	0.03125	0.02993	0.0302

(continued)

12.2 Methodology of the Knowledge-Based Urban Development … 167

Table 12.2 (continued)

Composite indicator	Indicator categories	Indicator sets	Indicators	Indicator descriptions	Equal category and indicator weightings	Equal category and variable indicator weightings	Variable category and indicator weightings
			Unemployment level	Ratio of unemployment	0.03125	0.02817	0.0260
			Category total		0.25000	0.25000	0.2476
	Spatial development	Sustainable urban development	Eco-city formation	International city ranking in eco-city	0.03125	0.02855	0.0291
			Sustainable transport use	Ratio of sustainable transport mode use for commuting	0.03125	0.03188	0.0313
			Environmental impact	CO_2 emissions in metric tons per capita	0.03125	0.02977	0.0258
			Urban form and density	Population density in persons per km^2	0.03125	0.03289	0.0327
		Quality of life and place	Quality of life	International city ranking in quality of life	0.03125	0.04042	0.0377

(continued)

Table 12.2 (continued)

Composite indicator	Indicator categories	Indicator sets	Indicators	Indicator descriptions	Equal category and indicator weightings	Equal category and variable indicator weightings	Variable category and indicator weightings
			Cost of living	International city ranking in cost of living	0.03125	0.02645	0.0229
			Housing affordability	Ratio between GPD per capita and median dwelling price	0.03125	0.02675	0.0249
			Personal safety	International city ranking in personal safety	0.03125	0.03329	0.0331
			Category total		0.25000	0.25000	0.2375
	Institutional development	Governance and planning	Government effectiveness	Level of government effectiveness	0.03125	0.03396	0.0367
			Electronic governance	International city ranking in e-government	0.03125	0.02613	0.0256
			Strategic planning	Level of KBUD strategies in strategic regional and local development plans	0.03125	0.03613	0.0340

(continued)

12.2 Methodology of the Knowledge-Based Urban Development …

Table 12.2 (continued)

Composite indicator	Indicator categories	Indicator sets	Indicators	Indicator descriptions	Equal category and indicator weightings	Equal category and variable indicator weightings	Variable category and indicator weightings
		Leadership and support	City branding	International city ranking in city branding	0.03125	0.03013	0.0320
			Effective leadership	Level of institutional and managerial leadership in overseeing KBUD	0.03125	0.02874	0.0293
			Strategic partnership and networking	Level of triple-helix and public-private-partnerships and global networking-global city ranking	0.03125	0.03258	0.0332
			Community engagement	Level of institutional mechanisms for community building and public participation opportunities	0.03125	0.02961	0.0303

(continued)

Table 12.2 (continued)

Composite indicator	Indicator categories	Indicator sets	Indicators	Indicator descriptions	Equal category and indicator weightings	Equal category and variable indicator weightings	Variable category and indicator weightings
			Social cohesion and equality	Level of income inequality (national gini coefficient)	0.03125	0.03273	0.0290
			Category total		0.25000	0.25000	0.2502
			Index total		1.00000	1.00000	1.0000

12.2 Methodology of the Knowledge-Based Urban Development ...

and planning, and leadership and support indicator sets, respectively. Then, indicator domain scores for each development domain are calculated, as specified by the following Eq. (12.3):

$$I_{EcoDev} = \sum_{i=1}^{n} \frac{EcoDev_i}{n}; I_{SocDev} = \sum_{i=1}^{n} \frac{SocDev_i}{n}; I_{EnvDev} = \sum_{i=1}^{n} \frac{SpaDev_i}{n};$$
$$I_{InsDev} = \sum_{i=1}^{n} \frac{InsDev_i}{n} \qquad (12.3)$$

In 12.3, I corresponds to the indicator score and *EcoDev, SocDev, SpaDev,* and *InsDev* subscripts represent economic, societal, spatial and institutional development indicator categories, respectively. Lastly, the composite indicator scores are calculated, as specified by the following formula (12.4):

$$I_{KBUD} = \sum_{i=1}^{n} \frac{KBUD_i}{n} \qquad (12.4)$$

In 12.4, I corresponds to the indicator score, *KBUD* corresponds to the KBUD composite indicator and $KBUD_i$ corresponds to each of the economic, societal, spatial and institutional development indicator category scores.

After calculation of the KBUD performance scores and rankings of the case cities, the methodology of the KBUD-AM includes one last step—i.e., clusters analysis. This statistical analysis is undertaken to understand how assessed cities are grouped together. The cluster analysis provides an additional angle in the interpretation of the results—revealing insights on better understanding the similarities and differences between cities and the gap between them.

12.3 Global Standing of Brisbane as an Emerging Knowledge City

The basis of prosperity and welfare of cities largely depends on their capacity to take advantage of opportunities for sustained employment growth and minimise the challenges of competitive economic conditions and urban population growth. Therefore, developing sound KBUD strategies is highly valuable for cities in managing growth and augmenting economic performances. In the era of a global knowledge economy, Brisbane seeks to increase its competitive edge, become a destination for talent and investment and provide prosperity and high quality of life to its inhabitants through effective KBUD strategies (Yigitcanlar and Velibeyoglu 2008). Nonetheless, knowing the standing of Brisbane on the global stage is important to adjust its KBUD policies to retain and improve its KBUD performances.

This section of the book applies KBUD-AM to measure, compare and evaluate KBUD performances of Brisbane against 10 global and emerging knowledge cities—namely Birmingham, Boston, Helsinki, Istanbul, Manchester, Melbourne, San Francisco, Sydney, Toronto, and Vancouver. These cities are selected purposely to represent a different level of knowledge city progress—i.e., established and emerging. Among these case cities, seven of them are considered as globally well-known knowledge cities—that are Boston, Helsinki, Melbourne, San Francisco, Sydney, Toronto, and Vancouver—and three of them are as emerging knowledge cities—that are Birmingham, Istanbul, and Manchester (see Garcia and Leal 2012). Some of the best practice knowledge cities investigated in an earlier section of this book have not been included in this quantitative analysis. The main reason for this exclusion is the data collection challenges due to data availability and compatibility, contextual differences or language barrier.

The city-regions of these global and emerging knowledge cities are considered in the analysis to take their metropolitan impacts in the account. The analysis compares the KBUD performances of the selected cities based on the data dating to the 2010–2011-time period—as these were the period that census conducted in the countries of the case cities. Raw data and equal weighted normalised index score values of the analyses are provided in Tables 12.3 and 12.4.

The results of the KBUD-AM global (or established) and emerging knowledge cities indexing exercise are listed in Table 12.5. The analysis ranks all case study cities based on their KBUD characteristics. In all alternative-weighting systems, Brisbane comes in the ninth place. In two out of three weighting systems Toronto and in one San Francisco comes as the top knowledge city. According to the cluster analysis Toronto, San Francisco, Boston, Vancouver, Helsinki, Melbourne and Sydney group closely together, where the ranks of these cities change in the differing weighing systems (see Fig. 12.1).

The ranking of the bottom four cities stays the same in all weighting options (i.e., Manchester, Brisbane, Birmingham, and Istanbul). However, Istanbul city shows a significantly lower performance than Manchester, Brisbane, and Birmingham, and thus not located in the second-tier cluster with the other three cities. The cluster analysis forms three significant city clusters according to their performances—see Clusters A, B, C in Fig. 12.1. Cities in Cluster A can be considered as 'established knowledge cities', Cluster B, including Brisbane, as 'progressive knowledge cities', and Cluster C as 'emerging knowledge cities'. Figure 12.2 illustrates ranking of the compared cities along with their population sizes.

In order to obtain a better understanding of KBUD performance of the city a number of spidergram scorecards are created presenting Brisbane's comparative performance in overall KBUD (Fig. 12.3) and individual KBUD domains of economic development (Fig. 12.4), societal development (Fig. 12.5), spatial development (Fig. 12.6), and institutional development (Fig. 12.7). As Fig. 12.3 indicates Brisbane's strengths are in the spatial and to a degree in societal development domains, however, the city is rather low performing in the economic and institutional development areas. The ranking of Brisbane in the main KBUD domains are as follows:

12.3 Global Standing of Brisbane as an Emerging Knowledge City

Table 12.3 Raw data (derived from Yigitcanlar 2014a, b)

Indicators	Birmingham	Boston	Helsinki	Istanbul	Manchester	Melbourne	San Francisco	Sydney	Toronto	Vancouver	Brisbane
Gross domestic product	31,421	64,568	47,672	13,498	31,532	38,141	76,727	41,891	40,394	38,675	**31,600**
Major international companies	0	5	1	1	0	4	10	3	9	0	**0**
Foreign direct investment	0.035	0.178	0.003	0.006	0.035	0.023	0.178	0.023	0.018	0.018	**0.023**
Urban competitiveness	83	13	16	136	36	22	9	31	11	47	**113**
Innovation economy	65	1	42	89	32	17	2	20	10	49	**85**
Research and development	0.018	0.027	0.035	0.007	0.018	0.021	0.027	0.021	0.020	0.020	**0.021**
Patent applications	37.7	581.0	529.0	15.5	58.4	113.7	676.1	154.6	247.6	159.1	**69.0**
Knowledge worker pool	0.275	0.463	0.494	0.252	0.275	0.351	0.453	0.370	0.444	0.432	**0.313**
Education investment	0.054	0.055	0.061	0.041	0.054	0.044	0.055	0.044	0.048	0.048	**0.044**
Professional skill base	0.280	0.390	0.348	0.075	0.310	0.336	0.404	0.346	0.408	0.390	**0.295**
University reputation	67	2	89	401	29	31	11	38	23	51	**65**

(continued)

Table 12.3 (continued)

Indicators	Birmingham	Boston	Helsinki	Istanbul	Manchester	Melbourne	San Francisco	Sydney	Toronto	Vancouver	**Brisbane**
Broadband access	0.316	0.276	0.286	0.104	0.316	0.242	0.276	0.242	0.298	0.298	**0.242**
Cultural diversity	0.169	0.168	0.086	0.004	0.106	0.289	0.300	0.317	0.499	0.396	**0.349**
Social tolerance	10	8	19	34	10	5	8	5	1	1	**5**
Socio-economic dependency	0.238	0.184	0.185	0.076	0.220	0.183	0.181	0.176	0.158	0.169	**0.192**
Unemployment level	0.083	0.041	0.051	0.155	0.064	0.043	0.044	0.039	0.069	0.042	**0.042**
Eco-city formation	55	25	3	121	47	25	32	46	39	13	**23**
Sustainable transport use	0.230	0.171	0.420	0.420	0.250	0.136	0.189	0.204	0.442	0.256	**0.129**
Environmental impact	8.5	17.9	10.6	4	8.5	18.6	17.9	18.6	16.3	16.3	**9.1**
Urban form and density	2864	2009	162	1409	385	495	474	357	921	776	**1007**
Quality of life	52	36	35	112	57	18	30	11	15	5	**34**
Cost of living	150	129	42	70	148	21	106	14	59	65	**22**
Housing affordability	0.127	0.199	0.107	0.251	0.133	0.065	0.128	0.061	0.089	0.057	**0.069**
Personal safety	53	53	2	187	44	25	53	25	17	17	**24**
Government effectiveness	0.910	0.900	0.990	0.650	0.910	0.960	0.900	0.960	0.970	0.970	**0.960**

(continued)

12.3 Global Standing of Brisbane as an Emerging Knowledge City

Table 12.3 (continued)

Indicators	Birmingham	Boston	Helsinki	Istanbul	Manchester	Melbourne	San Francisco	Sydney	Toronto	Vancouver	Brisbane
Electronic governance	4	2	19	69	4	8	2	8	3	3	**8**
Strategic planning	68.75	87.50	87.50	50.00	75.00	93.75	93.75	81.25	87.50	93.75	**75.00**
City branding	61	23	36	42	61	8	7	1	14	16	**99**
Effective leadership	62.50	81.25	87.50	50.00	76.25	91.25	87.50	77.50	81.25	90.00	**50.00**
Strategic partnership and networking	56.25	90.00	68.75	60.00	73.75	77.50	95.00	93.75	91.88	81.88	**60.00**
Community engagement	90.63	96.88	90.63	80.00	90.63	90.63	96.88	90.63	93.75	93.75	**50.00**
Social cohesion and equality	0.345	0.378	0.259	0.409	0.345	0.336	0.378	0.336	0.324	0.324	**0.394**

Table 12.4 Equal weighted normalised index scores (derived from Yigitcanlar 2014a, b)

Indicators	Birmingham	Boston	Helsinki	Istanbul	Manchester	Melbourne	San Francisco	Sydney	Toronto	Vancouver	Brisbane
Gross domestic product	0.008692	0.028500	0.020059	0.001584	0.008761	0.013213	0.030642	0.015936	0.014843	0.013596	**0.008803**
Major international companies	0.006445	0.022112	0.009138	0.009138	0.006445	0.018989	0.030377	0.015625	0.029669	0.006445	**0.006445**
Foreign direct investment	0.012976	0.030537	0.007449	0.007892	0.012976	0.010705	0.030537	0.010705	0.009783	0.009783	**0.010705**
Urban competitiveness	0.006471	0.024367	0.023720	0.000680	0.018704	0.022334	0.025177	0.020053	0.024779	0.015625	**0.002099**
Macro-economic foundations	*0.008646*	*0.026379*	*0.015092*	*0.004823*	*0.011721*	*0.016310*	*0.029183*	*0.015579*	*0.019769*	*0.011362*	***0.007013***
Innovation economy	0.005967	0.027388	0.013832	0.001590	0.017773	0.023186	0.027182	0.022195	0.025259	0.011156	**0.002050**
Research and development	0.009755	0.024752	0.030491	0.000625	0.009755	0.014936	0.024752	0.014936	0.013147	0.013147	**0.014936**
Patent applications	0.006216	0.028834	0.027692	0.005440	0.006997	0.009335	0.030177	0.011264	0.016013	0.011484	**0.007418**

(continued)

12.3 Global Standing of Brisbane as an Emerging Knowledge City 177

Table 12.4 (continued)

Indicators	Birmingham	Boston	Helsinki	Istanbul	Manchester	Melbourne	San Francisco	Sydney	Toronto	Vancouver	Brisbane
Knowledge worker pool	0.003927	0.026408	0.028593	0.002469	0.003927	0.012264	0.025501	0.014947	0.024601	0.023281	**0.007464**
Knowledge economy foundations	0.006466	0.026846	0.025152	0.002531	0.009613	0.014930	0.026903	0.015836	0.019755	0.014767	*0.007967*
Economic development	0.007556	0.026612	0.020122	0.003677	0.010667	0.015620	0.028043	0.015708	0.019762	0.013065	**0.007490**
Education investment	0.023308	0.024806	0.030048	0.002548	0.023308	0.005584	0.024806	0.005584	0.012088	0.012088	**0.005584**
Professional skill base	0.009801	0.023540	0.018564	0.000120	0.013560	0.016997	0.024932	0.018305	0.025300	0.023540	**0.011632**
University reputation	0.016335	0.023081	0.013885	0.000052	0.020450	0.020243	0.022241	0.019508	0.021062	0.018105	**0.016558**
Broadband access	0.025368	0.018267	0.020265	0.000118	0.025368	0.011269	0.018267	0.011269	0.022497	0.022497	**0.011269**
Human and social capitals	*0.018703*	*0.022423*	*0.020691*	*0.000709*	*0.020672*	*0.013523*	*0.022561*	*0.013667*	*0.020237*	*0.019058*	*0.011261*
Cultural diversity	0.009620	0.009546	0.004530	0.001686	0.005553	0.019336	0.020203	0.021496	0.029885	0.026436	**0.023730**
Social tolerance	0.015147	0.017764	0.005059	0.000160	0.015147	0.021482	0.017764	0.021482	0.025582	0.025582	**0.021482**

(continued)

Table 12.4 (continued)

Indicators	Birmingham	Boston	Helsinki	Istanbul	Manchester	Melbourne	San Francisco	Sydney	Toronto	Vancouver	Brisbane
Socio-economic dependency	0.002232	0.013904	0.013601	0.031064	0.004786	0.014208	0.014818	0.016349	0.021612	0.018468	**0.011525**
Unemployment level	0.008184	0.022572	0.019280	0.000096	0.014600	0.021950	0.021631	0.023174	0.012801	0.022264	**0.022264**
Diversity and independency	*0.008796*	*0.015947*	*0.010618*	*0.008251*	*0.010021*	*0.019244*	*0.018604*	*0.020625*	*0.022470*	*0.023187*	***0.019750***
Societal development	0.013749	0.019185	0.015654	0.004480	0.015347	0.016384	0.020583	0.017146	0.021353	0.021122	**0.015505**
Eco-city formation	0.009509	0.021030	0.027356	0.000136	0.012467	0.021030	0.018395	0.012855	0.015625	0.024917	**0.021741**
Sustainable transport use	0.012553	0.007000	0.028692	0.028692	0.014676	0.004510	0.008537	0.009934	0.029477	0.015321	**0.004094**
Environmental impact	0.025638	0.005930	0.021781	0.030068	0.025638	0.004864	0.005930	0.004864	0.008853	0.008853	**0.024651**
Urban form and density	0.030902	0.027918	0.004919	0.021759	0.007238	0.008576	0.008311	0.006918	0.014621	0.012453	**0.015926**

(continued)

12.3 Global Standing of Brisbane as an Emerging Knowledge City

Table 12.4 (continued)

Indicators	Birmingham	Boston	Helsinki	Istanbul	Manchester	Melbourne	San Francisco	Sydney	Toronto	Vancouver	Brisbane
Sustainable urban development	0.019651	0.015469	0.020687	0.020164	0.015005	0.009745	0.010293	0.008643	0.017144	0.015386	**0.016603**
Quality of life	0.009531	0.015968	0.016386	0.000180	0.007777	0.023013	0.018456	0.025224	0.024008	0.026796	**0.016804**
Cost of living	0.002185	0.004501	0.023214	0.016874	0.002356	0.026775	0.008475	0.027678	0.019513	0.018088	**0.026634**
Housing affordability	0.013647	0.002812	0.017602	0.000460	0.012388	0.025033	0.013368	0.025573	0.021105	0.026097	**0.024446**
Personal safety	0.013751	0.013751	0.025236	0.000073	0.015988	0.020585	0.013751	0.020585	0.022353	0.022353	**0.020813**
Quality of life and place	0.009779	0.009258	0.020609	0.004397	0.009627	0.023851	0.013513	0.024765	0.021745	0.023334	**0.022174**
Spatial development	0.014715	0.012364	0.020648	0.012280	0.012316	0.016798	0.011903	0.016704	0.019444	0.019360	**0.019389**
Government effectiveness	0.014782	0.013467	0.024474	0.000072	0.014782	0.021208	0.013467	0.021208	0.022368	0.022368	**0.021208**

(continued)

Table 12.4 (continued)

Indicators	Birmingham	Boston	Helsinki	Istanbul	Manchester	Melbourne	San Francisco	Sydney	Toronto	Vancouver	Brisbane
Electronic governance	0.020473	0.021623	0.011154	0.000055	0.020473	0.018040	0.021623	0.018040	0.021054	0.021054	**0.018040**
Strategic planning	0.005486	0.021234	0.021234	0.000308	0.010016	0.025764	0.025764	0.015625	0.021234	0.025764	**0.010016**
City branding	0.005638	0.019864	0.014574	0.012138	0.005638	0.025025	0.025310	0.026849	0.023146	0.022460	**0.000464**
Governance and planning	*0.011595*	*0.019047*	*0.017859*	*0.003143*	*0.012727*	*0.022509*	*0.021541*	*0.020431*	*0.021950*	*0.022912*	*0.012432*
Effective leadership	0.005859	0.019940	0.024325	0.001352	0.015906	0.026404	0.024325	0.016935	0.019940	0.025762	**0.001352**
Strategic partnership and networking	0.002338	0.025371	0.008785	0.003703	0.012722	0.015917	0.027829	0.027298	0.026400	0.019610	**0.003703**
Community engagement	0.018376	0.023619	0.018376	0.008806	0.018376	0.018376	0.023619	0.018376	0.021134	0.021134	**0.000071**

(continued)

Table 12.4 (continued)

Indicators	Birmingham	Boston	Helsinki	Istanbul	Manchester	Melbourne	San Francisco	Sydney	Toronto	Vancouver	Brisbane
Social cohesion and equality	0.016531	0.007298	0.030766	0.002175	0.016531	0.019201	0.007298	0.019201	0.022489	0.022489	**0.004137**
Leadership and support	*0.010776*	*0.019057*	*0.020563*	*0.004009*	*0.015884*	*0.019975*	*0.020768*	*0.020452*	*0.022491*	*0.022249*	***0.002316***
Institutional development	0.011185	0.019052	0.019211	0.003576	0.014306	0.021242	0.021154	0.020442	0.022221	0.022580	**0.007374**
Knowledge-based urban development	0.011801	0.019303	0.018909	0.006003	0.013159	0.017511	0.020421	0.017500	0.020695	0.019032	**0.012439**

Italic values are the sub-category scores

Table 12.5 Ranking of cities based on alternative weighting schemes (derived from Yigitcanlar 2014a, b)

Rank	City	Equal weighted score (equal category and indicator weighted score)	Cluster
1	Toronto	0.020695	A (established knowledge city)
2	San Francisco	0.020421	A (established knowledge city)
3	Boston	0.019303	A (established knowledge city)
4	Vancouver	0.019032	A (established knowledge city)
5	Helsinki	0.018909	A (established knowledge city)
6	Melbourne	0.017511	A (established knowledge city)
7	Sydney	0.017500	A (established knowledge city)
8	Manchester	0.013159	B (progressive knowledge city)
9	**Brisbane**	**0.012439**	**B (progressive knowledge city)**
10	Birmingham	0.011801	B (progressive knowledge city)
11	Istanbul	0.006003	C (emerging knowledge city)
Rank	City	Semi-equal weighted score (equal category and variable indicator weighted score)	Cluster
1	Toronto	0.020746	A (established knowledge city)
2	San Francisco	0.020456	A (established knowledge city)
3	Boston	0.019438	A (established knowledge city)
4	Vancouver	0.019072	A (established knowledge city)
5	Helsinki	0.019060	A (established knowledge city)

(continued)

12.3 Global Standing of Brisbane as an Emerging Knowledge City

Table 12.5 (continued)

Rank	City	Semi-equal weighted score (equal category and variable indicator weighted score)	Cluster
6	Sydney	0.017484	A (established knowledge city)
7	Melbourne	0.017470	A (established knowledge city)
8	Manchester	0.013145	B (progressive knowledge city)
9	**Brisbane**	**0.012303**	**B (progressive knowledge city)**
10	Birmingham	0.011805	B (progressive knowledge city)
11	Istanbul	0.005831	C (emerging knowledge city)
Rank	City	Variable weighted score (variable category and indicator weighted score)	Cluster
1	San Francisco	0.020730	A (established knowledge city)
2	Toronto	0.020720	A (established knowledge city)
3	Boston	0.019730	A (established knowledge city)
4	Helsinki	0.019097	A (established knowledge city)
5	Vancouver	0.018965	A (established knowledge city)
6	Sydney	0.017428	A (established knowledge city)
7	Melbourne	0.017392	A (established knowledge city)
8	Manchester	0.013091	B (progressive knowledge city)
9	**Brisbane**	**0.012105**	**B (progressive knowledge city)**
10	Birmingham	0.011728	B (progressive knowledge city)
11	Istanbul	0.005704	C (emerging knowledge city)

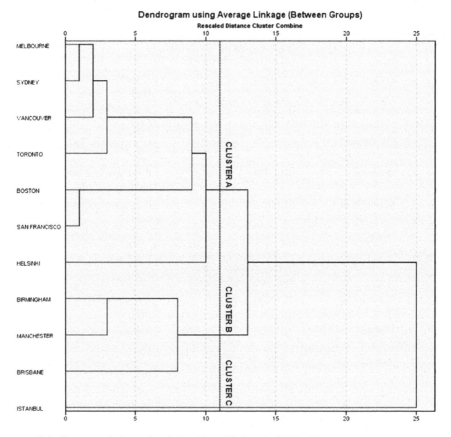

Fig. 12.1 Cluster analysis results (derived from Yigitcanlar 2014a, b)

- Among the compared cities Brisbane comes 10th (only ahead of Istanbul) when the economic development is concerned.
- In societal development domain, the city's performance is slightly better, coming 8th (in front of Manchester, Birmingham, and Istanbul).
- Brisbane receives its highest standing with an outstanding performance in the spatial development domain coming 3rd (just behind Helsinki and Toronto).
- In terms of institutional development, Brisbane showcases a weaker performance—similar to the economic development domain—and ranks 10th (only ahead of Istanbul).

12.3 Global Standing of Brisbane as an Emerging Knowledge City

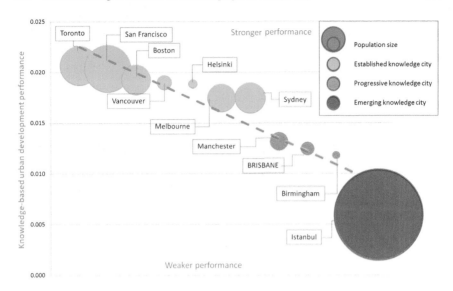

Fig. 12.2 Brisbane's knowledge-based urban development standing

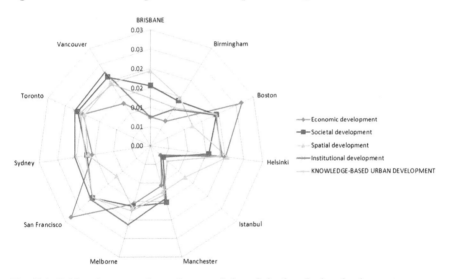

Fig. 12.3 Brisbane's comparative performance in knowledge-based urban development

186   12 Benchmarking City Performance

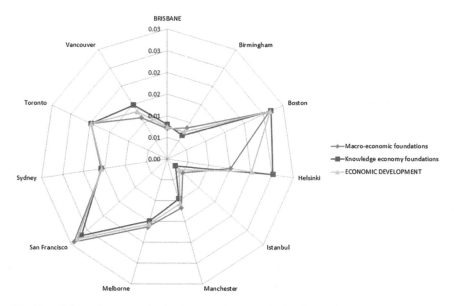

Fig. 12.4 Brisbane's comparative performance in economic development

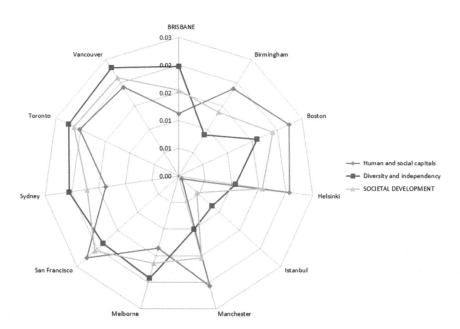

Fig. 12.5 Brisbane's comparative performance in societal development

12.3 Global Standing of Brisbane as an Emerging Knowledge City   187

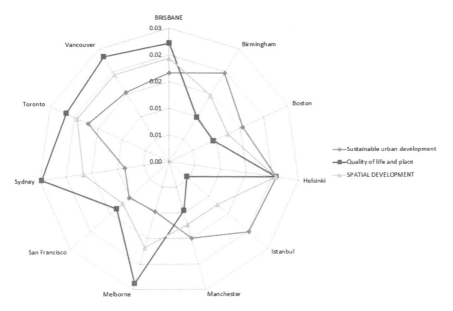

Fig. 12.6 Brisbane's comparative performance in spatial development

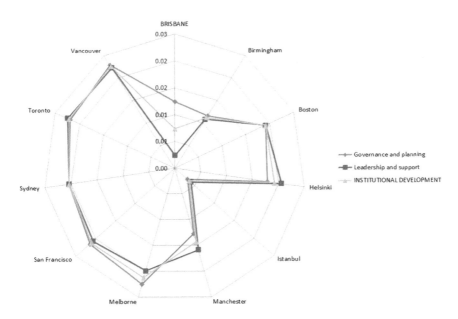

Fig. 12.7 Brisbane's comparative performance in institutional development

Beyond the overall rankings, it is also important to look into each indicator set and indicator results to understand the reasons behind Brisbane's weaker and stronger performances in specific KBUD areas.

The 10th place in the economic development domain (Fig. 12.4) is due to a weaker performance in establishing macroeconomic foundations (10th and only in front of Istanbul) and knowledge economy foundations (9th and in front of Birmingham and Istanbul). The ranking of Brisbane in specific economic development indicators are as follows:

- In gross domestic product (GDP) value the city is ranked 8th;
- In hosting major international companies, it is ranked 9th;
- In attracting foreign direct investment, the rank is 7th;
- In urban competitiveness, the city is ranked 10th;
- In innovation economy of the city is ranked 10th;
- In research and development activities it is ranked 6th;
- In patent applications the rank is 8th, and;
- In knowledge worker pool of the city is ranked 8th.

The 8th place in the societal development domain (Fig. 12.5) is due to weaker performance in establishing strong human and social capitals (10th and only in front of Istanbul), and stronger diversity and independency in the city (4th and just behind Vancouver, Toronto and Sydney). The ranking of Brisbane in specific societal development indicators are as follows:

- In education investment of the city is ranked 10th;
- In professional skill base, the city is ranked 9th;
- In university reputation, it is ranked 8th;
- In broadband access the rank is 10th;
- In cultural diversity, the city is ranked 3rd;
- In social tolerance, it is ranked 3rd;
- In socioeconomic dependency the rank is 8th, and;
- In unemployment level, the city is ranked 4th.

The 3rd place in the spatial development domain (Fig. 12.6) is where the performance of the city excels. This is due to strong performance in sustainable urban development (5th in front of Boston, Vancouver, Manchester, San Francisco, Melbourne and Sydney), and as strong in the quality of life and place (4th just behind Sydney, Melbourne, and Vancouver). The ranking of Brisbane in specific spatial development indicators are as follows:

12.3 Global Standing of Brisbane as an Emerging Knowledge City 189

- In eco-city formation, the city is ranked 3rd;
- In sustainable transport use the city is ranked last (11th);
- In environmental impact, it is ranked 4th;
- In urban form and density, the rank is 4th;
- In quality of life, the city is ranked 6th;
- In cost of living it is ranked 3rd;
- In housing affordability, the rank is 4th, and;
- In personal safety, the city is ranked 4th.

The 10th place in the institutional development domain (Fig. 12.7) is where the performance of the city weakest. This is due to low performance in governance and planning (9th in front of Birmingham and Istanbul), and weakest in leadership and support (11th behind all cities). The ranking of Brisbane in specific institutional development indicators are as follows:

- In government effectiveness, the city is ranked 4th;
- In electronic governance, the city is ranked 7th;
- In strategic planning, it is ranked 9th;
- In city branding the rank is the last (11th);
- Ineffective leadership, it is ranked 10th;
- In strategic partnership and networking the rank is 9th;
- In community engagement, the city is ranked the last (11th), and;
- In social cohesion and equality, the city is ranked 10th.

The empirical analysis has not drawn a so bright picture of Brisbane's KBUD performance. However, the findings of the analysis should be interpreted carefully by keeping in mind that most of the cities used for comparison are in the top global and knowledge cities and used as global benchmarks in many studies. Most of them are also the primate cities in their cities.

12.4 Strengths, Weaknesses, Opportunities and Threats of Brisbane

In the light of the empirical analysis, literature findings, and personal professional observations of the city, in this section, a strengths, weaknesses, opportunities, and threats (SWOT) analysis is conducted. SWOT analysis is a useful technique for understanding Brisbane's strengths and weaknesses, and for identifying both the opportunities open to the city and the threats the city faces. Table 12.6 presents the key findings of the SWOT analysis.

Table 12.6 SWOT analysis of Brisbane

Strengths	Weaknesses
• *Population growth*	• *Public transport accessibility and affordability*
The city has an increasing population growth, even though the increase is slowed down in recent years, this means the city will continue to expand its social capital	In spite of the increasing investment on the public transport system of the city due to public transport inaccessibility and unaffordability issues, the main mode of travel is private motor vehicles
• *Health infrastructure and research*	• *Second-tier city position*
Healthcare is equitable and health infrastructure is solid with numerous health precincts in the city not only providing quality health services but also undertaking cutting edge research—e.g., Translational Research Institute Australia	Brisbane lacks the business and social activity diversity and volume that of Sydney and Melbourne's due to its second-tier city disadvantages and lacks developing strong policies to narrow the gap between Sydney and Melbourne
• *Higher education institutions*	• *Housing affordability*
There are three reputable universities located in the city providing a high standard of tertiary education and training, along with research activities	Even though in the last year the inner-city rent has shown a drop, housing affordability, particularly for students and young people, remains to be a major challenge
• *Cultural heritage*	• *Walking and cycling*
The city has an increasingly multicultural nature with a preserved unique architectural characteristic (i.e., Queenslander, tin and timber houses) adding vibrancy and authenticity to the urban life	The walkability level of the city, besides some inner city central areas, is very low and further encouraging private motor vehicle dependency. Besides in spite of improved bicycle network, walking and cycling networks are disconnected
• *Development around transport hubs*	• *End of mining boom*
The city has the number of transport hubs assist increasing and establishing a denser residential population with easy access to transport options—i.e., transit-oriented development	After the end of the mining boom, unemployment level is increased, generating an urgent need for the provision of new employment opportunities ideally with knowledge-intensive jobs. This challenge has not been addressed yet
• *High density mixed use development and urban design*	• *International appeal*
City centre and inner-city suburbs are hosting higher density mixed use developments surrounded with quality urban design elements that are contributing to the appeal and vibrancy of the city	Brisbane lacks reputation and appeal to attract international knowledge industries' regional headquarters or businesses, which most of them are located in Sydney and Melbourne

(continued)

12.4 Strengths, Weaknesses, Opportunities and Threats of Brisbane

Table 12.6 (continued)

Strengths	Weaknesses
• *Youth population*	• *Lack of incentives for start-ups*
City centre and inner-city areas are hosting a higher number of younger student, professional and service sector worker populations	The city has limited resources to provide incentives for knowledge-intensive SMEs and start-ups, which makes them move to Sydney, Melbourne or overseas
• *Brownfield development*	• *Broadband access*
There are still a few good size brownfield areas in the city available for development particularly for new knowledge precincts and providing room for knowledge-based growth	Among the developed nations Australia's broadband internet (including 4G) is one of the slowest and expensive service. NBN coverage is even in the major metropolitan areas like Brisbane is very limited
• *Green space and recreation*	• *Retail and late-night trade*
Inner city green space offerings for recreation activities are excellent, free Wi-Fi service in these parks is a drawcard for visitors, and younger residents	Limited operating hours for stores in the city, reducing prospective sales and thus profit in the retail sector. Online shopping affects some of the local retail businesses negatively that do not offer online shopping option
• *Brisbane River*	• *Expensive tertiary education*
Brisbane river is a fundamental asset as a recreational, transport and aesthetic appeal, and stands as a pivotal centrepiece in the development of the city	The increasing cost of tertiary education fees are discouraging students from pursuing studies in universities and resulting in slowly thinning of the knowledge worker pool of the city
• *Airport and seaport*	• *Perception*
The ports with their existing capacity and potential for expansion provide international transport hubs for trade and other commercial purposes and helps in limiting the impact of the tyranny of distance	Queensland, and Brisbane by association is still perceived as the 'slow state' with antiquated and protectionist government policies. The general perception of the city as a 'country town' has still not completely changed, despite the branding efforts of the city as 'Australia's New World City'
• *Climate and lifestyle*	• *Red and green tapes and entrepreneur culture*
The subtropical climate allows residents and visitors to enjoy and active lifestyle and buildings to adopt innovative solutions. Good weather is a major drawcard for tourists and people that migrate to work or retire in the city	A significant amount of red and green tapes exists and prevents new economic development opportunities to happen in and around the city. Furthermore, the lack of strong entrepreneur culture hampers achieving the knowledge economy goals

(continued)

Table 12.6 (continued)

Strengths	Weaknesses
• *Geographic location*	• *Risk taking*
The geographic location of the city in Australia and internationally allows for strong connections with the fast-growing South-East Asia region	Governments play safe and avoid risk-taking in providing opportunities for new knowledge-intensive businesses, without strong financial support moving towards knowledge economy is quite impossible
• *Anticorruption practice*	• *Dedicated authority*
Both State and Local governments have effective systems and checks in place to prevent corruption and rotting across the system. This increases the trust in the governance system and foreign direct investment	There is no dedicated authority to oversee the knowledge-based urban development of the city and knowledge city transformation. Furthermore, disconnect or lack of strong collaboration between government departments is slowing the knowledge-based urban development progress
• *Policymaking*	• *Lifestyle and social scene*
There is a strong culture of strategic policymaking and planning in the city including policies on knowledge-based urban development—i.e., Smart State Strategy	The city has a growing but still limited art, café, sports and entertainment culture, besides inner-city areas there is not many sociocultural activities taking place
Opportunities	Threats
• *Lifestyle*	• *Financial crises*
Brisbane's outdoor lifestyle could be fostered, and the river could be embraced through urban design and pedestrian corridors supported by art houses, cafes, restaurants, bars, recreational facilities	The current global economy and outlook provide limitations to foreign investment. A potential future financial crisis would further limit the growth of the city and its economy
• *Cultural diversity*	• *Climate change*
Cultural integration efforts could enrich and empower the population and promote an accepting multicultural society, along with opportunity to ramp up Southbank and Fortitude Valley as classical and contemporary art and culture centres	Brisbane is one of the highest greenhouse gas emitters in the country. The predicted climate change has a likelihood of pushing environmental conditions beyond the ability to exist infrastructure to handle
• *Higher education institutions*	• *Governance challenges*
Major universities of the city could better engage with the industry, government and community (i.e., triple-helix model participation) to help in knowledge city transformation of the city	The shortcomings of the present urban governance system, such as lack of inter-departmental healthy collaboration, cooperation, participation, could provide a major threat for knowledge-based urban development

(continued)

12.4 Strengths, Weaknesses, Opportunities and Threats of Brisbane

Table 12.6 (continued)

Opportunities	Threats
• *Infill development and vertical streets*	• *End of resource-based economy*
Infill development and density increase in the city could provide the space needed for knowledge industries and social activities required to increase the attraction of the city	The end of the mining boom has already had a big hit on the economy and society of the city. In the case the move towards a knowledge economy cannot be achieved, the impact could be much dramatic
• *Modal shift*	• *Business climate*
Along with the new public transport projects, rideshare options could provide opportunities to lessen the private motor vehicle dependency	Continuum of not being able to create a business-friendly environment for the city would jeopardise its growth and employment generation
• *Cross-river rail*	• *Aging population*
The project is an opportunity to connect Brisbane CBD with other high-growth suburbs and generate employment hubs and knowledge industries around the stations	Aging population and out-migration of skilled young people of the city could place further stress on social and health services and increase competition for employment
• *Smart urban technology*	• *Decaying infrastructure*
The use of smart urban technology, such as LED lights, smart bins, smart surveillance system, smart transport systems, smart parking, could help efficiency and attractiveness of the city	Increase demand on the existing infrastructure that is aging and decaying could strain provision of quality urban amenities and services
• *Links to South-East Asia*	• *Community conflict*
Brisbane could use its proximity, as the closest major Australian city, to South-East Asia as an opportunity to increase economic relationships with the region	Community involvement is important in planning and policy-making processes, yet conflicting views and values of interest groups could limit growth and investment
• *Knowledge economy*	• *Housing affordability*
Transition to a knowledge economy has been a goal of Brisbane since the introduction of Smart State Strategy. The city could prepare another major strategy building on the past achievements and its potentials	Housing unaffordability could continue to increase particularly in the areas with close proximity to major employment and social infrastructure and services
• *Skilled workers*	• *Socioeconomic divides*
The city could foster and attract skilled workers by providing employment opportunities, quality and affordable living options with competent policies and investment	Further disadvantaging of the population without appropriate employment, affordable education opportunities and broadband access could deepen the socioeconomic and knowledge divide in the society

(continued)

Table 12.6 (continued)

Opportunities	Threats
• *Anticipation*	• *Sprawling development*
The government could develop more proactive policies as they create more opportunities than those reactionary policies to boost knowledge-based urban development. The end of the resource based booming economy could be an opportunity to do the right thing and invest in knowledge economy and society	Sprawling development trend could continue due to housing unaffordability and ineffectiveness of density increase attempts. This leads to an unsustainable development and highly costly infrastructure needs
• *Hosting international events*	• *Fossil fuel dependency*
The city could continue to host internationally significant events as have done in the past that places the city in the world map, along with the opportunity to scale up existing established festivals such as Big Sound	Continuum of the existing trend of heavy reliance on private motor vehicle use and road freight methods could create major problems if alternative energy resources and solutions are not implemented immediately
• *City branding and marketing*	• *Public values*
The recent branding and marketing efforts are promising but definitely not sufficient. The city could potentially further invest in these areas to turn itself into a popular destination for investment	New policies not properly communicated with the public and public views and values not being reflected in policies forming the development could result in no strong public support behind these policies. This risks their effectiveness
• *Governance reform*	• *Lifestyle change*
Possible governance reforms could help the decision and policy-making mechanisms to work more effective, efficient and address the critical developmental issues in a timely manner	The risk of residents not willing to change their lifestyles to adopt the change, for example adapting green alternatives of travel or lifelong learning, could risk the development
• *Knowledge city vision*	• *Disconnect between knowledge generators*
The city could develop a strong knowledge-based urban development vision to better engage all actors in focusing on the knowledge city transformation of the city	The potential risk of developing knowledge precincts in isolation from the rest of the other knowledge producers could result in a disconnect between them and jeopardise the success of knowledge-based urban development across the city
• *Health and knowledge precincts*	• *Poor urban governance and politics*
The city has been developing health and knowledge precincts, a continuum of this investment policy could help forming globally recognised knowledge and innovation clusters in the city	It is critical that three-tier government to jointly support the knowledge-based urban development agenda with support from industry, academia, and society. The party politics and poor governance could risk the development. Particularly acting slowly would result in falling behind in the tough global competition

12.5 Summary

As stated by the Committee for Sydney (2016, p. 2), "[c]ities that want to preserve and reinforce their best assets and address their key failings must benchmark their performance against appropriate targets and the cities with which they are in competition. The world's best cities, like the best companies, need to know how well they are doing and take action to progress further". This chapter of the book, hence, focused on providing an understanding on the international and national standing of Brisbane in the league of knowledge cities with a central question in mind. The question, how Brisbane stands as an emerging knowledge city internationally and nationally, is addressed through assessing Brisbane's KBUD achievement in comparison to eight international and two Australian knowledge cities with the help of the KBUD Assessment Model (KBUD-AM).

In Brisbane, the foundations of knowledge city aspiration were seeded with a number of government policies, including Smart State and City Strategies and most recently Advance Queensland innovation initiative, dating back as early as to the late 90 s. As part of these policies, some investment has been made to achieve KBUD through various urban and regional planning schemes in and around Brisbane. More recently, Brisbane's new city brand 'Australia's New World City' aims to support the knowledge city transformation of the city.

It is critical, however, to find out where the city stands globally in terms of its KBUD achievements, and at the same time be aware of how the other cities are performing. For that reason, a comparative KBUD analysis has been conducted, and Brisbane's performance is put into test in comparison to some of the world's best performing knowledge cities. These cities included Birmingham, Boston, Helsinki, Istanbul, Manchester, Melbourne, San Francisco, Sydney, Toronto, and Vancouver. The findings have shown that Brisbane is a progressive knowledge city with various potential. In both international and national contexts, the city's KBUD performance is progressive and promising. However, it is not adequate enough to compete with the top-tier knowledge cities.

Brisbane needs to do a lot to make a jump to the established knowledge cities category. The analysis highlighted two major weakness areas of the city. They are mostly concerning of economy and governance, which are in reality highly interrelated areas. This is to say in order to boost the economic base, the city needs to improve its governance mechanisms and approaches. Additionally, the city needs to strategize its development to be able to compete with the well-established knowledge cities. Perhaps one of the first steps in doing so is to form a knowledge city vision.

In interpreting the findings of the comparative study, the following limitations needs to be noted:

- For the practicality of data collection and convenience of similar governance practices mostly Anglophone country cities and the cities that the researcher is familiar with are included in the analysis. There are, however, many other suitable knowledge cities that Brisbane could learn from.

- Due to data availability, the 2010–2011-time period is selected for the analysis, this is to say the results are at least five years old and changes that have occurred since this date are not reflected in the results—such as the global financial crisis and end of the mining boom's economic and social effects.
- In total 32 indicators are selected for representing the key characteristics of KBUD. The data availability across 11 cities restricted the indicator selection criteria. This data limitation may have an impact on the results.
- There is a lack of 'perfect' indicators for capturing the intangible factors (such as effective leadership). Thus, in some cases, a choice had to be made to use somewhat low-quality data or remove relevant constructs from the analysis. In this study, the relevance criterion is chosen over the data quality criterion.
- Blurring boundaries between the greater city, city-region and the metropolitan area may have an impact on the selection of different administrative or regional boundaries in compared cases. Therefore, the impact of the different boundaries and population sizes should be carefully considered in analysing the results.

Lastly, as mentioned earlier the application of indicator-based comparative studies has certain limitations. However, these constraints should not prevent the performance of such studies for cities such as Brisbane. Rather, these analytical studies serve as a basis of curiosity, some level of scepticism, learning, and future development. Furthermore, presented cross-comparison studies are quite useful for cities not only for them to understand how they are performing in the global or national contexts, but also useful to monitor who their competitors are, and how they are progressing.

References

Carrillo, F. J., Yigitcanlar, T., García, B., & Lönnqvist, A. (2014). *Knowledge and the city: Concepts, applications and trends of knowledge-based urban development*. New York: Routledge.

Garcia, C., & Leal, M. (2012). The MAKCi index: Using logistic regression modelling for predicting most admired knowledge cities. *International Journal of Knowledge-Based Development, 3*(1), 83–99.

Greene, F., Tracey, P., & Cowling, M. (2007). Recasting the city into city-regions: Place promotion, competitiveness benchmarking and the quest for urban supremacy. *Growth and Change, 38*(1), 1–22.

Holloway, A., & Wajzer, C. (2008, October 7–10). *Improving city performance through benchmarking*. Sydney, Australia: International Cities Town Centres & Communities Society.

Huggins, R. (2010). Regional competitive intelligence. *Regional Studies, 44*(5), 639–658.

Luque-Martinez, T., & Munoz-Leiva, F. (2005). City benchmarking. *Cities, 22*(6), 411–423.

Malecki, E. (2007). Cities and regions competing in the global economy. *Environment and Planning C, 25*(1), 638–654.

Rondo-Brovetto, P., & Saliterer, I. (2007). Comparing regions, cities, and communities. *The Public Sector Innovation Journal, 12*(3), 1–18.

Stokie, T. (1999). Benchmarking Melbourne: Indicators of liveability and competitiveness. In L. Yuan, B. Yuen, & C. Low (Eds.), *Urban quality of life: Critical issues and options*. Singapore: National University of Singapore Press.

The Committee for Sydney. (2016). *The seven habits of highly successful cities: Advice for Sydney.* Sydney: The Committee for Sydney.

Yigitcanlar, T. (2014a). Position paper: Benchmarking the performance of global and emerging knowledge cities. *Expert Systems with Applications, 41*(12), 5549–5559.

Yigitcanlar, T. (2014b). *Valleys, alleys and roundabouts: innovating beyond a precinct. The Conversation.* Accessed on March 17, 2017 from http://theconversation.com/valleys-alleys-and-roundabouts-innovating-beyond-a-precinct-24290.

Yigitcanlar, T., & Lönnqvist, A. (2013). Benchmarking knowledge-based urban development performance: Results from the international comparison of Helsinki. *Cities, 31*(1), 357–369.

Yigitcanlar, T., & Velibeyoglu, K. (2008). Knowledge-based urban development: The local economic development path of Brisbane, Australia. *Local Economy, 23*(3), 195–207.

Chapter 13
Recommendations and Strategic Directions for Knowledge City Place Making

Abstract This chapter focuses on deriving insights and recommendations based on the empirical cross-city comparison conducted in the previous chapter of this book. The benchmarking study involved measuring Brisbane's performance, against Birmingham, Boston, Helsinki, Istanbul, Manchester, Melbourne, San Francisco, Sydney, Toronto, and Vancouver, through the Knowledge-Based Urban Development Assessment Model (KBUD-AM). The chapter provides strategic directions and recommendations in a structured way in the following areas: economic, societal, spatial and governance aspects along with a commentary for a long-term vision for 2050.

Keywords Benchmarking performance · Knowledge-based urban development
Knowledge city · Smart city · Sustainable urban development
Knowledge-Based Urban Development Assessment Model (KBUD-AM)
Brisbane

Reviewed academic and grey literature, investigated global best practice cases, the benchmarking exercise of Brisbane against global knowledge cities, and SWOT analysis of the city have generated invaluable findings. This chapter of the book aims to address the following critical question by providing insights, suggestions and potential strategic directions in light of these conceptual and practical findings.

- How can a holistic long-term knowledge city vision suitable for Brisbane be developed?

In order to address this question, the chapter proposes some recommendations and strategic directions concerning the key domains of KBUD for Brisbane. Following this, a potential vision for Brisbane 2050 is advocated.

13.1 Economic Aspects

The literature and international success stories underline the importance of having a sound economic base to spring up knowledge city development. This is to say basic economic foundations need to be in place in a city before aspiring knowledge city formation (Yigitcanlar et al. 2008a, b). The end of the mining boom impacted Queensland's economy negatively, and Brisbane as the State capital took a big hit. On top of the macroeconomic foundations, some basic knowledge economy foundations also need to be present in a city that is aspiring knowledge city recognition. Even though in the international comparison Brisbane does not stand out with its economic base, the city still has the required basic economic potential to support its knowledge city endeavours. Moreover, the end of the mining boom could be turned into an opportunity for the city to diversify its economic activities and invest more in the knowledge economy. Many of the now prosperous knowledge cities built their legacy on their past challenges—e.g., Austin, Helsinki and Singapore. The key issue for Brisbane is how to strategically prioritise the spending of public funds rather than whether the existing economic base is sufficient enough.

Brisbane had and has some successful past (e.g., Smart State Strategy) and present (e.g., Advance Queensland, Brisbane Economic Development Plan) initiatives in this direction. However, unlike the investigated global best practice, these efforts are rather not well coordinated or have limited resources allocated to fund knowledge generation and innovation activities and business start-ups. In the absence of private venture capital, Brisbane's public sector (State and Local governments and also with Commonwealth funds) needs to play the venture capitalist role to further support knowledge economy and society formation efforts by taking risks and investing capital in new knowledge-intensive projects and businesses. This might be a useful approach to help to form a thriving business climate in the city, particularly for knowledge-intensive start-ups and SMEs.

13.2 Societal Aspects

Brisbane has a promising global profile in terms of its societal development aspects. The city is home to three major Australian universities and a talented workforce. The global cosmopolite city characteristic of the city is developing day by day. The city is moving away from its past 'big country town' stigma and ironic nickname of 'Bris Vegas'—attributed to the city's nightlife, compact size of CBD, and perceived lack of sophistication similar to Las Vegas. While the city is developing its sophistication through time, housing affordability still remains to be a major issue and most likely contributes to the slowing of Brisbane's social sphere sophistication process.

High cost of living has been a major policy challenge not only for Brisbane, but also for all other Australian cities. This is an area that requires strong policy interventions from Federal, state and local governments for increasing the affordability conditions—particularly for the disadvantage groups and young people.

Moreover, strong branding and marketing campaigns would help in improving the image of the city; this also contributes to increasing attractiveness of the city particularly for international skilled migrants and university students. Affordable student accommodation solutions like those used in Amsterdam might be useful for Brisbane to investigate—e.g., recycled shipping containers to create student housing.

As the case of Singapore revealed, in order to attract knowledge-intensive businesses and industries a locality should invest in the human and social capital. Florida (2002) emphasised the importance of fostering and attracting not only human capital, but also the right kind of them—so-called creative class of knowledge workers. In order to attract and retain talent in the city, Brisbane needs to further enhance its human and social capital. For instance, making tertiary education more affordable could be an important step the Commonwealth government takes to increase the talent base in Australian cities. University-industry-government collaborations are critical as this type of collaboration (i.e., Triple helix model partnership) is rare in Queensland. This model of partnership requires and helps university research better align with industry needs in addressing real life problems. This could also help in generating jobs for new university graduates and students and assist in cost recovery of the tertiary education.

Additionally, as the case of Austin revealed developing a major festival or forum with international reputation helps in not only increasing the attractiveness of the city for visitors, but also in fostering the local creativity, innovation and human capital growth. Brisbane, therefore, should consider further investing on and expanding its existing successful festivals—such as Brisbane Festival, Big Sound Festival, and World Science Festival—or develop new ones in order to reinforce Brisbane's international reputation and provide opportunities for local talents.

13.3 Spatial Aspects

In the international comparison exercise, Brisbane's best achievement was in the spatial development domain. This is also evident from the State and Local governments' investments in major infrastructure projects to form world-class hard and soft infrastructures and amenities in Brisbane. A recently released State government strategy on cities and regions aims to deliver economic and community development outcomes through key infrastructure development. For example, cross-river rail innovation and economic development corridor, and health and knowledge precincts are among these infrastructures (Queensland Government 2016). However, it is important to invest in this type of KBUD projects as part of an umbrella knowledge city planning vision. This way each project will contribute achieving the bigger picture goal instead of a stand-alone project with lesser impact.

In order to attract the creative class of knowledge workers, creating liveable and attractive environments is considered as a central strategy for the knowledge economy growth. While Brisbane has a good international standing in this domain, there are some areas that still require more work to do. For instance, public transport availability and affordability are critical. Public transport fares in Brisbane (even after the recent fare reductions) are two or three times costlier than for the same distance travel in Helsinki and Singapore. Lower densities in Brisbane are discouraging hospitality services to locate in suburban areas besides some clustered nodes—that might also not be accessible by walking to many residents.

Additionally, as a result of the great climate factor (Sunshine State), Brisbane and its neighbouring coastal areas—i.e., Sunshine Coast and Gold Coast—have been popular as tourism, international immigration and interstate retirement destinations. Brisbane can improve its destination appeal further by better promoting natural assets and qualities—e.g., subtropical climate and coastal and outback areas—in its metropolitan region; and thus, benefit from visitor and destination power.

13.4 Governance Aspects

Literature and global best practice analysis indicates that governance is the key area of excellence of highly successful knowledge cities. Brisbane's performance is, unfortunately, the lowest in this domain compared to the prosperous knowledge cities investigated as part of this research. Therefore, the governance system and culture of Brisbane needs to be revisited. Closely examining successful knowledge cities common governance practices could be a good starting point for Brisbane. The following are among the key indicators of highly successful knowledge cities (see The Committee for Sydney 2016): (a) Prioritise identity, reputation, story and experience; (b) Investigate their own DNA: know their story deeply; (c) Benchmark: know others very well; (d) Brand and marketing alliances: obtain supporters for the city; (e) Make the city's promise personal; (f) Align the experience with the identity: renew the city offer, and; (g) Build up the next generation with integrity and consistency.

Successful knowledge-based development of a city requires integration of interdepartmental decision/policy making mechanisms to co-create effective and efficient policies. In Brisbane, knowledge/innovation and urban planning policies and practices are mostly detached from each other. For example, the Smart State Strategy, South East Regional Plan, and Brisbane City and Neighbourhood Plans—besides limited and ad hoc knowledge precinct developments—have not shared a common vision and not worked as pieces of an integrated planning mechanism (see Yigitcanlar and Velibeyoglu 2008; Hortz 2016) . Departments overseeing these two processes (i.e., knowledge generation and innovation, and urban planning) are not co-working and co-creating the knowledge city agenda of the city. Brisbane could highly benefit from either integrating these two processes or forming an institution (or government department) overseeing both and orchestrating KBUD and knowledge city progress of the city.

Manchester: Knowledge Capital, and Melbourne's Office of Knowledge Capital institutions can be a model for forming such an organisation for Brisbane. Besides, strong leadership around long-term knowledge city vision is also critical for Brisbane. Furthermore, continuity of planning, particularly at the strategic level, is highly important. Since the conclusion of the Smart State Strategy in 2015, there have not been any efforts to come up with a similar overarching strategy for the State and the city. Losing track, in a tough global rivalry for economic progress, could easily make cities highly vulnerable. Therefore, it is necessary to develop an umbrella strategy with a strong knowledge city vision for Brisbane—and Queensland.

Transformation of Brisbane into a prosperous knowledge city, in the guidance of such strategy/vision, will also require funding to initiate exemplar projects, upgrade or develop hard and soft infrastructure, and so on. This is an area that governance plays a critical role. Three-tier government collaboration along with partnership with the private sector and academia (in the form of triple helix model partnership) could help get projects going. In the past Brisbane benefited from the Better Cities Program that provided funding from the Commonwealth and Queensland governments in revitalising Brisbane's inner-city suburbs of Fortitude Valley, Teneriffe, New Farm, Newstead and Bowen Hills (see Bryant 2016).

Lastly, Brisbane City Council is the largest city council in Australia, and this provides council power and strength in dealing with spatial and socioeconomic challenges of the city. However, this also means a much larger area to manage, and the necessity to balance the delivery of investment and services across an extensive space. This translates into some of the collected council rates/tax from the CBD and inner suburbs (see Davis 2011—that states CBD rate revenue is about 10 times or so higher than the outer suburbs) being used for funding services or property rate offsetting of the other suburbs.

In other major Australian metropolitan cities—such as Sydney and Melbourne—the defragmented local government structure model, while having its own challenges such as collaboration with surrounding local councils for major projects, helps all collected council rates to be spent for the CBD and inner city. This has helped Sydney and Melbourne's inner-city to house world-class facilities and activities. Along with this, the conflicting views on the city's future between the Brisbane City Council and the State Government create duality for Brisbane's governance, and slow or hamper the decision-making process. The Council and the State should align their forces and in collaboration develop a vision and supporting policy actions for the knowledge city transformation of Brisbane and thus securing a better future for the Brisbanians.

13.5 Recommendations for a Vision

Brisbane has a long history in strategic plan making to use planning tools to combat the challenges the city and its metropolitan region is facing. Besides the Smart State Strategy 2005–2015 and Smart Cities Strategy 2007, for the city, a number of planning instruments are operationalized. These include, South East Regional Plan 2009–2031, Brisbane Economic Development Plan 2012–2031, Creative Brisbane Creative Economy 2013–2022, Brisbane City Plan, Brisbane City Centre Master Plan 2014, and Brisbane 2022 New World City Action Plan. This is to say Brisbane is not in shortage of economic and spatial development plans. However, the main issue or the lack thereof is a better alignment of those individual governmental/departmental approaches under a widely agreed and supported vision for a distant future.

Besides interdepartmental coordinated efforts, plans and policies of Brisbane targeting individual domains of KBUD—i.e., economy, society, space and governance—should be brought together under a common long-term vision—for example for 2050. This book suggests considering a 'knowledge city vision for Brisbane'. Despite another city brand gaining more popularity—i.e., smart city—in the recent years, knowledge city is a more comprehensive development perspective for a city. Interestingly the Smart State Strategy of Queensland was indeed a knowledge city strategy rather than literally a smart city strategy. Furthermore, cities like Barcelona, Melbourne, San Francisco, and Vienna have been investing in both knowledge city and smart city formation by mostly adopting citywide knowledge city policies, and precinct or infrastructure/amenity focused smart city solutions. These cities see the knowledge city strategy as a backbone of their growth and development, and the smart city strategy as a useful approach for developing high-quality urban spaces, delivering intelligent urban services, and fostering the growth of local smart technology firms (Yigitcanlar and Bulu 2018).

13.6 Summary

This chapter concentrated on producing some generic recommendations and strategic directions for developing a long-term focused vision for Brisbane. Developing such vision requires meticulous work with the involvement of all interested stakeholders and parties. Barcelona case could be useful to further investigate how the city managed to engage about 200 public sector departments and wide-range of stakeholders, including the public, private firms, and a large number of volunteers in the development and application of the knowledge city vision.

The recommendations made in this chapter are generic and aim to create an interest in the State government to consider incorporating a strong overarching knowledge city vision and strategy. However, in order to develop such vision more thorough investigations need to be conducted. These studies should include scrutinising Brisbane's knowledge assets, achievements, and potentials in a comprehensive way, while benchmarking against the other prosperous knowledge cities (see Velibeyoglu and Yigitcanlar 2010). Furthermore, the list of recommendations has also some generic nature and most likely to apply across the board to many emerging knowledge cities of the world. Thus, these recommendations could be considered, after a careful evaluation of the suitability to the local context, by the other emerging knowledge cities.

References

Bryant, L. (2016). *Investing in Australian cities: The legacy of the better cities program*. Brisbane: Property Council of Australia and Queensland University of Technology.
Davis, T. (2011, November 4). Brisbane City council rates revenue: All 191 suburbs listed. *The Courier Mail*.
Florida, R. (2002). *The rise of the creative class: And how it's transforming work, leisure, community and everyday life*. New York: Basic Books.
Hortz, T. (2016). The Smart State test: A critical review of the Smart State Strategy 2005–2015's knowledge-based urban development. *International Journal of Knowledge-Based Development, 7*(1), 75–101.
Queensland Government. (2016). *Advancing our cities and regions strategy*. Brisbane: Economic Development Queensland.
The Committee for Sydney. (2016). *First amongst equals? Benchmarking Sydney's performance*. Sydney: The Committee for Sydney.
Velibeyoglu, K., & Yigitcanlar, T. (2010). An evaluation methodology for tangible and intangible assets of city-regions: The 6K1C framework. *International Journal of Services, Technology and Management, 14*(4), 343–359.
Yigitcanlar, T., & Bulu, M. (Eds.). (2018). Urban knowledge and innovation spaces: Insights, inspirations and inclinations from global practices. New York, NY: Routledge.
Yigitcanlar, T., & Velibeyoglu, K. (2008). Knowledge-based urban development: The local economic development path of Brisbane, Australia. *Local Economy, 23*(3), 195–207.
Yigitcanlar, T., Velibeyoglu, K., & Baum, S. (Eds.). (2008a). Knowledge-based urban development: planning and applications in the information era. Hersey, PA: IGI Global.
Yigitcanlar, T., Velibeyoglu, K., & Baum, S. (Eds.). (2008b). Creative urban regions: Harnessing urban technologies to support knowledge city initiatives. Hersey, PA: IGI Global.

Chapter 14
Conclusion to Part III

Abstract This chapter provides a concluding remark to this part of the paper that placed Brisbane from Australia as an exemplar emerging knowledge city under the microscope of knowledge-based urban development. The part discussed the city benchmarking topic and presented an international city comparison study of Brisbane. Generated conclusions, in the light of the analysis reported in the earlier chapters, are not only useful for the Brisbane city but also invaluable for other emerging knowledge cities of the world that aspires a prosperous knowledge-based urban development and knowledge-city formation.

Keywords Benchmarking performance · Knowledge-based urban development Knowledge city · Smart city · Place making

The research results presented in this part of the book indicate a number of key strategies and actions that are critical for transforming Brisbane successfully into a prosperous knowledge city. This study, therefore, recommends the followings—complementary to the other recommendations made earlier—that are useful to be taken into consideration for implementation in Brisbane (and also in many other emerging knowledge cities with similar characteristics):

- Establishing strong political and societal will through effective governance, leadership, stewardship, and community engagement;
- Setting up a dedicated agency to orchestrate and promote knowledge-based urban development and knowledge city transformation through an enhanced intergovernmental and interdepartmental collaboration;
- Developing a strategic vision and dynamic and long-term development planning with a specific knowledge-based urban development and knowledge city focus;
- Stocktaking to understand the city's existing achievements and potential, and benchmarking to find out what other cities have been doing and capable of;
- Providing adequate financial support, incentives, and strategic investments to foster entrepreneurial culture, and empower citizens as knowledge creators and innovators;

- Forming urban innovation engines through knowledge and innovation space development (e.g., knowledge precincts) as an integrated part of the urban fabric to trigger, speed up, host, and facilitate innovation;
- Investing in and encouraging triple helix model (i.e., public-private-academic) partnerships in knowledge-intensive projects and investments;
- Supporting and closely working with universities and R&D institutions that are the primary sources of research excellence, innovation, growth, and prosperity;
- Utilising effective strategic city branding, marketing and alliance building activities to increase the international and domestic reputation to attract talent and investment;
- Creating and preserving the unique identity and image of the city based on endogenous assets to increase the appeal for attracting and retaining knowledge workers;
- Enhancing the international and multicultural character of the city by encouraging and increasing the societal openness and tolerance to diversity;
- Investing in the further development of human and social capitals to foster the talent base and intellectual wealth of the city, and create a cohesive knowledge society, and;
- Providing sustainable, high-quality, diverse and affordable housing and urban infrastructure, services and amenities to all citizens.

In the light of these recommendations, Brisbane has the opportunity to take a leadership role in developing an effective and efficient planning and development policy for the city and the region with a strong vision incorporating knowledge-based urban development principles and encouraging knowledge city transformation. This would help Brisbane, and many other emerging knowledge cities, to become a city that is:

- Connected and accessible—places a focus on mobility and the ability to be connected with the world;
- Attractive for talented people and businesses to move and invest in—thinking of long-term prosperity;
- Inclusive and for everyone—not for a narrow demographic or limited socioeconomic group;
- A place for people to live-work-play in—functional, operational and also enticing to live-work-play in, and;
- Sustainable—considers long-term social, economic, environmental, and governance considerations.

Lastly, as this chapter aimed to tackle a significant issue in a rather limited time frame with limited resources, it only provided generic and concise outputs to Brisbane. The study, therefore, points out for more detailed investigations on Brisbane to precisely understand its assets, achievements, potentials, and the most appropriate directions for the city to follow in its knowledge-based development journey—along with finding suitable ways to operationalise the recommendations put forward in this study.

Part IV
Place Making for Knowledge Precincts

Chapter 15
Introduction to Part IV

Abstract Knowledge and innovation spaces or in other words knowledge precincts are the nexus of knowledge-based urban development and knowledge economy. Planning the emergence of these innovation and creativity hubs is at the forefront of the urban agenda in many parts of the world. Place making is seen as an invaluable approach in their planning and emergence where place making is a multifaceted approach to the planning, design and management of spaces of public and private. This chapter provides an introduction to this part of the book that concentrates on the topic of place making for knowledge precincts that are essential elements of forming prosperous and sustainable cities of the global knowledge economy era.

Keywords Knowledge-based urban development · Knowledge city · Smart city
Sustainable urban development · Knowledge and innovation spaces
Knowledge precinct · Innovation district · Place making

In recent years, construction of 'knowledge precincts' has become a highly popular trend in many parts of the world for fostering, attracting or retaining innovation and knowledge generation activities, industries, and workers. Knowledge precincts are defined as integrated centres of knowledge generation, learning, commercialisation and lifestyle. In other words, they are high-growth knowledge industry and worker clusters, and distinguish the functional activity in an area, where an agglomeration of knowledge and technological activities has positive externalities for the rest of the city and the region as well as companies located there. Besides being spaces of innovation and knowledge generation, these precincts have been considered as a solution to complex political and economic issues in societies—e.g., regional industry problems, under-commercialisation of publicly financed research, shortage of new product development, productivity issues, and unemployment. Urban administrators often view these precincts as a significant policy instrument for promoting or supporting urban development and economic growth. Many of these administrators have pursued the idea of creating knowledge precincts that are able to attract global flows of ideas, knowledge, people, and capital, and become trendsetting knowledge and innovation hubs/clusters (Yigitcanlar and Dur 2013).

Even though the knowledge precinct concept is highly popular amongst the many cities of the world, only a small number of cities to date have managed to develop thriving knowledge precincts. For instance, 22@Barcelona (Barcelona), Arabianranta (Helsinki), Boston Innovation District (Boston), One-North (Singapore), Silicon Hills (Austin), Silicon Valley (San Francisco), and Strijp-S (Eindhoven) are a few of the most successful ones (Pancholi et al. 2015).

In the context of Brisbane, the foundations of knowledge precinct developments were seeded in the mid-1980s. The very first knowledge precinct development was the Brisbane Technology Park—a Queensland Government initiative established to foster the growth and collaboration of companies in knowledge-based industries in 1986. This is followed by the initiation of a number of knowledge precinct projects in South East Queensland—such as Dutton Park Knowledge Precinct, Herston Health Precinct, Kelvin Grove Urban Village, Health and Food Sciences Precinct, Sippy Downs Knowledge Precinct, and Gold Coast Health and Knowledge Precinct. Despite these knowledge precinct initiatives and over 30 years of experience and investment, Brisbane has not been entirely successful in establishing a vibrant venture capital environment, attracting major global companies, and placing any of these precincts on the global map.

This Part IV of the book concerns of knowledge precincts and their place making efforts. The part particularly focuses on generating insights on the potential pathways for cities and their knowledge hubs—for them to adopt an effective knowledge-based urban development approach and develop prosperous knowledge precincts. As working on a specific case city naturally makes generation of sound lessons and insights possible and easier, the part of the book concentrates its attention on a vibrant case study from a developed country context. Namely, Brisbane, capital city of the Queensland state of Australia—an emerging knowledge city—is selected as the case city for this investigation. The selection of the case study city was already justified in Part III of the book. In sum, this part of the book focuses on investigating a case city's (Brisbane, Australia) potential pathways and adopting suitable approaches for the city and the region to develop thriving knowledge precincts that would also help the knowledge city transformation of the city. This study also generates many insights for other emerging knowledge cities.

15.1 Rationale and Scope

Against the aforementioned backdrop, this part of the book aims to develop a thorough understanding of the underlining conceptual and practical issues on the planning, designing, and fostering existing and emerging knowledge precincts. The part intends to place Brisbane's knowledge precincts under the microscope scrutinising their potentials and progress from the lens of knowledge-based urban development and place making. In terms of the geographical boundary of the investigation area metropolitan region (South East Queensland) scale is selected. This Part IV of the book attempts to address the following research questions.

- What is the role of knowledge precincts in knowledge city formation?
- What are the key conditions and policy drivers to plan and develop thriving knowledge precincts?
- How can lessons learned from global knowledge precinct best practice be evaluated for their potential in Brisbane?
- How do Brisbane's knowledge precincts compare to the other reputable ones in Australia and overseas?
- What are the key issues concerning planning and design of Brisbane's knowledge precincts in a city-wide holistic and integrated perspective?

15.2 Methodology

The methodology of the study on knowledge precincts investigation contains the following systematic approaches:

- Review of academic and grey literature to determine the key conceptual and practical issues concerning the knowledge precinct concept.
- Global best practice analyses to determine lessons from the best practice knowledge precincts—i.e., 22@Barcelona (Spain), Macquarie Park (Australia), One-North (Singapore), and Silicon Valley (USA)—by focusing on the success and failure factors and analysing their suitability for Brisbane's knowledge precincts.
- Empirical analysis to compare Brisbane's knowledge precincts with major cities' precincts from Australia and overseas—i.e., Cambridge, Barcelona, Helsinki, Eindhoven, Dublin, Sydney, and Melbourne—by focusing on the key contextual and place making aspects relevant to knowledge precincts.
- Recommendations and strategic directions to provide insights, suggestions and potential directions in the light of the findings from conceptual and practical analyses targeting both hard and soft factors and measures of knowledge precincts.

References

Pancholi, S., Yigitcanlar, T., & Guaralda, M. (2015). Place making facilitators of knowledge and innovation spaces: Insights from European best practices. *International Journal of Knowledge-Based Development, 6*(3), 215–240.

Yigitcanlar, T., & Dur, F. (2013). Making space and place for knowledge communities: Lessons for Australian practice. *Australasian Journal of Regional Studies, 19*(1), 36–63.

Chapter 16
Theory and Practice of Knowledge Precincts

Abstract Global knowledge economy placed innovation at the heart of urban development as innovation clusters of cities significantly contribute to the economic and societal development of these locations. A knowledge precinct is in most cases an urban cluster that knowledge is generated, exchanged and marketed by talented knowledge workers. The contemporary practice moves from work focused knowledge precincts—e.g., science and technology parks, innovation parks—multi-activity focused knowledge community precincts—i.e., work, live, play, and cyber. The term is used interchangeable to the popular innovation districts or knowledge or creativity hubs. This chapter concentrates on the theory and practice of knowledge precincts and generates useful information to improve our understanding on this new land use form.

Keywords Knowledge-based urban development · Knowledge city · Smart city · Sustainable urban development · Knowledge and innovation spaces · Knowledge precinct · Innovation district · Place making

The expansion of knowledge economy, globalisation, and economic competitiveness has imparted the importance of knowledge and innovation in local economies worldwide (Yun et al. 2017, 2018). As a result, integrating knowledge generation and innovation considerations in urban planning and development processes have become an important agenda for establishing sustainable growth and long-term competitiveness of contemporary cities (Yigitcanlar and Kamruzzaman 2015; Ioppolo et al. 2018). Today making space and place that concentrate on knowledge generation and innovation is a priority for many cities all across the globe. This paved the way for the establishment of 'urban knowledge and innovation spaces'. Such spaces are branded as technology, science, innovation, creative, and knowledge hubs or precincts or districts and the like (Yigitcanlar and Bulu 2016).

These knowledge and innovation spaces are established with two primary objectives in mind. The first objective is to be a seedbed and an enclave for knowledge and technology and to play an incubator role nurturing the development and growth of new, small, high-technology firms, facilitating the transfer of university know-how to tenant companies, encouraging faculty-based spin-offs, and stimulating innovative products and processes. The second objective is to act as a catalyst for regional economic development that promotes economic growth and contributes to the development of the city as a 'knowledge or innovative city' (Yigitcanlar and Bulu 2016).

Against this background, this chapter aims to address the following two critical questions by determining the key conceptual and practical issues based on the review of the academic and grey literature and offering essential background information on urban knowledge and innovation spaces— in this book, generally the term 'knowledge precinct' is used when referring to these spaces.

- What is the role of knowledge precincts in knowledge city formation?
- What are the key conditions and policy drivers to plan and develop thriving knowledge precincts?

16.1 Evolution of Knowledge Precincts

The effects of globalisation, urbanisation, and deindustrialisation, particularly in the 21st century, are rapidly changing the contemporary local economies, forcing cities across the world to adopt advanced information and communication technologies (ICTs), and pushing them to become more innovative and thus competitive (Bulu 2011; Lee et al. 2014; Yigitcanlar and Lee 2014). In this era of global rivalry, the key driver of penetration in global knowledge markets is excelling in the knowledge economy by adopting innovative mechanisms for knowledge generation (Bulu et al. 2014; Pancholi et al. 2014). For this reason, presently many cities around the globe are targeting to outperform in this type of economy by pursuing a knowledge-based urban development. As a popular development approach KBUD aims to bring economic prosperity, environmental sustainability, a just socio-spatial order and good governance to cities, and produce a city purposefully designed to encourage the production and circulation of knowledge in an environmentally conserved, economically secure, socially just and well-governed human setting—a knowledge city (Bulu 2014; Yigitcanlar 2014a). This is to say, following a robust KBUD pathway may create cities an opportunity to address economic, social, spatial and institutional needs of their growth (Carrillo et al. 2014; Yigitcanlar and Bulu 2016).

KBUD as a policy not only applied to the formation of knowledge cities, but also shaping the economic geography of talent by fostering, attracting and retaining high-skilled workers and innovative companies (Yigitcanlar and Bulu 2015). As knowledge industries and workers are the driving forces of the knowledge economy, fostering, attracting and retaining them is an important factor for achieving a flourishing economy and sustainable city growth. Cities are the natural hosts of these industries, workers and their economic and socio-cultural activities, which make urban spaces a key factor for them to emerge or locate (Yigitcanlar 2010). In other words, knowledge and innovation spaces, particularly in the urban context, are the natural hosts of talent and investment that generate the added value that is critical for success in the knowledge economy (Yigitcanlar and Lonnqvist 2013).

For instance, international best practice experience indicates that knowledge and innovation spaces—such as clusters of innovative industries (Boschma 1999)—make an effective contribution to the development of regional innovation systems by encouraging knowledge transfer between academic institutions and knowledge-intensive establishments, thereby resulting in start-ups and growth in innovation industries (Cooke 2001; Mudambi 2008; Inkinen and Suorsa 2010). Knowledge generation and innovation systems—e.g., network of organisations, enterprises, and individuals focused on bringing new products, new processes, and new forms of organisation into economic use—are spatially interlinked (Bunnell and Coe 2001; Scott 2001; Coenen et al. 2004; Asheim and Coenen 2005; Cooke 2008; Asheim et al. 2011), and commodification of science, technology and arts takes place in knowledge and innovation spaces in the form of generated knowledge and innovation that has a high market value (Makkonen and Inkinen 2014). Hence, these spaces demonstrate the potential to enhance economic growth in a city or region (Hommen et al. 2006; Ratinho and Henriques 2010; Yigitcanlar and Bulu 2016).

The knowledge cluster concept, founded by Michael Porter, has rapidly become the focus of economic competitiveness theory and policy. According to Huggins (2008), the underlying tenet of the cluster theory is that national or urban competitiveness is determined by the strength of key concentrations of specific industries within a nation or city. Porter (1998) defines a knowledge cluster as a geographically proximate group of interconnected knowledge-intensive companies and associated institutions such as universities, linked by commonalities and complementarities. Knowledge clusters possess a socioeconomic business culture linking certain fundamental conditions that are the drivers of economic growth within nations or cities. These clusters are considered as offering a means for creating higher value-added by tapping into and distributing the potential of local strengths as a whole rather than as a series of fragmented companies (Huggins 2008). In his seminal article Robert Huggins (2008, p. 277) writes that knowledge clusters can "positively influence economic growth in three core ways: by increasing the productivity of companies based in an area; by driving the direction and pace of innovation, underpinning future productivity growth; and by stimulating the formation of new businesses, expanding and strengthening the cluster itself".

Urban knowledge and innovation spaces, as knowledge clusters, are integrated centres of knowledge generation, learning, commercialisation and lifestyle. In other words, they are high-growth knowledge industry and worker clusters, and distinguish the functional activity in an area, where an agglomeration of knowledge and technological activities has positive externalities for the rest of the city as well as firms located there. Besides being spaces of innovation and knowledge generation, these spaces have been considered as a solution to complex political and economic issues in societies—e.g., regional industry problems, under-commercialisation of publicly financed research, shortage of new product development, productivity issues, and unemployment (Link and Scott 2003; Carrillo et al. 2014; Yigitcanlar et al. 2015).

Urban administrators often view knowledge and innovation spaces as a significant policy instrument for promoting or supporting urban development and economic growth (Scott 2006; Fikirkoca and Saritas 2012). Many have pursued the ideal of creating a knowledge and innovation space that is able to attract global flows of ideas, knowledge, people, and capital, and become a trend-setting innovation hub (Anttiroiko 2009; Florida 2012; Makkonen and Inkinen 2013). However, developing knowledge and innovation spaces and catering to the needs of innovative industries have been a major challenge for most of the cities—due to a low level of science and technology expenditure, a high level of government involvement in financing and undertaking research, a low level of private sector R&D, lack of collaboration amongst firms, and exceptionally high dependence on foreign technology (Dodgson et al. 2011; Yigitcanlar and Bulu 2016).

The past few decades have witnessed an increased interest from local and national government in the development of knowledge and innovation spaces—particularly in Europe, North America, and Australasia (Cooke 2002). Consequently, today, cities around the world possess the model knowledge and innovation spaces (Anttiroiko 2004; Yigitcanlar 2009; Katz and Wagner 2014). Pioneering examples of knowledge and innovation spaces include industry, science or technology parks—e.g., Stanford Industrial Park (USA), Cambridge Science Park (UK), Sophia Antipolis Technology Park (France). Following the success of Silicon Valley, which literally emerged from the Stanford Industrial Park, the 'silicon' tag has become a ubiquitous trademark for cities positioning themselves as the next major centre of innovation—e.g., Silicon Hills Austin (USA), Silicon Alley New York (USA), Silicon Roundabout London (England), Silicon Glen (Scotland).

Some knowledge and innovation spaces created identities based on their industrial heritage—e.g., Arabianranta Helsinki (Finland), Strijp-S Eindhoven (The Netherlands), and Ørestad Copenhagen (Denmark). Many others used more descriptive labels to indicate the specialisation area—e.g., Brisbane Creative Industries and Ecosciences Precincts (Australia), Biotech Bavaria (Germany), and Solid State Pharmaceutical Cluster (Ireland). Others seek to integrate R&D, technology and medical activity with top universities in an urban setting including those Innovation and Knowledge Districts in Boston, New York, San Francisco, Pittsburgh, Providence (USA), and 22@Barcelona (Spain) (Yigitcanlar and Bulu 2016).

Knowledge and innovation spaces take on many different forms and highlight different industries, which vary according to local contextual features. Whilst there are some common features amongst all, many of them are typologically dissimilar from each other. For example, they are founded in different locational settings (i.e., urban, suburban, exurban), they may host different anchor industries, possess different physical layouts, size, developmental constraints or objectives. They may be driven by different regional industries, policy and planning regimes and institutional characteristics.

Although some evolved organically or without government assistance, in most cases, their development is planned or incentivised by government—e.g., One-North, Singapore; Ørestad, Denmark; Hsinchu Science and Industrial Park, Taiwan (Hu 2008: Yigitcanlar et al. 2008b). In both organic and planned knowledge and innovation spaces local contextual features play a key role in determining their shape and outcomes (Baum et al. 2009; Grodach 2012) According to the literature, human, physical and institutional capital (Kozak 2011; Lonnqvist et al. 2014), economic development policy-making (Etzkowitz and Leydesdorff 1995; Leydesdorff 2000; Cooke and Leydesdorff 2006; Huggins and Strakova 2012), and planning systems (Knight 1995; Yigitcanlar et al. 2008a; Grodach 2013) play an indispensable role in the formation of knowledge and innovation spaces (Pancholi et al. 2015; Yigitcanlar and Bulu 2016).

In recent years, as the new generation urban knowledge and innovation spaces 'knowledge precincts' or 'knowledge community precincts' have gained popularity. Knowledge precincts denote groups of interconnected knowledge-intensive firms, highly-skilled labour, and private and public knowledge institutions, specialised within a few and related high value-added economic activity areas, and are located together in particular urban areas (Lorenzen and Mahnke 2002). They are seen to form the nucleus of KBUD as they offer a home for knowledge industries and their workers—and some also for their families.

Knowledge precincts are also seen as mixed-use environments of housing, business, education and leisure in an urban-like setting to gather the creative class of knowledge workers and innovative knowledge industries (Yigitcanlar et al. 2008b). The relation between creativity, human capital and firm growth rates shows fostering, attracting and retaining knowledge workers to knowledge precincts is a force for achieving successful economic and sustainable development (Boschma and Ter Wal 2007; Arbolino et al. 2018). Likewise, during the last decade, the importance of attracting talented workers to innovate and generate new knowledge especially in knowledge precincts has been highlighted in numerous urban and regional development policies (Verdich 2010). While knowledge workers are highly-mobile and their choice of place are influenced by different economic, social, environmental and individual factors, a comprehensive understanding of these factors and correlation between them remain to be the challenging issues. The recent literature (e.g., Yigitcanlar et al. 2007; Yigitcanlar and Dur 2013) elaborates some of these challenges and the key role of knowledge workers in developing knowledge precincts (Yigitcanlar and Bulu 2016).

16.2 Challenges for Urban Planning and Development Practice

The socio-spatial development of knowledge precincts is a recent hot topic in the fields of urban development and planning. The main reason for this is that because knowledge workers ought to live in creative urban regions, the role of planning and the built environment are back on the urban and economic policy agenda (Hospers and Van Dalm 2005). There is an urgent necessity of researching on this agenda and providing a broad policy framework that informs the weaving policy to knit together the quite different urban land uses with, at times, conflicting requirements. Such policy will need to account for the substantial spatio-technical parameters of the different land uses. Equally important, however, is the wider social context of public policy intervention. Such parameters are considered as the structure of governance, urban politics, the urban land market and the role of corporate partners. Along with these parameters, high-tech economic development and the close connection between the political economy of urban development and revitalisation are worth taking into consideration (O'Mara 2005; Carrillo et al. 2014).

During the last decade, this new community type—knowledge community clustered in knowledge precincts—has been the subject of considerable international writing (e.g., Florida 2005; Nathan and Urwin 2006; Peck 2005; Storper and Manville 2006). Interest has been sharpened by the suggestion that a knowledge community is not merely integrated by knowledge workers but constitutes an important element in the dynamics of knowledge production itself. The community, in classic terms, is a factor of production. Here we see a common motif in the literature on the knowledge economy/society or community/precinct—the juxtaposition, if not joining, of activities previously seen as quite separate—in this case, 'life' and 'work' (Boddy 2000; Yigitcanlar and Dur 2013; Carrillo et al. 2014).

According to Felsenstein (1994), knowledge precincts are generally established with two primary objectives in mind. The first objective of a knowledge precinct is to be a seedbed and an enclave for technology and to play an incubator role, nurturing the development and growth of new, small high-tech firms. It facilitates the transfer of university knowhow to tenant companies, encouraging the development of faculty-based spin-offs and stimulating the development of innovative products and processes. The second objective is to act as a catalyst for regional economic development or revitalisation and to promote economic growth, while providing work, play, and living spaces to knowledge workers (Henry and Pinch 2000). Conceptually, knowledge precincts do not foster innovation itself, but applied innovation—i.e., innovative knowledge production (Carrillo et al. 2014).

The implicit assumption that the social form of the knowledge precinct is an aggregation of creative individuals/class is worth examining. The pattern of contemporary knowledge production argues against this essentially an innocent formulation. The literature suggests that the 'culture' of the knowledge community be seen as complementary to material conditions of their work (see Nathan 2008). Such culture would be a mirror image, as it were, of conditions of production. Thus, if the abstract work of knowledge production increasingly marginalises the human body, the spatio-culture of knowledge workers requires opportunities for its display. Similarly, if the industrial organisation of knowledge production on a global scale puts more and more time pressure on workers, compensatory cultural dynamics can be expected in the knowledge community. The literature also notes the desire manifest in existing knowledge communities for 24/7 urban development and, further, the demand for art spaces, museums, and authentic forms and experiences (Baum et al. 2007; Yigitcanlar et al. 2007; Yigitcanlar and Dur 2013; Carrillo et al. 2014).

Like Florida (2005), Jacobs (2000, 2004) also insisted on the importance of KBUD in knowledge precincts for socioeconomic development. However, unlike Florida, Jacobs (2004) sees such developments as the unplanned, but the systematic chaos of the urban environment that is the driving force behind our welfare and well-being. Silicon Valley and Route 128 from the US are good examples of organically developed world-class knowledge precincts—that is, developed without direct government and planning intervention (Henton 2002). The 'One-North' knowledge precinct in Singapore is an example that falsifies her point on unplanned nature of such precincts. Clearly, the concept and design of One-North as a knowledge precinct has been influenced by Florida's understanding of the importance of attracting creative talent with, and retaining it in, a total knowledge environment to become the 'knowledge capital' of South East Asia (Baum et al. 2007).

The success of the One-North knowledge precinct comes from its uniqueness of not choosing to replicate other global best practices (i.e., Silicon Valley). In her book on cities of knowledge, O'Mara (2005) discussed many failed efforts to imitate Silicon Valley. She scrutinises the relationship between knowledge production and metropolitan development including debates about urban and regional planning. O'Mara concludes that knowledge precincts are not simply accidental market creations, but planned communities of knowledge production that were shaped and subsidised by venture capital. The previously mentioned examples reveal that knowledge precinct formation can occur, in some cases, organically while, in others, it can occur as a result of planned and orchestrated efforts. However, either at the conception or during the expansion stages of knowledge precinct formation, an engineered approach is utilised to speed up the development progress of these knowledge precincts (Yigitcanlar et al. 2008b; Ergazakis and Metaxiotis 2011; Yigitcanlar and Dur 2013; Carrillo et al. 2014).

Baum et al. (2007, p. 69) portrayed what an ideal knowledge precinct would be like. Briefly, they described it as "a networked space of many places—a stimulating, disjunctive environment that both echoes the multiplicity of contemporary knowledge production while, in its physical quality(s), compensating for the abstract nature of such work". They also pointed out the most critical aspect of such development as the constant change: "Change is fundamental to this environment: it continually refreshes experience—it is the medium of the new. If the nature of the contemporary knowledge precinct can be summed up in a phrase like this: excessive change. The challenge for policy-makers is to provide for such change in space (the juxtaposition of multiple land uses and a myriad style) and through time (the task is never complete)" (Baum et al. 2007, p. 69). They also stressed that creating and managing 'dynamism and change' are the two major challenges for the socio-spatial development of knowledge precincts (Carrillo et al. 2014).

The main challenge urban planning and development discipline has been facing today is being unable to address the needs of KBUD as its nexus knowledge precincts—for example, knowledge precincts. This challenge can only be met with either the discipline reforming and becoming more dynamic in nature to cope and manage the changes and challenges the era of global knowledge economy brought to us. Alternatively, we need to look for alternative options including the formation of new paradigms and perhaps disciplines. Knight (1995), in his pioneering article on KBUD, suggested reforming the city planning discipline as 'city development', thus equipping planners with the skills of urban developers and creating a better factor in knowledge-based development in cities' future. Many other scholars have suggested a direction for KBUD, even though it requires further theorisation (e.g., Kunzmann

2008; Yigitcanlar and Lönnqvist 2013) and employing suitable tools and approaches for practical KBUD implementation (Carrillo et al. 2014).

The next two sections of the chapter some useful insights into the essential elements for success in knowledge precincts along with the emergence and development of planned (or induced) knowledge precincts, and how planning could play a potentially effective role.

16.3 Essential Success Elements of Knowledge Precincts

During the last two decades, issues relating to knowledge generation and innovation have taken the central stage all across the world. The crucial role of knowledge—in various types such as tacit, codifiable, scientific, technological, cultural, aesthetic, expressive, and symbolic—in underpinning economic competitiveness have been recognised more widely, whether this be at the level of the company, cluster, city, region, or nation (Pinch et al. 2003). Consequently, the emergence of the global knowledge economy, with an increased spatial focus, has provided opportunities to develop innovative systems and models of work, business interaction, and production that some cities through their knowledge precincts are taking more advantage of.

Huggins (2008, p. 287), however, points out that most knowledge precinct development "policies across the globe have focused far too much on the structural products of development—especially hard infrastructure—to the detriment of the functional processes, for example, the networks and the value and supply chains underlying successful growth". According to Martin and Sunley (2003), this is due to policy makers becoming seduced by the knowledge precinct concept without sufficient understanding of its fundamental underpinnings.

Hard factors or infrastructures are the large physical structures that serve as the nexus of buildings and institutions. They encompass networks necessary for the functioning of a contemporary city. They are, thus, also critical for the emergence and survival of knowledge precincts. The urban physical components of interrelated systems provide commodities and services essential to enable, sustain, or enhance societal living conditions (Fulmer 2009). In other words, they are built constructions based on the rational needs of a city, which then has irrational consequences on the economy, the environment, and people's lives on the ground (Atlantis 2013). The main hard factors of a city that are required for the successful functioning of a knowledge precinct include, but not limited to, the followings (Van Winden 2010; Carrillo et al. 2014):

- Municipal utilities and services—to supply water, sewers, waste collection and treatment, greenspaces and so on for supporting urban lifestyles, and enhancing liveability, attractiveness, and quality of life and place of the area;

16.3 Essential Success Elements of Knowledge Precincts

- Energy infrastructure—to provide power through generation plants, electrical grid, local distribution network, ideally in smart and sustainable forms, for running businesses and daily urban services and activities;
- Transport infrastructure and services—to boost local, regional, national and international level accessibility and mobility of the people and goods;
- Advanced ICT infrastructure and services—to improve connectivity, networking, and communication capabilities, and business reach and productivity;
- Higher education and R&D institutes—to foster knowledge workers, local know-how and knowledge-intensive companies, ideally located in a close proximity to the knowledge precinct, and;
- Advance healthcare system—to provide accessible and affordable quality health services, and specialised health professionals delivering compassionate high-quality care to every patient, every time;

Soft factors or infrastructures encompass the system of associative structures and social or business networks, connections, and human interactions that underpins and encourages the flow of ideas between individuals and institutions (Landry 2012).

The concept of soft factors indicates that knowledge precincts should not simply be conceived of in terms of economic benefits, but also be contextualised as nodes of interaction in communities, linked to interpersonal as well as virtual terms (Chen 2012). Soft factors refer to social, political, and cultural institutions and services supporting a community or city to provide core services to a culture such as health, public safety, emergency services, and education (Atlantis 2013). Soft factors of a city that are required for healthy functioning of a knowledge precinct include, but not limited to, the followings (Carrillo et al. 2014):

- Economic infrastructure—such as financial systems and institutions, markets, venture capital companies, business logistics facilities and systems, manufacturing systems that make business activity possible and lucrative;
- Social infrastructure—is the interdependent mix of facilities, places, spaces, programs, projects, services and networks that maintain and improve the standard of living and quality of life in a community; examples include assets that accommodate social services such as education, training, healthcare, housing, social welfare systems, corrections and justice systems, and recreational facilities;
- Cultural infrastructure—in the forms of people, technology, cultural collections, and buildings support the development of cultural products and activities, serve as civic identity markers, and help in community building and cohesion; examples include concert halls, museums, sports venues and clubs, libraries, theatres, studios, and specialised training facilities, business travel and tourism infrastructure, and;
- Governance infrastructure—such as political, legislative, law enforcement, and justice systems, and planning processes as well as specialised facilities as entities to manage the processes of interaction and policy-making among the actors involved in a collective problem to make binding decisions including those relate to knowledge precincts.

Table 16.1 Lifecycle of knowledge precincts (derived from Huggins 2008)

Stage of lifecycle	Precinct force	Key actors	Network types	Network ties
Genesis	Institutional trigger	Institutional trigger	Local linkage, social networks	High trust, strong ties
Development	Centrifugal forces	New entrepreneurs and spin-offs	Informal networks	Dense ties, frequent contacts
Growth	Centripetal forces	New knowledge investment, venture capital, business service firms	Formal networks, strategic alliances	Global ties
Renewal (or demise)	Knowledge trajectories	New skilled workers creating new products and markets	Integrated new modes of interaction	Indirect ties

As soft factors, Huggins (2008) also investigates the role of 'network types and ties' in knowledge precincts in order to highlight the essentiality of soft factors in their success. His four-stage lifecycle process of a knowledge precinct includes the following stage of evolution: (a) Genesis; (b) Development; (c) Growth, and; (d) Renewal (or demise). These stages are elaborated below and illustrated in Table 16.1 (Huggins 2008):

- *Genesis*: The creation of knowledge precincts is inevitably related to an institutional trigger. Institutional triggers act as an initial magnet for attracting talent, and although they may not necessarily be a singular institution there is a high degree of correlation with the existence of specific universities and research institutes. Triggers are institutional in the sense of there being a set of pre-existing relationships or associations. The primary feature of institutional triggers consists of the existence of social networks, high trust, strong ties, and localised linkages.
- *Development*: The initial developmental phase of a knowledge precinct is based on forces that spin-off knowledge from the institutional trigger, which remains localised. The existing relationships held by the institutional trigger—based on informal networks with frequent contact and dense ties—are proactively utilised to push out new entrepreneurs and spin-off firms that create new relationships and collaborations between both themselves and the existing institutions.
- *Growth*: Knowledge precincts that successfully develop a critical mass act as a centripetal magnet for new capital inputs in the form of inward investing knowledge-based firms, venture capital organisations, and other firms specialising in specific business and professional service activities. At this stage, the scope of existing relationships evolves with the creation of more formalised networks and strategic alliances, particularly through the generation of new ties beyond the precinct.

16.3 Essential Success Elements of Knowledge Precincts

- *Renewal (or demise)*: The final stage of a knowledge precinct's initial lifecycle is dependent on the technological trajectory or path of its product and process base. Precincts able to adapt to disruptive knowledge shifts through the creative destruction associated with new product and market development will survive and grow, whereas those clusters that have become overly path dependent will eventually die. Perhaps the most important feature is the requirement for the continual development and mobilisation of human capital. At the renewal stage, this is strongly related to the capacity to renew networks and create new modes of interaction, often with actors who are one step removed—indirect ties—from existing associations.

In addition to the abovementioned hard and soft factors, the following urban or regional conditions or contextual and knowledge precinct conditions are also among the essential elements of successful emergence and development of knowledge precincts (Royal Haskoning 2015).

- Essential urban or regional conditions or contextual elements:
  - A well-functioning network of innovative/creative businesses and institutions: Strong, specialising economies with a good regional or local innovative ecosystem form a sound basis for successful knowledge precincts;
  - A well-functioning job market of knowledge workers: Technology and knowledge companies are even more dependent than other economic sectors on well-educated, creative workers;
  - An attractive residential and living environment: An essential condition for attracting highly educated people and retaining knowledge workers already living in the area;
  - The presence of tertiary education, universities and other knowledge institutions: Being in the immediate vicinity of universities or other key knowledge institutions stimulates informal contacts;
  - Available sources of financing: Such as specific financial arrangements, provided by cooperating banks, as well as access to other sources of financing, such as innovation funds.

- Essential knowledge precinct elements:
  - Embedding: The development must be explicitly embedded in a regional or local innovation strategy to promote economic development;
  - Market, vision and strategy: The development must be based on a thorough market survey. Working from this basis, the target group can be clearly delineated and a distinctive concept can be conceived of. This is set down in a clear vision and a long-term strategy for attaining the goals that have been formulated;
  - Strong management: The precinct must be professionally managed by a managing body and a single (or overarching) body corporation. A great deal of attention should be devoted to the set-up and management of incubator/accelerator centres, in the thinking that this is where the future 'residents' of the precinct will be coming from;

- Broad package of services: These services include access to sources of financing, management, and marketing advice, development of innovation strategies, fostering of contact between entrepreneurs, organisation of seminars, training, informal meetings, events, concerts, and exhibitions;
- Clear choice of target group: A specialisation of the precinct is critical to ensure a clear profile—such as creative industries or biotechnology—relevant complementary businesses can also take up residence, such as consultancy firms and other technical service providers;
- A strong urban planning and design concept: An overall urban planning and design concept is needed to construct a work environment that stimulates creativity and innovation, form suitable public interaction spaces/uses, and control architecture of the individual buildings. Furthermore, in view of the long-term planning, flexibility in the concept is of great importance;
- A unique, distinctive identity: The unique identity of the precinct should be fleshed out in, for instance, the name, logo, precinct design and common activities must be distinctive and consistently expressed externally by all the parties involved;
- Facilitating role for the government: The state and local governments' commitment is important as they (in cooperation with other government agencies) can play a role in attracting businesses and improving national and international accessibility. Government subsidies will often be necessary, certainly in the starting phase. Government loans are also attractive because the term of these loans is often longer than those provided by financial institutions.

16.4 Emergence and Development of Planned Knowledge Precincts

One of the most frequently applied approach to increase the interactions between knowledge generation and the economic progress of cities is the development of knowledge precincts (Yigitcanlar and Bulu 2016). In many cities across the globe, therefore, planned knowledge precincts are seen as a type of 'silver bullet' with the capability of dramatically improving a city's ability to compete in the global knowledge economy. However, too many of these precincts are turned into high-tech fantasies or pure real-estate businesses that are unable to deliver up to their original knowledge and innovation generation promises. According to many scholars (e.g., de Van Winden et al. 2012; Yigitcanlar and Bulu 2016; Carvalho and Van Winden 2017), the key reason for such failure include: (a) Issues associated with grandiose promises being detached from a city's socioeconomic context; (b) Policy-makers being failed to adjust the precinct's key concept overtime, thus making it unfit to changing economic and political contexts, and; (c) Avoiding to undertake continuous spatial and time dynamics analyses, and determining necessary and timely actions to change the vision and design of the precincts in reaction to multiple pressures.

16.4 Emergence and Development of Planned Knowledge Precincts

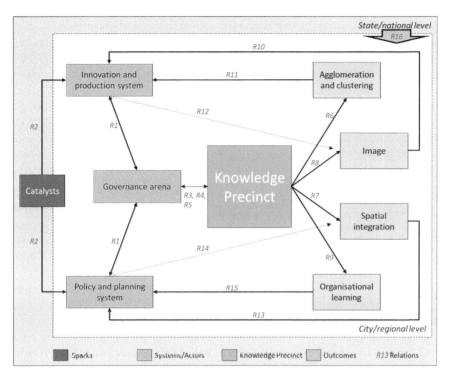

Fig. 16.1 Framework of emergence of knowledge precincts (derived from de Van Winden et al. 2012)

In order to avoid these critical failure points and provide the best nurturing opportunity for planned knowledge precincts' emergence and development, de Van Winden et al. (2012) developed a conceptual approach. Figure 16.1 illustrates the conceptual framework of emergence and development of knowledge precincts overtime. This approach pinpoints the triggering factors, as sparks of planning knowledge precincts, coming both inside and outside the local context. It highlights the dominant role of both innovation (innovation and production system) and planning (policy and local planning system) systems and their fine-tuning in the collective action (governance) arena for planning actions for the emergence and further development of knowledge precincts. 'Innovation and production system' is underlined by three combined sets of mechanisms: (a) Labour mobility; (b) Knowledge spill-overs, and; (c) Spin-offs. As for 'policy and planning system' the mechanism involves a set of elements, structures and institutional environments: (a) Policy-makers and related agents; (b) Formal political and administrative skeleton, and; (c) Specific formal and informal institutions (Carvalho and Van Winden 2017).

According to Carvalho and Van Winden (2017, p. 49), the "emergence of [planned knowledge precinct] is resulting from the coupling of interests of individuals and organisations [as the fundamental actors], which are deeply influenced by the character of the [city's] production dynamics and planning traditions; thus, ... it is misleading to think of [knowledge precincts] as the sole result of the action and free will of visionary (or short-sighted) decision-makers... Visions and designs for [knowledge precincts] are not once and for all, but evolve along multiple rounds of decision-making. Again, those changes do not necessarily result from good or bad planning, but from the coupling of: (a) External-to-the-location changes and actor's self-organisation with; (b) The progressive co-evolution between the location and its spatial-economic context". This approach, illustrated in Fig. 16.1, also lists several expected outcomes resulting from the development of a knowledge precinct. These outcomes include the followings (de Van Winden et al. 2012):

- *Agglomeration and clustering*: Successful knowledge precincts are magnets for knowledge companies. The colocation of companies generates positive external economies of agglomeration to be appropriated by the tenants and to the general benefit of the local economy—such as specialised resource and equipment sharing, knowledge spill-overs, formal and informal co-operative and competitive links, and maintaining face-to-face contact.
- *Image*: Successful knowledge precincts generate images and perceptions for inside and outside the city, becoming intertwined with a city and/or with a particular sector or cluster of activities. For instance, such knowledge precincts become unique selling points for certain activities, signalling vibrant atmospheres, buzz and places for the acquisition of the necessary knowledge resources: the 'place-to-be' effect. Often knowledge precincts become the 'face of the new economy' in the city that hosts them, and this may attract new tenants.
- *Spatial integration*: Successful knowledge precincts may contribute to the cleaning up of former derelict areas and improvements in the environmental quality of the place. However, a special attention should be paid here to not to turn them into a detached urban 'enclave', dominated by urban élites, a gentrified area without the human scale of a sense of identity. They may also steer the renaissance of deprived urban districts into more lively places. Knowledge precincts become rather diverse and vibrant urban areas and create a new added-value for the city. They also turn into excellent testbeds and experimentation laboratories of new urban living concepts and associated technologies.
- *Organisational learning*: The development of a knowledge precinct is a complex project that requires a large number of resources, skills and organising capacity. It involves many people within the public administration sphere, requiring new routines and organisation models, eventually more flexible to cope with the project requirements. Therefore, the development of knowledge precincts generates organisational learning effects in the organisations responsible for its development, namely within organisational structures of regional or local governments. These learning effects are dynamic and may turn into new organisational competences.

Van Willem et al. (2012, pp. 233–236) also put forward the following propositions to elaborate the process of emergence and development of planned knowledge precincts (see Fig. 16.1 to spot the visual linkages of these relations, e.g., R1, R2):

16.4 Emergence and Development of Planned Knowledge Precincts

- R1: The emergence of a planned knowledge precinct results from a dynamic governance process in which actors from two distinct and localised systems strategically engage: (a) Innovation and production system, and; (b) Policy and planning systems.
- R2: Governance dynamics are triggered by the need to anticipate or respond to challenges that can be both external and internal to the localised systems.
- R3: The design of a knowledge precinct depends on the power exerted by the actors of each system in governance arenas. However: (a) In governance arenas with more uneven power distributions and/or less wide participation, the knowledge precinct's concept design will emerge faster, but is likely to be more volatile, and; (b) In governance arenas with more uneven power distributions and/or less generalised participation, the knowledge precinct's concept design will be more unbalanced.
- R4: Governance arenas co-evolve with the knowledge precinct, as the precinct (as an agent and as an institution) becomes part of both localised systems over time.
- R5: The vision for a knowledge precinct evolves over time, over multiple rounds of decision-making.
- R6: The development of a knowledge precinct provokes new spatial agglomeration and clustering effects.
- R7: There is a degree of urban-spatial integration associated with the development of a knowledge precinct.
- R8: There is a degree of image effects associated with the development of a knowledge precinct.
- R9: There is a degree of organisational learning associated with the development of a knowledge precinct.
- R10: Over time, the image of a location affects the local innovation and production system.
- R11: The history of the localised system of innovation and production moderates the agglomeration and clustering potential of knowledge precincts.
- R12: Over time, the capacity of a knowledge precinct to attract new tenants and steer clustering effects feeds back into the localised system of innovation and production.
- R13: The history of the localised system of policy and local planning moderates the urban-spatial integration potential of knowledge precincts.
- R14: Over time, the capacity of a knowledge precinct to steer urban-spatial integration dynamics feeds back into the local system of policy and planning.
- R15: Over time, organisational learning effects of a location affect the local system of policy and planning.
- R16: The whole process underlying the emergence and development of knowledge precincts is influenced by higher-level political economy institutions and planning systems—of the State and/or National level.

Complementary to the abovementioned framework, Wasim (2014) suggested an approach to the planning and development of knowledge precincts—comprising four

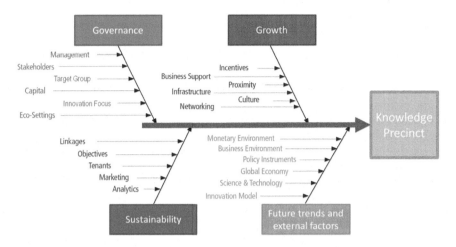

Fig. 16.2 Key factors of knowledge precinct planning (derived from Wasim 2014)

key domains of Governance, Growth, Sustainability, and Future Trends and External Factors (see Fig. 16.2).

The first domain 'governance' involves planning and setting up an environment for autonomous firms that are to be located in the precinct and are to be engaged in explicit or implicit contracts to coordinate and safeguard knowledge exchanges. Key factors that are related to planning and setup of such environment in this domain include (Wasim 2014):

- *Management*: A team of individuals that governs planning, development, administration, and operation of a knowledge precinct—with a clear vision and mission that helps in fulfilment of the objectives that aid in knowledge precinct development and growth.
- *Stakeholders*: An entity or an individual that has an investment, share, or interest in a knowledge precinct—such as management team, tenant companies, R&D organisations, local, regional and national development authorities, and the private sector.
- *Target group*: A group of customers towards whom a knowledge precinct decides to aim its services—such as start-ups, spin-offs, small and medium enterprises (SMEs), privately or publicly owned R&D centres, business support agencies (e.g., law firms), and venture capital companies.
- *Capital*: Wealth in the form of money or other assets that is required by a knowledge precinct during development and operations. Seed capital can be acquired from grants, loans, leasing and so on, whereas, working capital can be generated by offered services or incentives at a knowledge precinct.
- *Innovation focus*: The new marketable knowledge generation and innovation focus helps knowledge precincts predominantly oriented towards specific economy sectors, companies, or functions of companies with innovation outcome in mind.

16.4 Emergence and Development of Planned Knowledge Precincts

- *Eco-settings*: The eco-friendly relationship between a knowledge precinct and environment—such as initiatives for improving the environment by taking proactive measures like use of cleaner and renewable energy sources, and development of green buildings and infrastructures.

The second domain 'growth' is enabled by the services knowledge precinct offers to its target group. Commonly these services include: networking support throughout its value chain, infrastructure for a desirable quality of life and work ambience, access to business opportunities that lie inside/outside a knowledge precinct, economic incentives, access to eminent organisations, and culture with a general familiarity with entrepreneurial behaviour and ethics. Key factors in this domain include (Wasim 2014):

- *Incentives*: Financial or nonfinancial benefits offered to tenants of a knowledge precinct—such as access to university assets such as R&D equipment, skilled human resource, and employ training and counselling activities, access to funds and grants, tax waivers, subsidies on R&D activities, and single window clearance.
- *Business support*: Services provided in relation to business operations at a knowledge precinct—such as incubation, business development and training, intellectual property rights management, technology transfer, financial support, settlement management, and investor consulting.
- *Proximity/location*: The geographical proximity of a knowledge precinct to organisations and infrastructure prompting trade and commerce—such as industries, R&D organisations, government agencies, universities, city centre, airport, seaport, and railway.
- *Infrastructure*: Top-notch physical and social infrastructure is critical for knowledge precincts. Availability of space and land to accommodate the expected needs of potential tenants and the availability of urban infrastructure services—such as streets, parking facilities, water distribution system and sewers, an uninterrupted power and fast broadband supply. Social facilities help to uplift knowledge precinct's brand value—such as sports ground, medical centre, emergency response centre, gymnasium, kindergarten, shopping centre, food court, clubhouse, and multiplex dormitories and apartments.
- *Culture*: A vibrant environment that fosters social relation development and interaction, and therefore contributes to knowledge exchange while meticulously adhering to intellectual property rights.
- *Networking*: Knowledge precincts are built through collaboration and shared resources, and thus require strong networking practices—such as linkages between different knowledge precincts, between knowledge precinct management and tenants, and among tenants themselves, and if the university is one of the stakeholders then linkages between university constituents and tenant companies in a knowledge precinct.
- *Anchor tenants*: The first and the leading tenant in a knowledge precinct, whose prestige and name recognition attracts other tenants, and thus plays a key role in the growth of a knowledge precinct.

The third domain 'sustainability' refers to the ability of a knowledge precinct to gauge its performance against given criterion—performance appraisal in reference to its objectives, linkages, and tenants. Key factors in this domain include (Wasim 2014):

- *Evaluating linkages*: Network analysis in reference to linkages between tenants and organisations outside the knowledge precinct, between tenants and associated university, and amongst tenants is vital for the survivability of a knowledge precinct.
- *Objectives evaluation*: Objectives give a clear picture of knowledge precincts standings and its progress towards sustainability. Evaluations are separately performed on both short-term and long-term objectives of a knowledge precinct. Short-term objectives revolve around established firms moving to a knowledge precinct, new high-tech businesses fostered, turnover/profitability of firms, rental income from firms, patents by firms, increase in patents, and joint publications. Long-term objectives consider on-site employment quantity and quality, the value of purchases from firms in the region, increase in city's balance of trade, increase in gross domestic product (GDP) for the city, and change in relative unemployment level.
- *Evaluating tenants*: Helps in minimising the risk of failures for tenants of a knowledge precinct—such as applying a filter to international standards for evaluation of start-ups in light of the following criteria: What is innovative about this idea? What is entrepreneurial about it? What kind of partnership is included that will expand the asset base? To what degree is it an international project?
- *Marketing and analytics*: A sustainable brand image of a knowledge precinct depends upon publicly available periodicals on its performance reviews—such as releasing an annual index based on indicators that measure the socioeconomic strength and highlighting challenges and provides an analytical foundation for leadership and decision making at a knowledge precinct.

The final domain focuses on the 'external factors and future trends' that may influence planning and development of a knowledge precinct. It is essential for a knowledge precinct to closely monitor changing conditions and future trends to develop its resilience by preparing and adopting itself to those changes. Key factors in this domain include (Wasim 2014):

- *Monetary environment*: The emergence and progress of a knowledge precinct depend on a number of factors that have effects on the monetary environment—such as stock markets, foreign direct investment, foreign exchange rates and informal economy.
- *Business environment*: The vibrancy of the business environment and the willingness of the local industry to invest in new ventures, that would take place in a knowledge precinct, have critical importance.
- *Policy instruments*: Fiscal policies and investment climate policies along with urban development policies have significant impacts on the knowledge precincts emergence and progress.

- *Global economy*: The stability of the global economy is critical for knowledge precincts as on-going economic recession influences financial endurance of knowledge precinct tenants.
- *Science and technology*: Estimating the future trends in science and technology is critical for knowledge precincts. For example, from synthetic genomics to stem cell therapy, biology in collaboration with other sciences will become a central source of scientific and technological breakthroughs in future.
- *Innovation models*: Following from the above example, research at the intersection of biology, informatics, and nanotechnology will require transdisciplinary skillsets and a new model of innovation, i.e., open innovation, among tenants of knowledge precincts.

16.5 Summary

This chapter of the book focused on providing an understanding of the knowledge precinct concepts and their evolution—with the following two central questions in mind.

The first question focusing on what the role of knowledge precincts are in knowledge city formation, is addressed through elaborating the literature on KBUD, and underlining the linkages between knowledge cities, KBUD, and knowledge precincts. In the contemporary literature, knowledge precincts are seen as the nexus of KBUD that are only showing healthy emerge in the flourishing best practice knowledge cities (e.g., de Van Winden et al. 2012; Yigitcanlar and Dur 2013; Carrillo et al. 2014; Yigitcanlar et al. 2016; Carvalho and Van Winden 2017).

The second one, what the key conditions and policy drivers are to plan and develop thriving knowledge precincts, has been a subject of extended discussion in the literature. Despite the countless contemporary attempts taking place all around the world to plan new knowledge precincts from scratch, knowledge precincts do not emerge in a vacuum (Carvalho and Campos 2013; Carvalho 2015). As stated by Carvalho and Van Winden (2017, p. 49), they are "influenced by the socioeconomic and institutional characteristics of the [cities] in which they are planned. It can thus be expected that the way knowledge precincts emerge and unfold—i.e., their envisioned knowledge specialisations, physical profile and development dynamics—to be strongly influenced by their specific spatial-economic context". This context encompasses both innovation and production, and policy and planning systems (with their actors, networks and institutions) of cities. It also tends to be path-dependent, changes slowly, and be highly localised in space (see Cooke 2001).

Lastly, the findings suggest, as evidenced by the literature (e.g., de Van Winden et al. 2012), prosperous knowledge precincts emerge and develop in solid socioeconomic urban landscapes with strong history and texture. This provides clear insights into why the most successful knowledge precincts (e.g., 22@Barcelona, Arabianranta, Cambridge Science Park, Digital Hub, Macquarie Park, Monash Employment Cluster, One-North, Silicon Valley, and Strijp-S) happened to be emerged and flourished in the most prosperous and liveable cities of the world. These prosperous knowledge precincts of the world are placed under the microscope in the following two chapters of this book.

References

Anttiroiko, A. V. (2004). Science cities: Their characteristics and future challenges. *International Journal of Technology Management, 28*(3–6), 395–418.

Anttiroiko, A. V. (2009). Making of an Asia-pacific high-technology hub: Reflections on the large-scale business site development projects of the Osaka City and the Osaka Prefecture. *Regional Studies, 43*(5), 759–769.

Arbolino, R., Simone, L., Carlucci, F., Yigitcanlar, T., & Ioppolo, G. (2018). Towards a sustainable industrial ecology: implementation of a novel approach in the performance evaluation of Italian regions. *Journal of Cleaner Production, 178*(1), 220–236.

Asheim, B. T., & Coenen, L. (2005). Knowledge bases and regional innovation systems: Comparing Nordic clusters. *Research Policy, 34*(8), 1173–1190.

Asheim, B. T., Moodysson, J., & Tödtling, F. (2011). Constructing regional advantage: Towards state-of-the-art regional innovation system policies in Europe? *European Planning Studies, 19*(7), 1133–1139.

Atlantis (2013). Definition box: Hard infrastructure, soft infrastructure, network and social domain. Accessed on 8 March 2017 from https://atlantistudelft.wordpress.com/2013/12/17/definition-box-hard-infrastructure-soft-infrastructure-network-social-domain/.

Baum, S., O'Connor, K., & Yigitcanlar, T. (2009). The implications of creative industries for regional outcomes. *International Journal of Foresight and Innovation Policy, 5*(1–3), 44–64.

Baum, S., Yigitcanlar, T., Horton, S., Velibeyoglu, K., & Gleeson, B. (2007). *The role of community and lifestyle in the making of a knowledge city*. Brisbane: Griffith University.

Boddy, M. (2000). Technology, innovation, and regional economic development in the state of Victoria. *Environment and Planning C, 18*(3), 301–319.

Boschma, R. A. (1999). The rise of clusters of innovative industries in Belgium during the industrial epoch. *Research Policy, 28*(8), 853–871.

Boschma, R. A., & Ter Wal, A. L. (2007). Knowledge networks and innovative performance in an industrial district: The case of a footwear district in the South of Italy. *Industry and Innovation, 14*(2), 177–199.

Bulu, M. (2011). Measuring competitiveness of cities: Turkish experience. *International Journal of Knowledge-Based Development, 2*(3), 267–281.

Bulu, M. (2014). Upgrading a city via technology. *Technological Forecasting and Social Change, 89*(1), 63–67.

Bulu, M., Önder, M. A., & Aksakalli, V. (2014). Algorithm-embedded IT applications for an emerging knowledge city: Istanbul, Turkey. *Expert Systems with Applications, 41*(12), 5625–5635.

Bunnell, T. G., & Coe, N. M. (2001). Spaces and scales of innovation. *Progress in Human Geography, 25*(4), 569–589.

Carrillo, F. J., Yigitcanlar, T., García, B., & Lönnqvist, A. (2014). *Knowledge and the city: Concepts, applications and trends of knowledge-based urban development*. New York: Routledge.

Carvalho, L. (2015). Smart cities from scratch? A socio-technical perspective. *Cambridge Journal of Regions, Economy and Society, 8*(1), 43–60.

Carvalho, L., & Campos, J. B. (2013). Developing the PlanIT Valley: A view on the governance and societal embedding of u-eco city pilots. *International Journal of Knowledge-Based Development, 4*(2), 109–125.

Carvalho, L., & Van Winden, W. (2017). Planned knowledge locations in cities: Studying emergence and change. *International Journal of Knowledge-Based Development, 8*(1), 47–67.

References

Chen, Y. (2012). Making Shanghai a creative city: Exploring the creative cluster strategy from a Chinese perspective. In M. Van Geenhuizen & P. Nijkamp (Eds.), *Creative knowledge cities: Myth, visions and realities* (pp. 437–464). Northampton: Edward Elgar.

Coenen, L., Moodysson, J., & Asheim, B. T. (2004). Nodes, networks and proximities: On the knowledge dynamics of the Medicon Valley biotech cluster. *European Planning Studies, 12*(7), 1003–1018.

Cooke, P. (2001). Regional innovation systems, clusters, and the knowledge economy. *Industrial and Corporate Change, 10*(4), 945–974.

Cooke, P. (2002). *Knowledge economies: Clusters, learning and cooperative advantage*. London: Routledge.

Cooke, P. (2008). Distinctive proximities: Between implicit and explicit knowledge in ICT and biotechnology innovation. *Revue d'Économie Régionale & Urbaine, 3*, 381–409.

Cooke, P., & Leydesdorff, L. (2006). Regional development in the knowledge-based economy: The construction of advantage. *The Journal of Technology Transfer, 31*(1), 5–15.

de Van Winden, W., Carvalho, L., Van Tuijl, E., Van Haaren, J., & Van den Berg, L. (2012). *Creating knowledge locations in cities: Innovation and integration challenges*. London: Routledge.

Dodgson, M., Hughes, A., Foster, J., & Metcalfe, S. (2011). Systems thinking, market failure, and the development of innovation policy: The case of Australia. *Research Policy, 40*(9), 1145–1156.

Ergazakis, K., & Metaxiotis, K. (2011). The knowledge-based development agenda: A perspective for 2010–2020. *VINE, 41*(3), 358–377.

Etzkowitz, H., & Leydesdorff, L. (1995). The triple helix—University-industry-government relations: A laboratory for knowledge-based economic development. *EASST Review, 14*(1), 14–19.

Felsenstein, D. (1994). University-related science parks: "Seedbeds" or enclaves of innovation? *Technovation, 14*(2), 93–110.

Fikirkoca, A., & Saritas, O. (2012). Foresight for science parks: The case of Ankara University. *Technology Analysis & Strategic Management, 24*(10), 1071–1085.

Florida, R. (2005). *Cities and the creative class*. New York: Routledge.

Florida, R. (2012). *The rise of the creative class: Revisited*. New York: Basic Books.

Fulmer, J. (2009). What in the world is infrastructure. *Pei Infrastructure Investor, 1*(4), 30–32.

Grodach, C. (2012). Before and after the creative city: The politics of urban cultural policy in Austin, Texas. *Journal of Urban Affairs, 34*(1), 81–97.

Grodach, C. (2013). Cultural economy planning in creative cities: Discourse and practice. *International Journal of Urban and Regional Research, 37*(5), 1747–1765.

Henry, N., & Pinch, S. (2000). Spatialising knowledge: Placing the knowledge community of Motor Sport Valley. *Geoforum, 31*(2), 191–208.

Henton, D. (2002). Lessons from Silicon Valley: Governance in a global city region. In A. Scott (Ed.), *Global city-regions: Trends, theory, policy* (pp. 391–400). Oxford: Oxford University Press.

Hommen, L., Doloreux, D., & Larsson, E. (2006). Emergence and growth of Mjärdevi Science Park in Linköping, Sweden. *European Planning Studies, 14*(10), 1331–1361.

Hospers, G., & Van Dalm, R. (2005). How to create a creative city: The viewpoints of Richard Florida and Jane Jacobs. *Foresight, 7*(4), 8–12.

Hu, T. S. (2008). Interaction among high-tech talent and its impact on innovation performance: A comparison of Taiwanese science parks at different stages of development. *European Planning Studies, 16*(2), 163–187.

Huggins, R. (2008). The evolution of knowledge clusters: Progress and policy. *Economic Development Quarterly, 22*(4), 277–289.

Huggins, R., & Strakova, L. (2012). Knowledge-based economic development in emerging regions: Policy issues and implications in the Balkan Peninsula. *Regional Studies, 46*(7), 961–975.

Inkinen, T., & Suorsa, K. (2010). Intermediaries in Regional Innovation Systems: High-technology enterprise survey from northern Finland. *European Planning Studies, 18*(2), 169–187.

Ioppolo, G., Cucurachi, S., Salomone, R., Shi, L., & Yigitcanlar, T. (2018). Strategic environmental assessment and material flow accounting: A novel approach for moving towards sustainable urban futures. *International Journal of Life Cycle Assessment*, https://doi.org/10.1007/s11367-018-1494-0.

Jacobs, J. (2000). *The nature of economies*. New York: Vintage Books.

Jacobs, J. (2004). *Dark age ahead*. Toronto: Random House of Canada.

Katz, B., & Wagner, J. (2014). *The rise of innovation districts: A new geography of innovation in America*. Washington, DC: Brookings Institute.

Knight, R. V. (1995). Knowledge-based development: Policy and planning implications for cities. *Urban studies, 32*(2), 225–260.

Kozak, M. (2011). Strategic approach to intellectual capital development in regions. *International Journal of Learning and Intellectual Capital, 8*(1), 76–93.

Kunzmann, K. (2008). Spatial dimensions of knowledge production. In T. Yigitcanlar, K. Velibeyoglu, & S. Baum (Eds.), *Knowledge-based urban development: Planning and applications in the information era* (pp. 296–300). Hershey, PA: IGI Global.

Landry, C. (2012). *The creative city: A toolkit for urban innovators*. London: Earthscan.

Lee, J. H., Hancock, M. G., & Hu, M. C. (2014). Towards an effective framework for building smart cities: Lessons from Seoul and San Francisco. *Technological Forecasting and Social Change, 89,* 80–99.

Leydesdorff, L. (2000). The triple helix: An evolutionary model of innovations. *Research Policy, 29*(2), 243–255.

Link, A. N., & Scott, J. T. (2003). US science parks: The diffusion of an innovation and its effects on the academic missions of universities. *International Journal of Industrial Organization, 21*(9), 1323–1356.

Lönnqvist, A., Käpylä, J., Salonius, H., & Yigitcanlar, T. (2014). Knowledge that matters: Identifying regional knowledge assets of the Tampere region. *European Planning Studies, 22*(10), 2011–2029.

Lorenzen, M., & Mahnke, V. (2002). *Global strategy and the acquisition of local knowledge: How MNCs enter regional knowledge clusters* (DRUID Working Paper Series (2002-8)). Aalborg and Copenhagen: Danish Research Unit of Industrial Dynamics.

Makkonen, T., & Inkinen, T. (2013). Innovative capacity, educational attainment and economic development in the European Union: Causal relations and geographical variations. *European Planning Studies, 21*(12), 1958–1976.

Makkonen, T., & Inkinen, T. (2014). Innovation quality in knowledge cities: Empirical evidence of innovation award competitions in Finland. *Expert Systems with Applications, 41*(12), 5597–5604.

Martin, R., & Sunley, P. (2003). Deconstructing clusters: Chaotic concept or policy panacea? *Journal of Economic Geography, 3*(1), 5–35.

Mudambi, R. (2008). Location, control and innovation in knowledge-intensive industries. *Journal of Economic Geography, 8*(5), 699–725.

Nathan, M. (2008). Creative class theory and economic performance in UK cities. In T. Yigitcanlar, K. Velibeyoglu, & S. Baum (Eds.), *Creative urban regions: Harnessing urban technologies to support knowledge city initiatives* (pp. 80–93). Hershey, PA: IGI Global.

Nathan, M., & Urwin, C. (2006). *City people: City centre living in the UK*. London: IPPR Centre for Cities.

O'Mara, M. (2005). *Cities of knowledge: Cold war science and the search for the next Silicon Valley*. Princeton, NJ: Princeton University Press.

Pancholi, S., Yigitcanlar, T., & Guaralda, M. (2014). Urban knowledge and innovation spaces: Concepts, conditions, and contexts. *Asia Pacific Journal of Innovation and Entrepreneurship, 8*(1), 15–38.

Pancholi, S., Yigitcanlar, T., & Guaralda, M. (2015). Public space design of knowledge and innovation spaces: Learnings from Kelvin Grove Urban Village, Brisbane. *Journal of Open Innovation: Technology, Market, and Complexity, 1*(1), 13.

Peck, J. (2005). Struggling with the creative class. *International Journal of Urban and Regional Research, 29*(4), 740–770.

References

Pinch, S., Henry, N., Jenkins, M., & Tallman, S. (2003). From 'industrial districts' to 'knowledge clusters': A model of knowledge dissemination and competitive advantage in industrial agglomerations. *Journal of Economic Geography, 3*(4), 373–388.
Porter, M. E. (1998). Clusters and the new economics of competition. *Harvard Business Review, 76*(6), 77–90.
Ratinho, T., & Henriques, E. (2010). The role of science parks and business incubators in converging countries: Evidence from Portugal. *Technovation, 30*(4), 278–290.
Royal Haskoning. (2015). *Success factors for science and technology parks*. Accessed on 8 March 2017 from https://www.royalhaskoningdhv.com/en-gb/blog/urban/success-factors-for-science-and-technology-parks/5154.
Scott, A. J. (Ed.). (2001). *Global city-regions: Trends, theory, policy*. Oxford: Oxford University Press.
Scott, A. J. (2006). Entrepreneurship, innovation and industrial development: Geography and the creative field revisited. *Small Business Economics, 26*(1), 1–24.
Storper, M., & Manville, M. (2006). Behaviour, preferences and cities. *Urban Studies, 43*(8), 1275–1300.
Van Winden, W. (2010). Knowledge and the European city. *Tijdschrift Voor Economische En Sociale Geografie, 101*(1), 100–106.
Verdich, M. (2010). Creative migration? The attraction and retention of the 'creative class' in Launceston, Tasmania. *Australian Geographer, 41*(1), 129–140.
Wasim, M. U. (2014). Factors for science park planning. *World Technopolis Review, 3*(2), 97–108.
Yigitcanlar, T. (2009). Planning for knowledge-based urban development: Global perspectives. *Journal of Knowledge Management, 13*(5), 228–242.
Yigitcanlar, T. (2010). Making space and place for the knowledge economy: Knowledge-based development of Australian cities. *European Planning Studies, 18*(11), 1769–1786.
Yigitcanlar, T. (2014). Position paper: Benchmarking the performance of global and emerging knowledge cities. *Expert Systems with Applications, 41*(12), 5549–5559.
Yigitcanlar, T., Baum, S., & Horton, S. (2007). Attracting and retaining knowledge workers in knowledge cities. *Journal of Knowledge Management, 11*(5), 6–17.
Yigitcanlar, T., & Bulu, M. (2015). Dubaization of Istanbul: Insights from the knowledge-based urban development journey of an emerging local economy. *Environment and Planning A, 47*(1), 89–107.
Yigitcanlar, T., & Bulu, M. (2016). From the guest editors: Urban knowledge and innovation spaces. *Journal of Urban Technology, 23*(1), 1–9.
Yigitcanlar, T., & Dur, F. (2013). Making space and place for knowledge communities: Lessons for Australian practice. *Australasian Journal of Regional Studies, 19*(1), 36–63.
Yigitcanlar, T., Guaralda, M., Taboada, M., & Pancholi, S. (2016). Place making for knowledge generation and innovation: Planning and branding Brisbane's knowledge community precincts. *Journal of Urban Technology, 23*(1), 115–146.
Yigitcanlar, T., Inkinen, T., & Makkonen, T. (2015). Does size matter? Knowledge-based development of second-order city-regions in Finland. *disP-The Planning Review, 51*(3), 62–77.
Yigitcanlar, T., & Lee, S. H. (2014). Korean ubiquitous-eco-city: A smart-sustainable urban form or a branding hoax? *Technological Forecasting and Social Change, 89,* 100–114.
Yigitcanlar, T., & Lönnqvist, A. (2013). Benchmarking knowledge-based urban development performance: Results from the international comparison of Helsinki. *Cities, 31*(1), 357–369.
Yigitcanlar, T., O'Connor, K., & Westerman, C. (2008a). The making of knowledge cities: Melbourne's knowledge-based urban development experience. *Cities, 25*(2), 63–72.
Yigitcanlar, T., Velibeyoglu, K., & Baum, S. (Eds.). (2008b). *Knowledge-based urban development: Planning and applications in the information era: Planning and applications in the information era*. Hersey: IGI Global.
Yigitcanlar, T., & Kamruzzaman, M. (2015). Planning, development and management of sustainable cities: A commentary from the guest editors. *Sustainability, 7*(11), 14677–14688.

Yun, J., Jung, K., & Yigitcanlar, T. (2018). Open innovation of James Watt and Steve Jobs: Insights for sustainability of economic growth. *Sustainability, 10*(5), 1553.

Yun, J., Won, D., Jeong, E., Park, K., Lee, D., & Yigitcanlar, T. (2017). Dismantling of the inverted U-curve of open innovation. *Sustainability, 9*(8), 1423.

Chapter 17
Global Knowledge Precinct Best Practice

Abstract Best practice analysis is a common method to learn from the success and also failure factors of the leaders of the field and develop customised policies to apply in another context. This also applies to the knowledge and innovation space development planning. The chapter focuses on placing global knowledge precinct best practice under the place making microscope and drawing lessons for other cities. The investigated cases include 22@Barcelona from Spain, Silicon Valley from the USA, One-North from Singapore, and Macquarie Park from Australia. The lessons generated are discussed in the context of for Brisbane from Australia to consider. However, the findings are also relevant to many other cities with emerging knowledge precincts.

Keywords Knowledge-based urban development · Knowledge city · Smart city · Sustainable urban development · Knowledge and innovation spaces · Knowledge precinct · Innovation district · Place making · 22@Barcelona · Silicon Valley · One-North · Macquarie Park

In order to develop effective and efficient policies for developing thriving knowledge precincts, it is essential to learn from the high-performing knowledge precincts' experiences. A best practice analysis is a commonly used method or technique to determine success factors and produce lessons from the outstanding knowledge precinct practices that set the standards. Findings of such best practice analysis are useful to form ideas, policies, and guidelines that represent the most efficient or prudent course of action. However, careful translation of the best practice lessons to the local context and meticulous tailoring of the potential actions that have worked in the investigated cases are of the utmost importance.

Against this background, this chapter of the book aims to address the following critical question through drawing lessons from the best practice knowledge precincts by focusing on the success and failure factors and analysing their suitability for Brisbane's knowledge precincts. The best practice analysis involved the following knowledge precincts: 22@Barcelona (Spain), Silicon Valley (USA), One-North (Singapore), and Macquarie Park (Australia).

- How can lessons learned from global knowledge precinct best practice be evaluated for their potential in Brisbane?

17.1 A Best Practice Knowledge Precinct from Europe: 22@Barcelona

As discussed in Part III of the book, Barcelona showcases a remarkable KBUD performance. The city archives this performance by engaging various stakeholders in the determination of strategies to form a development pathway for a sustainable and knowledge-based future. Being recognised as one of the global best practice for both knowledge city and smart city development efforts; Barcelona is also home to the world's first innovation district—22@Barcelona—located at the inner city El Poblenou district.

Until the mid-nineteenth century, El Poblenou district—situated northeast of the historical centre so-called Ciutat Vella—was an agricultural land with a number of salt swamps, whereas sporadically appearing buildings occupied long and narrow plots. Following the industrial revolution, the district gained an industrial characteristic—mostly specialising in textile industries; and by the mid-twentieth century, almost the entire area was transformed into an industrial district, where about 90% of the buildings were allocated for industrial and warehousing uses. The 50s, 60s and 70s brought along the implementation of several social estates in the eastern part of the district, which partially replaced the slums. The political, technological and logistic condition-dependent changes at the end of the twentieth century resulted in the demise or relocation of numerous local manufacturing plants and institutions (Gyurkovich 2012).

In 1985, a decision was taken to redevelop the entire area following to the success of the bid of Barcelona to host the 1992 Summer Olympic Games. The games provided an opportunity for the district to upgrade its urban infrastructure with an extended system of public transport, renewed residential districts, prestigious objects, and public spaces. A new district with the prevailing residential functions, opening to the sea, originally acting as the Olympic Village, was planned in El Poblenou. Another opportunity for the revitalisation of deteriorated post-industrial grounds of the district was the organisation of the World Forum of Cultures in 2004. The forum helped to introduce considerable structural changes (Gyurkovich 2012).

In 1998, Spain in general and Barcelona, in particular, were embarked upon a property boom that lasted a whole decade. The reforms of land and planning policy, shifts in banking behaviour at the national level, and deeper regional monetary integration with the introduction of the Euro allowed for the maintenance of low to negative real interest rates for a sustained period. This has generated a general trend towards capital switching and the financialising urban assets at the global level that made land and real estate a highly attractive investment opportunity in Barcelona (Charnock et al. 2014).

17.1 A Best Practice Knowledge Precinct from Europe: 22@Barcelona 241

Fig. 17.1 Snapshot of a creative industry at 22@Barcelona

In 2000, the City of Barcelona adopted a new urban strategy, which aimed at transforming the parts of the post-industrial district of El Poblenou that had not been revitalised. The former industrial and storage function is being replaced with knowledge-based activities. The name of the project '22@' replaced the traditional planning zone code of '22a', which refers to mono-functional industrial areas (Gyurkovich 2012). The plan for the district established "the criteria and terms for the conversion of the old industrial areas into a sector suited to the new forms of productive activity based on information and knowledge technology, and on a new balance of urban, residential, productive and service functions, in which all are integrated into a hybrid fabric, constructed around the historical morphology of the sector" (Clos 2004, p. 193).

The 22@Barcelona project, with a top–down redevelopment strategy—has managed to transform 200 ha of industrial land into an innovative district offering modern spaces for the strategic concentration of intensive knowledge-based activities (Casellas and Pallares-Barbera 2009). This project also produced an urban refurbishment model for Barcelona providing a response to the challenges posed by the knowledge-based society. The district has generated over 130,000 jobs, over 11 ha green space, almost 15 ha land for economic facilities, and over 4000 housing units—on top of the existing over 4500 dwellings (see Fig. 17.1).

According to Gyurkovich (2012), the 22@Barcelona model realised by the public and private partnership supported by a number of studies, plans, and special legal regulations prepared in collaboration with the local, metropolitan, regional and national government organisations. This broad coalition makes it possible to change the intensity of land development, offer more public spaces of diverse character, and attract investment. The district coexists with the historical architectural and infrastructural heritage, which builds the identity of the place together with some ultramodern structures.

The district is specifically planned to facilitate innovation and enhance competitiveness in the global economy. This is done by: (a) Primarily through the promotion of clusters—in the specific areas of media, ICT, medical technologies, and energy and design; (b) Establishment of technology and R&D centres in collaboration with the main local universities, and; (c) Implementation of a new special infrastructure plan oriented to meet sustainable development goals and to enhance digital telecommunications provision throughout the district. The district is also designed to promote the quality of work and life in the district. This is achieved by: (a) Aiding entrepreneurs and supporting vocational training and labour market entry; (b) Promoting knowledge-based learning in schools; (c) Making ICT more accessible to the elderly, and; (d) Making the most of the district's proximity to the city's famed beaches and entertainment centres so as to attract and retain that all-important talent (Charnock and Ribera-Furnaz 2011).

The district supports the formation of urban research and facilitates a new working space among the city administration, companies and research institutes. Moreover, it intends to foster research activities about the smart management of the urban space and e-services. The success of this development model comes from it being laid the foundation for dialogue on territory, economy and society, creating its own unique model, which puts universities, companies and the administration at the heart of economic and technological transformations. The accomplishments of the 22@Barcelona model have made it a benchmark of urban, economic and social transformation; and it is studied and followed by many cities around the world aiming to create their knowledge and innovation districts (Pareja-Eastaway and Piqué 2011; Yigitcanlar 2016).

The district, along with technology retrofitting in heritage buildings, is also advanced in smart building design. A good example is the Media-Tic building of the 22@Barcelona. It is a 40-m high transparent cube, clad in a warped mesh of steel cables and ethylene-tetrafluoroethylene (EFTE) bubbles (Australian Design Review 2010). The building hosts companies and institutions in ICTs and the media and audio-visual sectors. It was designed as a communication hub and meeting point for these businesses with the ambitious architecture of Barcelona. The facade of the building is striking and at the same time functional. The translucent and innovative covering material EFTE acts as an external covering and a mobile sunscreen. The use of ETFE not only lends the building a quality of transparency, but also serves as the basis for two very unique environmental controls that are perhaps the project's most remarkable features. The covering is activated using pneumatic mechanisms. These 'luxometer sensors' regulate levels of sunlight and temperature automatically and independently (Manville et al. 2014). The district also adopts advanced smart urban technology solutions particularly in the smart mobility area—e.g., smart parking (see Yigitcanlar and Kamruzzaman 2018; Yigitcanlar 2016).

The district is home to a number of knowledge-based businesses specialised in various areas ranging from ICT to multimedia and from creative industries to biomedical. Some of the well-known and businesses and establishments include: Barcelona Activa, Torre Agbar, Telecommunications Market Commission, Barcelona Media Park, Indra, Spanish National Radio, Media-Tic, Barcelona TV, The International Business Centre of Catalonia, Telefonica, International Convention Centre, The Blood and Tissue Bank, and Barcelona Biomedical Research Park (see http://www.22barcelona.com). Furthermore, all nine universities of the city have strong ties with companies located in the district.

As highlighted by Leon (2008), the 22@Barcelona district showcases a successful combination of knowledge-based and sustainable urban development agendas by promoting a culturally appealing, innovative and dynamic urban milieu that takes active measures against socio-spatial segregation and reinvigorates local self-governance. As much as the internal qualities of the district, contextual opportunities of the City of Barcelona—with its culture, architecture, food, football team, beach, and lifestyle—contributes to the international reputation and appeal of the district.

However, global financial crisis and high competition among European cities to capture economic activities with high value added, currently present Barcelona's public-led knowledge strategy with a number of uncertainties. According to Casellas and Pallares-Barbera (2009), a top–down high-tech economic development policy for a district with a significant proportion of public intervention and investment brings with it a range of challenges. The major one is that from an economic perspective, whether this large public investment would be able to capture sustainable high-value economic firms in the area. Time will tell the success of this top–down approach. Further info on this district is provided in the next chapter of this book.

17.2 A Best Practice Knowledge Precinct from North America: Silicon Valley

It would not be wrong to call Silicon Valley of California, the USA as the most popular knowledge and innovation space of the world—others from the US context include Boston's Route 128, and Research Triangle Park in North Carolina (Chatterji et al. 2014). Located in the southern part of the San Francisco Bay area, Silicon Valley—formerly an agricultural valley—is the result of an organic growth that emerged from a rich compost of government investment in science and technology (Engel 2014). While benefiting from its initial foundation as an industry cluster, the valley has evolved into a high technology cluster that spreads across different industries and sectors. Today, it has become the archetype of a cluster of innovation.

The convergence of talent, technology and capital has created a vibrant and dynamic ecosystem that is unceasingly creating and exploiting new ideas as they arise, which rapidly take the form of a new company and drive markets. Silicon Valley roots are in thousands of venturesome individuals around the world that came to California because of the Gold Rush seized by the opportunities that gold offered during the mid-nineteenth century.

In the second half of the nineteenth century, the interactions between enterprises were augmented with the creation of two academic institutions, the University of California at Berkeley in 1868 and the Stanford University in Palo Alto in 1891, which provided new ideas, inventions, engineers and entrepreneurs. These universities took the lead in the establishment of university-industry collaborations, fuelled in great part, by the commercialisation of telephone, electronics and computer technologies. A community was rapidly created, driving further innovation and business creation.

In response to the demand for industrial land near the university resources and the emerging electronics industry, in 1951 the Stanford Industrial Park was created—now Stanford Research Park. A series of small industrial units were put available for companies to be rented at low cost for the development of new technologies. Soon after, large corporations such as Eastman Kodak, General Electric, Hewlett-Packard, IBM and NASA showed their interest and opened their R&D centres there. In 1954 the Honors Cooperative Program was created, allowing the employees of these companies to obtain a university degree by undertaking part-time studies. These companies signed five-year agreements, paying a double tuition fee per student to cover expenses. In the mid-50s, the Valley was in a nascent but growing stage, and in the early 70s the number of corporations engaged in the production of semiconductors increased considerably, companies which in turn, supplied their counterpart firms that were working in the computer field area. This growth was driven by the parallel development of the venture capital industry.

In 1980, the Patent and Trademark Law Amendments Act changed the technology transfer system by enabling universities, small businesses and non-profit organisations to retain title to inventions made under federally-funded research programs and take the lead in patenting and licensing discoveries. The enactment of this law originated a new wave of venture capital investment. Venture capitalists become professional investment managers, investing large pools of other people's money and leveraging their own capital and expertise for a significant multiplier effect (Engel 2014). Today Silicon Valley is home to many of the world's largest corporations and thousands of start-up companies in the fields of electronics, software, biotechnology and other high-tech fields (see Fig. 17.2).

The growth of Silicon Valley as a knowledge cluster (or a giant knowledge precinct) is an example of the powerful influence of the key clusters of innovation components of university expertise, government investment, and a pool of eager entrepreneurial individuals. Together they created a vital cluster, attracting investors and specialised service providers and together evolving the behaviours and culture that characterise an innovation cluster—mobile resources, entrepreneurial process, aligned interests, global scope.

Fig. 17.2 Google bikes at Mountain View headquarters

In the beginning of the twenty-first century, however, Silicon Valley began to experience a new relocation phase (Engel and Forster 2014). Although it has always been somewhat geographically dispersed, with the generation of consumer-facing internet ventures and the explosion of social network ventures, more and more companies are increasingly being founded or relocated to the urban core of San Francisco—the most vibrant and dynamic urban area of the Bay Area. This migration trend is not unique to San Francisco, as the urban tech model seems to fit particularly well in the city's downtown area.

During the last two decades, two main booms—both driven by the internet industry—have shaken the economic landscape of San Francisco. By the late 1990s, with the dot-com bubble, a large number of entrepreneurs and start-up companies in the technological sphere invigorated the economy of the city, followed by professionals in other sectors. Demands for new housing and office space ignited a construction wave of new buildings, gentrifying once-poorer neighbourhoods such as the South of Market district. In 2001, the bubble burst, and 94,000 jobs disappeared. The market fell and many dot-com industries closed their doors. However, entrepreneurs and high technology remained as a mainstay of the local economy. A boom in social media and the advent of social mobile technology took place in the mid-2000s.

While some years ago San Francisco was a popular place to live for people employed in Silicon Valley companies, in the recent years, companies have started seeing the city as the place to locate. San Francisco provides the urbanity and lively street culture that bring people together, encouraging serendipitous interactions that have the potential to be transformed into powerful networks and durable bonds. It is filled with the dense and trendy districts where young techies prefer to live and work. The city becomes a recruiting tool by itself. The thrilling atmosphere of its inner-city neighbourhoods is a huge magnet for millennials, a generation with a strong propensity for walkable areas that enable them to live close to their friends and the amenities they need, rather than living in cookie-cutter isolated houses in the suburbs. In order to capture these workers, some companies such as Google run daily shuttle buses between the city and their suburban campuses in the Silicon Valley. However, recent moves from Palo Alto to San Francisco are evidencing a centre of gravity shifting away from suburban Silicon Valley to urban San Francisco (Florida 2012).

In addition to the inherent urban attractions for the skilled workforce and the supportive role of major corporations, the municipal government has played a significant role in stimulating the regeneration of San Francisco's urban core. A major city infrastructure project and the building of new skyscrapers have made the city an attractive place. In 2011, the administration of Mayor Ed Lee embarked upon a plan to foster a technology hub along the Market Street corridor. Tax exclusion was designed. Firms with an annual payroll of greater than US$250,000 would have their payroll taxes waived for six years. Twitter anchored the move and established its headquarters in a formerly vacant 1937 Art Deco landmark in Mid-Market Street, increasing rent rates by 60%. Just a year before the area was known for drug deals and homeless people. Soon, many major social media, software and technology companies relocated or expanded in the area including Airbnb, Dolby, Foursquare, LinkedIn, Salesforce, Trulia, Uber, Pinterest, Yahoo, Yelp, and Zynga among others. The presence of these high-tech companies also augmented the smart city endeavours of the city and its region—including Silicon Valley (Yigitcanlar 2016).

This trend is not new; a similar pattern was observed before the dot-com bust (1995–2000) when many companies settled in the South of Market neighbourhood (SOMA), a former warehouse and industrial district. From 1996 to 2003 over 700,000 m^2 of office space were built (San Francisco Planning Department 2014). With each economic cycle—such as the dot-com bust—the city and its community became stronger and more resilient to survive, adapt, and grow. Today, SOMA is the home of a comprehensive cultural offer, including convention centres, museums, art galleries and theatre companies. As of July 2014, over 600,000 m^2 of office space were approved or under construction, with over one million more pending or in the pipeline (Stehlin 2016). Due to its older, smaller office spaces and greater opportunities to convert former industrial spaces into live-work lofts San Francisco costs are advantageous for would-be entrepreneurs.

Another aspect of that characterises the metropolitan area is the concentration of highly ranked universities, research centres and hospitals. The dense urban core of the city facilitates the close collaboration of these institutions with businesses, resulting in cutting edge developments and inventions with a markedly practical application in the marketplace, which in turn, attracts more talent, firms and investments. A report from Citylab in 2013 listed San Francisco as the leader in venture capital investments with US$4.4 billion (Florida 2013)—approximately a third of the Bay Area total—which represent a 16% of total venture investment nationally. According to the Entrepreneur VC 100 list of top investors in early-stage start-ups, 12 venture capital firms are located in San Francisco. The city is also the home of nine of the Fortune 500 companies in 2015 (i.e., McKesson, Wells Fargo, PG&E, Gap, Charles Schwab, Salesforce.com, Levi Strauss, Williams-Sonoma, and Big Heart Pet Brands). It hosts the Federal Reserve Bank as well as the United States Mint, and more than 60 foreign banks have their offices in San Francisco.

Undoubtedly, San Francisco is the new 'Silicon Valley' for start-ups. It offers training programs, access to capital for new business, and start-up accelerator programs and shared workspaces. The city is also immersed in the provision of value-added services to its citizens. For instance, as part of the smart city initiative, using intelligent analytical tools based on global positioning system (GPS) technology, the city has implemented networked metering in water, electricity and gas using low-frequency radio frequency sensors. Urban maps also use sensing data from different sources. Such services open up new economic opportunities in the ICT domain.

All these ingredients, along with an appropriate transportation network make the city attractive to people, and consequently, to firms. However, the shift to urban tech is not without its problems. The influx of skilled workers and new companies is driving up housing prices and rental costs. San Francisco's neighbourhood transformation and gentrification have aggravated social inequality. The creative class is strongly concentrated in the city's central core, while the blue-collar and lower-wage service workers are displaced in the suburban areas, furthering the class divide (Ling and Dale 2011; Florida 2013). Creative strategies to deal with these challenges must be developed for San Francisco's urban core to continue to prosper and grow.

17.3 A Best Practice Knowledge Precinct from South East Asia: One-North

As discussed in Part III of the book, Singapore showcases a trendsetting KBUD performance as technologically advanced and culturally vibrant global city. The city-state archives this performance by heavily investing on the fostering, attracting, and retaining innovative and knowledge-intensive businesses, and highly talented workforce through effective KBUD strategies (Yigitcanlar 2009). Effective strategies have turned Singapore into a major global R&D hub (Philips and Yeung 2003). Being seen as one of the global best practice for both knowledge city development efforts; Singapore is home to a large number of knowledge precincts—e.g., Singapore Science Parks 1 and 2, Changi Business Park, International Business Park, One-North. Among those highly successful precincts, One-North stands out as the most famous one.

The planning process for One-North may be traced at least as far back as 1991, when its current site was designated as a 'science habitat' or 'business park' in the 'technology corridor concept plan' for the southwestern part of Singapore. The development of the precinct that originally was known as Science Hub initiated in 1998. In 2001, it was renamed to One-North, which symbolises Singapore's unique geographical location in the world and its aspirations to be connected to the region, the world, and be innovative. At the planning stage, the precinct was explicitly described in state discourse as not only as the icon of the new economy in Singapore, but also an important national project to transport Singapore's economy into the knowledge age (Wong and Bunnell 2006).

According to the JTC (2001)—a statutory board under the Ministry of Trade and Industry in charge of the precinct development—the masterplan of One-North sought to create an intellectually stimulating and creative physical environment where a critical mass of talents, entrepreneurs, scientists, and researchers would congregate, exchange ideas and interact. As stated by Wong and Bunnell (2006, p. 76), this goal was achieved by "locating key institutions, such as university campuses and government research agencies, in close proximity to private firms in high-technology industries". This goal is further supported by: (a) Providing residential options such as home offices to create a work-live-play-learn environment; (b) Fostering a vibrant cultural scene with art galleries, restaurants, pubs, and cafes, and; (c) Facilitating pedestrian movement within and across the different developments in One-North. These planning strategies result in a compact, mixed-use, pedestrian-centric urban form, which ideally fosters the kinds of face-to-face interactions that are deemed important for sustaining the innovative processes.

Today, One-North is a 200-ha development strategically positioned in the heart of Singapore. It is located 10-min drive to Singapore's CBD and well served by the One-North and Buona Vista metro stations. The proximity to the National University of Singapore, the Singapore Science Parks 1 and 2, the National University Hospital; the industry positioning of biomedical, infocomm technology and media companies, business schools, global leadership training campus. The possibility of technology collaborations amongst public research institutions, global multinational corporations, SMEs and start-ups have created a world-class home to technology and innovative global and growth companies in One-North. Today, it is now home to a community of over 18,000 knowledge workers, working in: public research institutions such as the Agency for Science, Technology and Research; business schools like INSEAD, ESSEC Business School; government agencies including Ministry of Education, Media Development Authority, SPRING Singapore; and companies like Autodesk, Fujitsu, Lucasfilm, Fox International Channels, Discovery Networks, Infinite Studios, Globecast, Bandai Namco, Garena, Canon, Oracle, Novartis, MSD, GlaxoSmithKline, Takeda, Electrolux, Llyod's Register, National Healthcare Group, and so on.

One-North has a number of specialised sub-precincts. The Fusionopolis Precinct houses various research organisations, high-tech companies, government agencies, retail outlets, and serviced apartments. The Biopolis Precinct is an international R&D centre for biomedical sciences and houses various international and Singaporean biomedical research institutes and companies. The Mediapolis Precinct is home to infocomm, media, physical sciences, and engineering sectors, and provides

17.3 A Best Practice Knowledge Precinct from South East Asia: One-North

Fig. 17.3 Snapshot from One-North

a synergistic business environment for the finest in the industry to thrive and grow. The Ayer Rajah Precinct is home to companies specialised in the hard disks and hybrid drives, and big data and social media analytics. The Nepal Hill Precinct is the education campus providing training centre and a global hub for located companies' leadership development. The Launchpad Precinct supplies collaborative spaces for entrepreneurs and start-ups to foster networking, and also hosts various cafes and restaurants. The Wessex Precinct provides accommodation, recreation, retail, and education facilities. The Vista Precinct offers accommodation, recreation, retail, healthcare and arts and culture facilities (see Fig. 17.3).

One-North's is one of the rare knowledge precinct examples that provides an exquisite work-live-play-learn environment for the creative class of knowledge workers. Clearly, the concept and design of One-North as a knowledge precinct has been influenced by Florida's (2012) understanding of the importance of attracting creative talent with, and retaining it in, a total knowledge ecosystem. In the planning of the precinct, it has been stressed that over and above key industrial infrastructure "a vibrant cultural scene, such as pubs, clubs and coffee houses is … crucial for facilitating networking among artists, designers, entrepreneurs, scientists, and venture capitalists, hence enhancing the milieu of innovation" (Wong and Bunnell 2006, p. 76). This creative milieu is seen as vital to the ongoing success of the precinct insofar as it retains local talent, trains new talent and attracts and retains migrant talent (Baum et al. 2007).

From the conception to development and from there to functioning stages the notable One-North project shows, in the broadest sweeps, how a very successful precinct intending to compete in the global knowledge economy could be. It takes from tradition by having a long-term plan and actualises it with almost a committed two decades of investment and development. The precinct focuses on three apparently separate industries that are, however, on closer inspection evermore interrelated. Certainly, ICT is linked to both bioscience and media production. This convergence, it could be maintained, reflects the fact technology itself rather than human need is increasingly the prime driver of knowledge production. The total One-North environment suggests a futuristic urban knowledge village. On an island where development is pinched for space, One-North plans to wrap through the built environment broad ribbons of green space—of 'nature'. The development, however, is not pastoral as the intensity of the built form surround ranges from tower blocks, through medium density housing to (relatively) low density detached housing. Through all the spaces—be they of work, of play, of home, of recreation, of retail—One-North targets for one major constant: 'mix'. For One-North a mix of experience, of stimulation, of activity is vital both for the production of new knowledge and the satisfaction of the desire of knowledge workers (Baum et al. 2007).

Besides the 'mix' or quality of urban development, other success factors' of the One-North project are worth elaborating. For example, Singapore in general and One-North, in particular, were highly sensitive to the global private sector's needs and requirements and saw them as the main drivers in the development of the precinct—influencing the characteristics and requirements based on their current and future business needs. Therefore, the planning and development activities of One-North were supported and guided by the private sector to ensure it has high-level 'end-user' buy-in. Another factor was the size of the precinct. One-North is an area large enough to be branded and recognised by a wide section of global business, but small enough to operate as a single entity. The historical evolution of the project and its region over the last three decades has significantly contributed the idea and practice of an internationally significant knowledge precinct to emerge. One-North was instigated as part of a technology corridor initiative and became a science hub, and then designed as a contemporary knowledge community precinct. Proximity to other key research, knowledge and innovation institutions—such as the National University of Singapore and Singapore Science Parks 1 and 2—helped in creating the right environment for new ideas, companies and knowledge transfer to flourish. Furthermore, One-North is also a test-bed for advanced smart urban technologies—including automated vehicles.

Lastly, several organisations that have played critical contributions to the success of One-North are worth mentioning. These organisations include (A*STAR 2013):

- The Agency for Science, Technology and Research (A*STAR)—the lead agency for fostering world-class scientific research and talent for a vibrant knowledge-based and innovation-driven Singapore. It oversees a number of biomedical sci-

ences and physical sciences and engineering research institutes, consortia, and centres, located in Biopolis and Fusionopolis as well as their immediate vicinity. A*STAR supports Singapore's key economic clusters by providing intellectual, human and industrial capital to its partners in industry. It also supports extramural research in the universities, and with other local and international partners.
- The Singapore Economic Development Board (EDB)—the lead government agency for planning and executing strategies to enhance Singapore's position as a global business centre. EDB dreams, designs and delivers solutions that create value for investors and companies in Singapore. EDB's mission is to create for Singapore, sustainable economic growth with vibrant business and good job opportunities. EDB's 'Host to Home' strategy articulates how we are positioning Singapore for the future. It is about extending Singapore's value proposition to businesses not only to help them improve their bottom line, but also to help them grow their top line through establishing and deepening strategic activities in Singapore to drive their business, innovation and talent objectives in Asia and globally.
- The JTC Corporation (JTC)—leading industrial infrastructure specialist spearheading the planning, promotion and development of a dynamic industrial landscape, including playing the key developer role for the One-North project. JTC has played a key role in the growth of the Singapore's economy by pioneering cutting-edge industrial real estate solutions. Some of its landmark projects include CleanTech Park, a chemicals hub on Jurong Island, Jurong Rock Caverns, Seletar Aerospace Park, and Tukang Innovation Park. In JTC's portfolio are other key projects including wafer fabrication parks, business parks, Biopolis and Fusionopolis at One-North, biomedical parks as well as logistics hubs for various industries. These industrial and business parks are now home to renowned global companies and promising local enterprises. As Singapore transforms itself for the future, JTC, as the industrial infrastructure innovator, will partner with its customers to understand their evolving needs and develop appropriate 'future-ready' infrastructure solutions. JTC will continue to break new ground and expand its innovation capacity to offer infrastructure facilities of a calibre that sets the city-state apart as an investment location.

17.4 A Best Practice Knowledge Precinct from Oceania: Macquarie Park

Macquarie Park—also referred as the Silicon Valley of Australia—with its flourishing economic output is a successful name on the innovation-scape of Australia as well as globally. The Park recently rebranded as 'Macquarie Park Innovation District'. Macquarie Park, located 12 km north-west of the Sydney central business district (CBD) in the local government area of the City of Ryde, is Australia's largest, research and business hub. The precinct has access to the M2 Hills Motor-

Fig. 17.4 Snapshot from Macquarie Park

way that runs through the northern part of the suburb. It is serviced by Busways, Forest Coach Lines, Hillsbus, Sydney Buses and Transdev NSW bus services (see Fig. 17.4). However, due to high car dependency during peak hours, the whole area gets very congested with traffic. It houses the Macquarie University, which is one of Australia's largest universities. The university campus features a 126-ha park-like campus, besides a high-technology corridor. The establishment of the university in 1964 was the defining moment for the precinct. Inspired by the Stanford model—anticipating collaboration and knowledge exchange between university and businesses at close proximity—the area adjunct to the university was identified as a principal industry growth location (Pancholi et al. 2018).

Soon after its establishment, in the 70s the precinct managed to attract numerous prestigious businesses. The opening of the shopping mall (Macquarie Centre) in the early 80s and continuing infrastructural investments to the area, such as Chatswood to Epping rail line, helped the precinct attract more business. From the 90s to 00s the increasing property prices and parking problems in the Sydney CBD boosted the popularity of the precinct. Macquarie Park had a total of 38,627 jobs in 2011, making it the fourth largest concentration of jobs in NSW after Sydney CBD, North Sydney, and Parramatta. The area has 32% more managers and 54% more professionals as compared to NSW average (ABS 2017).

17.4 A Best Practice Knowledge Precinct from Oceania: Macquarie Park

Socially, the presence of young and culturally diverse population due to its proximity to the university has been advantageous in the growth of knowledge-based industries. Demographically, the area represents a population with a continuous growth rate in last decade boasting a younger demographic profile with a large proportion, i.e., 46%, of the population aged 20–34—out of which 26% of the population consist of a younger workforce. Over 40% of people hold tertiary qualifications as compared to Greater Sydney's average of 24%, depicting a well-educated population with higher-qualifications. Around 48% of people are overseas born as compared to Greater Sydney's average of 20% reflecting a multi-culturally rich and tolerant society (ABS 2017).

The precinct is home to the headquarters of many Australian top-100 businesses. Large businesses of the precinct include: AC Neilsen, Avaya, AstraZeneca, BOC, Canon, CA, Compuware, Australian Radio Network: 101.7 WSFM and Edge 96.1, CSC, Kyocera, Fujitsu, Fuji Xerox, Ford Motor Co/Premiere Automotive Group (Aston Martin, Jaguar, Volvo, Land Rover), Foxtel, General Motors (Holden & Saab), George Weston Foods, Hitachi, Hyundai, Johnson & Johnson, Kenwood, Kimberly-Clark, Lucent Technologies, Metcash, Microsoft, Nestle Purina, Nortel Networks, Novartis, Bilfinger Berger, Optus, Oracle Corporation, Orix, Philips, Raytheon, Sanofi-Aventis, Siemens, Memjet, Sony, Toshiba, TPG Telecom, Warner Music Group, Wesfarmers Industrial and Safety. Macquarie Park is also the location for the Lane Cove River Tourist Park, where the Australian soap opera 'Home and Away' is partly filmed.

Macquarie Park had a real competitive advantage over the city (Sydney CBD) until recently, because of the availability of relatively cheaper and larger blocks of land with ample space for parking. However, the changing preferences of knowledge workers, few of the businesses find it challenging to attract people to work in their Macquarie Park office. Herewith the main reasons are the congestion and lack of a vibrant environment. Traffic congestion is identified as having a major impact on the productivity of employees by about 95% of the businesses surveyed in the area. Surprisingly, despite the time taken by bus to be about 50% less than the time taken by the car for the same journey, 70% of the people travel by car (NSW Government 2017).

Henceforth, to act as a key platform to address the key issues of the precinct, a formal group so-called 'Connect Macquarie' is established—as a joint initiative funded by the NSW government, City of Ryde and local businesses. Its major aim is to solve the transportation issues as faced by knowledge workers, businesses, students and the residents of Macquarie Park by tailoring out the best possible customised solutions on the basis of each company's and people's goal, current commuting pattern and bottlenecks. Recent initiatives for employees and society include co-hop carpooling, setting up of bike committee, purpose-built tools and apps, trip planners, centralised transport information, special discounts, and free safety equipment. In addition—while earlier attraction factors centred on the economy—planners are currently focusing on the societal aspects of the precinct to strengthen its appeal. In Macquarie Park, the Herring Road and the neighbouring north Ryde have been chosen as priority precincts—previously called urban activation precincts (NSW Government 2014). With an aim to develop work-live-play-learn community, the proposal will enhance housing in an economically, socially and environmentally sustainable manner by delivering up to 5800 new homes by 2031. Societally, this includes the redevelopment of Ivanhoe Estate—an on-site social housing estate—enhancing the current 259 existing social dwellings into a mix-housing estate neighbourhood with at least 556 social housing dwellings. This will develop the site into 'a true knowledge community precinct' with desired social and public amenities.

Originally it was a typical business park characterised by low-scale developments, however, the form of the precinct is constantly evolving. A look at the map of Macquarie Park reveals a mix of building forms. The low-scale development and its location adjacent to a national park give an open environment and opportunity for people to cycle or walk. However, there are few pitfalls (Pancholi et al. 2018). For example, the lack of collaborative spaces and connectivity, low density, strictly zoned land uses and—even more importantly—a vibrant public realm are the primary ones out of them. Large isolated corporate campuses behind a boom gate—with their own cafes and restaurants—exist as a current norm due to lack of ineffective implementation of initial planning.

Additionally, the undulating topography and lack of effective connections pose a challenge to the walkability of the site. It is not an inviting environment to explore as it has minimal shade, narrow surpass, traffic dominant streets, no pedestrian crossings. In the past, a conflict of opinion existed regarding height control between government and planners. Though architects considered it a drawback spatially in terms of density of the area, government's point of view was appreciably driven in the direction of keeping the societal assurance. In order to make the precinct more people-oriented and enhance its vibrancy, recently proposed plans aim to address these issues. The key objectives of Herring Road Urban Activation Precinct proposal (HRUAP) are to provide (NSW Government 2014):

- Activating the precinct by re-zoning to permit mixed land-uses with greater height;
- Establishing a higher density urban community;
- Providing access to soft and hard infrastructure;
- Providing safe, convenient and accessible pedestrian-friendly environment, and;
- Strengthening activities, landscaping, amenities, community facilities, green spaces and public places.

Herewith, new proposals are on board for developing an innovation district in university, entertainment precinct, and high-density housing with a mix of affordable housing. The university is also strongly putting new plans in place simultaneously aimed to create something more energising and to create transparency and accessibility across the university. In order to make the university more integrated, accessible, and people-oriented, as well as ensuring knowledge exchange into a creative environment along with developing a dynamic and collaborative arena few of the key physical initiatives laid down by university are as follows: (a) Land use re-zoning to mixed use; (b) More commercial enterprises close to the boundaries; (c) Visual display of creativity around the fringes, and; (d) Providing more pedestrian-friendly campus by clearly identifiable entry statements and engaging spaces (Pancholi et al. 2018).

17.4 A Best Practice Knowledge Precinct from Oceania: Macquarie Park

Despite its booming success, the precinct did not see much of collaboration happening until past few years. It is a maturing process that Macquarie Park is more of a case of collocation rather than genuine clustering or collaboration. Macquarie University, as a relatively young university, has so far failed to establish a collaborative innovation ecosystem in the precinct. However, there are some new developments. Recently, with the university opening its doors and few strong initiatives from the local council, collaboration levels are growing rapidly—reflecting the potential to be strengthened further. The university in recent years has changed its approach from what can you do for us to what can we do for you? The corporate team of the university organised a network giving a common platform to companies (Pancholi et al. 2018).

Like formal networks, equally crucial is the strengthening of informal networks within the community—including local community and knowledge workers—for developing a sense of place and exchange of knowledge and ideas. Community participation and engagement are the key drivers. University integrates them by providing access to amenities such as sports centres, swimming pools, playgrounds and so on as well as organising a number of joint programs like sports events, public workshops, festivals and seminars. In absence of a direct common interactive platform between businesses and community, programs like Program for After Class Enrichment (PACE) run by the university gives an opportunity to students to work not only with industry partners, but also the community groups. However, apart from university-led networking initiatives, due to the lack of a common body corporate management for the entire precinct, the area lacks organisation of common events or concerts that involve the local community (Pancholi et al. 2018).

Analysing knowledge worker perceptions, green environment and availability of shopping centre surface as major positives. However, issues like commuting time, housing affordability, high rentals, lack of common events, as well as day-to-day activities emerge as some of the key concerns. Some of the companies pay bonuses to their knowledge workers as it takes at least an hour or so to get to work. The local community has a great sense of pride in belonging to the area owing to its reputation. Transparency, effective communication, and public participation are few of the key aspects that help developing trust amongst the local community. In spite of its success in attracting and retaining businesses, Macquarie Park is still in need of becoming a desirable location for knowledge workers (Pancholi et al. 2018). Only then Macquarie Park will become a truly successful knowledge precinct. Further info on this knowledge park is provided at the next chapter of this book.

17.5 Lessons from the Global Knowledge Precinct Best Practice

Previous sections of this chapter placed four prosperous knowledge precincts from different country and continent contexts—namely, 22@Barcelona, Silicon Valley, One-North, and Macquarie Park—under the KBUD microscope. All of the investigated best practices showcase the capacity to spur KBUD and a commitment to enhance economic competitiveness through innovation. While there are some commonalities between them—e.g., developed as a result of government/university action and private initiatives, the ability to attract investment and talent, innovation and knowledge economy vision, financial support mechanisms—all four of these cities have built their achievements on their unique strength area bases.

In the light of the global knowledge precinct best practice analysis, the following generic lessons could be useful for Brisbane to consider—also relevant to many other cities with emerging knowledge precincts:

- Hosting international sports or cultural events/games and establishing international festivals to place the name of the city/knowledge precinct in the global map, and fund major infrastructure—Brisbane could learn from Barcelona's experiences;
- Orchestrating the emergence and development of knowledge precincts through strong leadership and a dedicated planning/development authority or bridging institutions—Brisbane could learn from Singapore and Barcelona's experiences;
- Providing strong government and planning support along with the existence of committed champions in the emergence and development of knowledge precincts—Brisbane could learn from Singapore and Barcelona's experiences;
- Developing brownfield knowledge precinct projects through urban renewal and economic revitalisation programs—Brisbane could learn from Singapore and Barcelona's experiences;
- Preserving local architectural and cultural characteristics of knowledge precincts through urban renewal and high-tech retrofitting projects—Brisbane could learn from Barcelona's experiences;
- Trialling and adopting smart urban technologies, such as driverless cars, smart parking, in knowledge precincts—Brisbane could learn from Barcelona, Singapore and San Francisco's experiences;
- Locating knowledge precincts adjoining or in close proximity to a major university—Brisbane could learn from Barcelona, Sydney, Singapore and San Francisco's experiences;
- Making start-up training, financial incentives, and venture capital funding opportunities available for knowledge precinct firms—Brisbane could learn from San Francisco and Singapore's experiences;
- Providing global/world city quality of life and place offerings to knowledge precincts through their host city—Brisbane could learn from Barcelona, Sydney, Singapore and San Francisco's experiences;
- Having long-term KBUD planning practice and experience of the city to support knowledge precincts' continuous evolution process—Brisbane could learn from Barcelona, Sydney, Singapore and San Francisco's experiences;
- Offering a vibrant entrepreneurial culture and environment to the companies and talented individuals located in knowledge precincts—Brisbane could learn from Singapore and San Francisco's experiences;

17.5 Lessons from the Global Knowledge Precinct Best Practice

- Forming knowledge precincts, and their sub-precincts, with highly specialised knowledge area/activities—Brisbane could learn from Sydney, Singapore and San Francisco's experiences;
- Establishing strong formal and informal networks for increasing communication, collaboration and coordination in knowledge precincts and their relevant industries—Brisbane could learn from Sydney, Singapore and San Francisco's experiences;
- Locating knowledge precincts in cities that capture the positive human/social capital (or soft infrastructure) built over a long-time of public investment—Brisbane could learn from Barcelona, Sydney, Singapore and San Francisco's experiences;
- Investing on local knowledge community emergence and development in knowledge precincts by investing in knowledge workers and their social needs—Brisbane could learn from Singapore's experiences;
- Showcasing smart and sustainable development, good architectural and urban design practices, with vibrant and diverse cultural activities in knowledge precincts—Brisbane could learn from Barcelona, Singapore and San Francisco's experiences;
- Developing effective metrics to help management set clear goals and, over time, gauge the effectiveness of the knowledge precinct—Brisbane could learn from Singapore's experiences.

Lastly, while closely following the success and failure factors in international best practice is a highly useful approach to form strategies and actions in order to learn these leading exemplar knowledge precinct cases, it is crucial for any city to build their knowledge precincts by preserving and consolidating their own uniqueness and strengths. Brisbane, therefore, needs to form a development pathway through carefully tailored knowledge precinct strategies that are suitable to its unique characteristics—such as population, culture, governance structure, and business climate.

17.6 Summary

Influenced by the successful global best practice—e.g., Silicon Valley—in recent years the establishment of knowledge precincts has been one of the features of national, regional and local strategies for promoting knowledge generation and innovation activities. This chapter of the book focused on providing lessons for Brisbane drawn from the prosperous knowledge precinct best practices with a central research question in mind—how lessons learned from global knowledge precinct best practice can be evaluated for their potential in Brisbane.

The analysis has revealed some potentially useful generic lessons for Brisbane to consider in its knowledge precinct development journey. However as mentioned earlier, these lessons need to be carefully considered for adoption to the local context. The potential actions also need to be meticulously tailored for Brisbane to make sure that they would work effective and efficiently to provide desired outcomes achieved from the knowledge precinct developments in and around the city. Similarly, generated lessons and insights have a value for many other cities with similar ambitions to Brisbane, but a careful evaluation and customisation of these suggestions is needed.

References

A*STAR (Agency for Science, Technology and Research). (2013). *Singapore's Biopolis: A success story*. Accessed February 21, 2017 from https://www.acnnewswire.com/press-release/english/14502/singapore's-biopolis:-a-success-story.

ABS (Australian Bureau of Statistics). (2017). *2011 census*. Canberra: ABS.

Australian Design Review. (2010). *The ambitious architecture of Barcelona*. Accessed August 26, 2016 from http://www.australiandesignreview.com/architecture/1538-media-tic.

Baum, S., Yigitcanlar, T., Horton, S., Velibeyoglu, K., & Gleeson, B. (2007). *The role of community and lifestyle in the making of a knowledge city*. Brisbane: Griffith University.

Casellas, A., & Pallares-Barbera, M. (2009). Public-sector intervention in embodying the new economy in inner urban areas: The Barcelona experience. *Urban Studies, 46*(5–6), 1137–1155.

Charnock, G., & Ribera-Fumaz, R. (2011). A new space for knowledge and people? Henri Lefebvre, representations of space, and the production of 22@Barcelona. *Environment and Planning D, 29*(4), 613–632.

Charnock, G., Purcell, T. F., & Ribera-Fumaz, R. (2014). City of rents: The limits to the Barcelona model of urban competitiveness. *International Journal of Urban and Regional Research, 38*(1), 198–217.

Chatterji, A., Glaeser, E., & Kerr, W. (2014). Clusters of entrepreneurship and innovation. *Innovation Policy and the Economy, 14*(1), 129–166.

Clos, O. (2004). The transformation of Poblenou: The new 22@ district. In T. Marshall (Ed.), *Transforming Barcelona* (pp. 191–202). London: Routledge.

Engel, J. S. (2014). USA: Clusters of innovation: Final thoughts. In J. S. Engel (Ed.), *Global clusters of innovation: Entrepreneurial engines of economic growth around the world* (pp. 378–390). Massachusetts: Edward Elgar.

Engel, J. S., & Forster, F. (2014). USA: Silicon Valley, the archetypal cluster of innovation. In J. S. Engel (Ed.), *Global clusters of innovation: Entrepreneurial engines of economic growth around the world* (pp. 41–94). Massachusetts: Edward Elgar.

Florida, R. (2012). *The rise of the creative class: Revisited*. New York: Basic Books.

Florida, R. (2013). Why San Francisco may be the new Silicon Valley. *Citylab*. Accessed January 10, 2017 from http://www.citylab.com/work/2013/08/why-san-francisco-may-be-new-silicon-valley/6295/.

Gyurkovich, M. (2012). 22@Barcelona—The city of knowledge civilization. *Technical Transactions, 13*(109), 25–56.

JTC. (2001). *JTC corporation press release 5 June 2001*. Accessed March 28, 2007 from http://www.jtc.gov.sg/Corporate/media+room/5june2001.asp.

Leon, N. (2008). Attract and connect: The 22@Barcelona innovation district and the internationalisation of Barcelona business. *Innovation: Management, Policy and Practice, 10*(2), 235–246.

Ling, C., & Dale, A. (2011). Nature, place and the creative class: Three Canadian case studies. *Landscape and Urban Planning, 99*(3), 239–247.

Manville, C., Cochrane, G., Cave, J., Millard, J., Pederson, J. K., Thaarup, R. K., et al. (2014). *Mapping smart cities in the EU*. Brussels, Belgium: European Union.

NSW Government. (2014). *Herring road Macquarie Park urban activation precinct planning report*. Department of Planning and Environment. Accessed April 7, 2016 from www.planning.nsw.gov.au.

References

NSW Government. (2017). *Macquarie Park*. Department of Planning and Environment. Accessed 7 April 2016 from http://www.planning.nsw.gov.au/macquariepark.

Pancholi, S., Yigitcanlar, T., & Guaralda, M. (2018). Societal integration that matters: Place making experience of Macquarie Park Innovation District, Sydney. *City, Culture and Society, 13*(1), 13–21.

Pareja-Eastaway, M., & Piqué, J. M. (2011). Urban regeneration and the creative knowledge economy: The case of 22@ in Barcelona. *Journal of Urban Regeneration & Renewal, 4*(4), 319–327.

Phillips, S. A. M., & Yeung, H. W. C. (2003). A place for R&D? The Singapore science park. *Urban Studies, 40*(4), 707–732.

San Francisco Planning Department. (2014). *Office development annual limitation program*. San Francisco: San Francisco Planning Department.

Stehlin, J. (2016). The post-industrial shop floor: Emerging forms of gentrification in San Francisco's innovation economy. *Antipode, 48*(2), 474–493.

Wong, K. W., & Bunnell, T. (2006). New economy discourse and spaces in Singapore: A case study of one-north. *Environment and Planning A, 38*(1), 69–83.

Yigitcanlar, T. (2009). Planning for knowledge-based urban development: Global perspectives. *Journal of Knowledge Management, 13*(5), 228–242.

Yigitcanlar, T. (2016). *Technology and the city: Systems, applications and implications*. New York: Routledge.

Yigitcanlar, T., & Kamruzzaman, M. (2018). Smart cities and mobility: Does the smartness of Australian cities lead to sustainable commuting patterns? *Journal of Urban Technology*, https://doi.org/10.1080/10630732.2018.1476794.

Chapter 18
Benchmarking Knowledge Precincts

Abstract Benchmarking is a powerful technique to determine the standing of a city or cluster against its competitors. The chapter places a knowledge precinct from Brisbane in a benchmarking exercise and evaluates its performance against the reputable knowledge precinct best practices from Cambridge, Barcelona, Helsinki, Eindhoven, Dublin, Sydney and Melbourne. The chapter employs a place making framework for the benchmarking exercise that is developed to be used in knowledge and innovation spaces. The findings not only provide insights for the further development of the Dutton Park Knowledge Precinct of Brisbane, but also generate overall lessons for other cities aiming to develop prosperous and sustainable knowledge precincts.

Keywords Knowledge-based urban development · Knowledge city · Smart city · Sustainable urban development · Knowledge and innovation spaces · Knowledge precinct · Innovation district · Place making · Brisbane · Cambridge · Barcelona · Helsinki · Eindhoven · Dublin · Sydney · Melbourne

Benchmarking against competitors or trend setting practices is recognised as an essential tool for continuous improvement of quality; hence it is widely used for progressive city planning, management, and development (Hong et al. 2012; Yigitcanlar 2014a). Benchmarking a knowledge precinct against the thriving examples helps the local practice to learn from achievements and strengths of the benchmarked precincts. This may also help in thoroughly investigating the success factors and in consequence adopting tailored strategies to emulate or even surpass the achievements of the benchmark. In other words, benchmarking as the process of identifying, understanding, and adapting outstanding practices from successful knowledge precincts anywhere in the world could help a knowledge precinct improve its performance (Kumar et al. 2006).

Against this background, this chapter of the book aims to address the critical question, listed below, through comparing one of Brisbane's knowledge precincts with major cities' precincts by focusing on the key contextual and place making aspects relevant to knowledge precincts. The benchmarking exercise involved the following cities' selective knowledge precincts—to compare with Brisbane's Dutton

Park Knowledge Precinct—Cambridge, Barcelona, Helsinki, Eindhoven, Dublin, Sydney, and Melbourne.

- How do Brisbane's knowledge precincts compare to the other reputable ones in Australia and overseas?

18.1 Place Making Facilitators of Knowledge Precincts

Knowledge precincts are environments that are designed for nurturing creativity, innovation and value knowledge that differ significantly from those that were developed for commodity-based services. They, therefore, call for different development strategies that have to be taken into account in their design and planning phases. It has to be highlighted in the literature that organisational factors are also strategic drivers for the development of knowledge precincts. In this section, the book discusses the conditions for place making in knowledge precincts from the lens of organisational, economic, spatial and social qualities as these factors need to be considered in order to inform good governance, business, spatial and people climates in knowledge precincts (Fernandez-Maldonado and Romein 2010; Yigitcanlar 2011; Pancholi et al. 2015).

In organisational terms, it is strategic to have a favourable policy and developmental context, and effective organisational structure for the successful development of knowledge precincts (Van Winden et al. 2012; Yigitcanlar and Dur 2013; Adams et al. 2014). The requirements include: (a) Stability of and strong support from the leadership possessing a prescient vision for development; (b) Strategic vision and development plans aiming for long-term sustainability; (c) Incentivising the development and knowledge-based activities; (d) Facilitation of diffusion of innovation in the form of technology and communication in their development, and; (e) Integration and balance among all domains of knowledge precincts at sectoral, horizontal and vertical levels (Maynard 2008; Van Winden et al. 2007; Yigitcanlar et al. 2008). Ergazakis et al. (2004) advocate that management plays one of the important roles in developing strong internal and external links and creating synergy and trust between all social actors—i.e., government, university, industry, and society (Pancholi et al. 2015).

In economic terms, Ergazakis et al. (2004) suggest that the aim of KBUD can only be achieved through continuous interactions amongst citizens. More and more firms are embracing the model of open innovation as their trajectory towards growth and have realised its potential in order to keep up with the pace of competition in the progressively dynamic marketplace (Chesbrough 2003; Chesbrough et al. 2006). Favouring the concept, Saxenian (1994) compares examples of Route 128 (Boston), and Silicon Valley (California)—the most popular knowledge precinct practices. She argues that Silicon Valley firms more quickly adapted to changing technologies and markets because firms were more innovative, open to collaboration with outside firms, and the region exhibits strong internal and external networks. Van Winden et al. (2012) consider 'cognitive proximity' between the firms as playing a major role in the success of knowledge precincts. Thus, increasingly externally focussed organisational structure is emphasised (Pancholi et al. 2015).

In societal terms, knowledge precincts aim to stimulate good people climate by providing social equity and inclusion achieved through strong social and human capitals, diversity and independency and promoting connectedness (Fernandez-Maldonado and Romein 2010; Yigitcanlar 2011). People in such knowledge precincts could develop the capacity to create and share knowledge resulting in a social ecosystem of learning (Berkes 2009). Explanations have promoted the need for democratisation of developed knowledge by making it accessible to all. Moreover, recent works have shown evidence advocating public participation as the factor of social change and one that brings innovation (Smith 1995; Gonzalez and Carrillo 2012; Pancholi et al. 2015).

In spatial terms, literature points out the role of physical environment and good design in initiating and encouraging ideas, creativity and innovation. Environmental structures that we design or the places that we create are an extension of the cognitive processes giving birth to innovation, thus an integral part of the knowledge creation process (Gottdiener 2000; McCoy and Evans 2002; Peschl and Fundneider 2012; Oksanen and Ståhle 2013). Musterd and Zoltan (2013), on the basis of the research carried out in Europe, draw attention to the equal role of soft as well as hard factors in the success of such locations. In a recent study, Pancholi et al. (2014, 2015) outline three major conditions that shape the concept of knowledge precincts in the light of context: (a) Policy in terms of supporting political and economic processes; (b) Place understood as produced space, and; (c) People as individual and society.

As evident from the above discussion, it is quite clear that the role of place making extends itself from the stage of conception to development and then to the fully-functioning of the knowledge precinct. This provides us with four major facilitators of place making in knowledge precincts—that form the basic elements of our framework. These are: (a) Management; (b) Firms; (c) Space, and; (d) People. Organisationally, 'management' acts as a facilitator by providing policy, financial and amenities support for knowledge-based activities and develop internal and external networks. Economically, 'firms and institutions' act as the second facilitator of place making that facilitate the vitality in the environment by generating processes and networks. Spatially 'space' acts as the third facilitator of place making by performing as a medium to disseminate knowledge and having a role in inspiring creativity and innovation. Socially, 'people' act as the facilitator of place making in knowledge precincts by being stakeholders/actors and producing the place in terms of the meanings that they associate with it (Pancholi et al. 2015).

18.2 Comparison of a Knowledge Precinct from Brisbane with Global Best Practices

The methodology adopted in this research rests on a case-study based approach to analyse knowledge precinct best practice cases with the aid of a framework derived from facilitators of place making. The major purpose of the best practice analysis is to assess place making facilitator in contemporary knowledge precincts. The definition of facilitating elements is based on place making in its physical manifestation, meanings attached as well as in the form of processes, i.e., organisational, economic and social layers. The four facilitators are considered in the light of the wider sphere of their 'context', where they are located. In brief, the context here is referred at four levels, i.e., governance, economic, spatial, and societal. Each of these facilitators is analysed considering—but not limited to—three major attributes. A brief description of the key facilitators of place making—forming the four dimensions of the analysis framework (Fig. 18.1)—with their major attributes is as follows (Pancholi et al. 2015).

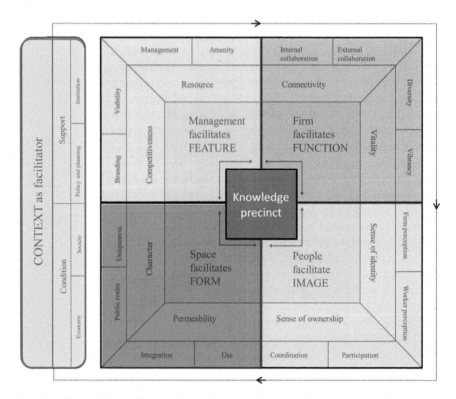

Fig. 18.1 Place making facilitators (derived from Pancholi et al. 2015)

Management: (a) Vision and brand—e.g., one that drives the shaping up of development; (b) Partnership—e.g., extent of participation of various actors determining their role in the decision-making process; (c) Managing body—e.g., formal, informal or semi-formal nature that determines access to support.

Firm: (a) Anchor—e.g., role played by the main anchor in encouraging inter-firm collaborations; (b) Diversity—e.g., in terms of industrial base and size of firms present in knowledge precincts; (c) Collaboration—e.g., presence of institutional mechanisms, physical and digital collaborating platforms in knowledge precincts.

Space: (a) Functional use—e.g., variety and interweave determining the vitality; (b) Design and uniqueness—e.g., integration of arts, creativity and heritage that encourage participation and drive the sense of place and identity; (c) Public realm—e.g., the presence of spaces for interactions.

People: (a) Image and perception—e.g., image to which people identify and produced by historical context or conceived image of produced space driven by objectives; (b) Networks and participation—e.g., informal and formal mechanisms in terms of institutions or digital networks that enhance the connectivity between people; (c) Attractions—e.g., events and functions that make a place vibrant and attractive (Pancholi et al. 2015).

18.3 Cambridge Science Park, United Kingdom

Context: Cambridge Science Park was set up in 1970 with the aim to attract existing science-based industry. This was fulfilled by academic entrepreneurship based on funded research. It gave rise to an ecosystem in the form of new science and technology-based business that has been growing ever since, with a strong acceleration in the 1990s. Knowledge arising from the research conducted in Cambridge is of world-class and path-breaking, meaning that the alumni and staff have an initial advantage in publishing, being awarded Nobel Prizes and, for academic entrepreneurs, being early or first in the market with commercial innovations, innovative processes or new business models. Globally important discoveries have been made ranging from the splitting of the atom, to the discovery of DNA and the science of low energy microprocessor design software and cybersecurity. There is plenty of outsourcing to global leader firms like Apple, Samsung, Google, Microsoft and so on, as testified by the existence of research institutes of many of these being found in Cambridge. Trinity College of the University of Cambridge historically formulated the development scheme that took shape as Cambridge Science Park (CSP). The plan included the establishment of science-based industries in the proximity of university area accompanied by other science-based organisations such as Medical Research Council. Apart from the proximity, there were numerous other strong reasons for the selection of site location like its proximity to the residential areas, strong accessibility with other parts of the city and presence into an area where unemployment rates were high. The major planning intervention was the publication of the 1969 Mott Report that gave extensive subsidised planning permissions (Bradfield 1981).

Management: Cambridge is not state-managed but is a collaborative enterprise complex. It has a high rate of networking among technology entrepreneurs, university researchers and government or military representatives and clients. It is not a top-down hierarchical system in any meaningful way. Government intervention has been more indirect than direct although the UK has been slow until recently in promoting the idea of 'innovation systems' of any kind. The project is privately owned by the university, Trinity College of the University of Cambridge, and does not depend on government grants. However, there was always 'arm's length' government funding for scientific research supplemented by industry and charitable trust-funding, assisted until 2016 (Pre-Brexit) by the EU Framework Program (Horizon 2020). It was set up after the realisation of the UK government's extended need for exchange between industry and academia of ideas and people. The vision for CSP was to develop it as a 'Science Park'. Unlike the recent knowledge precincts with a lot of fancy advertisements and branding, CSP has a simplistic approach aiming not at the private profit motive but at encouraging the educational science and research activities further. There is no formal organisation as the managing body. An informal institutional management set-up is run by a group of professionals from the local companies.

Firm: Due to the restriction in the type of companies permitted access to the park is limited to scientific research and development or related activities. The park boasts of a rich diversity in terms of knowledge bases of firms with the share distributed equally between bio-medical and IT/telecom related businesses (see www.cambridgescience park.co.uk). Non-technical firms also occupy the space, a considerable portion being SMEs. The University of Cambridge acts as a bridge between scientists and the companies by providing an easily accessible system, being the source of information and also a physical location to develop informal contacts that lead to the innovations.

Space: Low-density park-like setting characterises the space of CSP. Land use is majorly limited to R&D activities, light industrial production under the regular collaboration from university and the related ancillary economic activities. In order to give a conducive and peaceful environment for research activities to prosper, the design is introvert in nature secluding the inside from outside. A major thrust has been given towards the integration of nature in the design. The central common green area that comprises the lake and the natural green spaces and landscaping rich with a range of cycling and jogging tracks, accounts to more than 12% of into the total site. Apart from this 35–40% of the area under each plot is reserved for green areas. Heritage protection is also a major part of the project as seen from the considerable amount of expenditure over the cause (Fig. 18.2).

People: The development is quite people-oriented and efforts are made to create the opportunities to let the individuals collaborate and work together. Social and conference facilities provide support to the smaller companies. The facilities are provided by the Trinity Centre for people to eat and meet. The presence of social clubs further strengthens the informal networks. The simplistic informal institutional management set-up giving direct access to solutions to the park tenants has been a

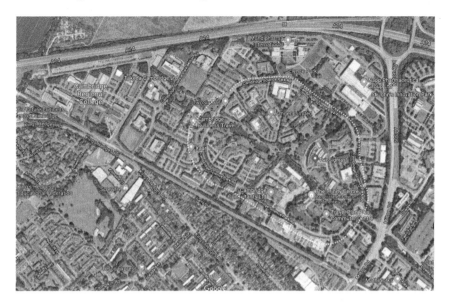

Fig. 18.2 Cambridge Science Park (Google Earth)

major factor contributing to the environment of collaboration. Thus the place making has been done in this knowledge precinct by keeping the administrational activities informal and accessible to a considerable extent, and keeping its tenants and workers informed and connected (Pancholi et al. 2015).

18.4 22@Barcelona, Spain

Context: The Barcelona model has been recognised worldwide as a successful example of social, economic and cultural revitalisation (Yigitcanlar 2009). The Olympics-driven urban transformation that led to collaborative economic environment together with cultural regeneration bias of policies laid the background that shaped the forthcoming success of Barcelona model (Martí-Costa and Miquel 2014). 22@Barcelona is located in the Poblenou district, a deindustrialised site used by the transport businesses for a short period during the 1970s; the site later was chosen by the artistic groups as their workspace. The conversion into a creative cluster led to the transformation of old industrial areas into new artistic uses, the emergence of diverse functions and the establishment of close-knit networks between the artistic groups. Martí-Costa and Miquel (2014) assert that these factors laid the creative context in the development of the creative economy in 22@Barcelona (or 22@ for short).

Management: 22@ is a government initiative through a major land-use reform with two major objectives in mind: (a) To establish the city in the global knowledge economy as a leading centre for technology-based production, and; (b) Revitalisation and integration of the deindustrialised downtown area as the knowledge core. In order to achieve these objectives, 22@bcn a private municipal corporation was formed in the form of a public-private initiative. 22@network is the collaborating organisation created as a group of companies and institutions in the proximity.

Firm: While the planning of the area focused on the redevelopment of brownfield site, economic development strategy concentrated on the development of five clusters centred on design, technology, biomedicine, audio-visual sector and energy. Based on the positive externalities from clustering like profitable collaborations and shared ideas, facilities and innovations, the firms of the similar sector are clustered together. Apart from the institutes in proximity, the firms receive fresh talent from programs like 22@staying in a company that connects regional institutes with the knowledge precinct (see www.22barcelona.com). 22@network acts as a common network bringing together the community, associations and private firms.

Space: An integration of industrial, commercial, residential and leisure activities, 22@ is characterised by the diversity in types and styles of architecture, compactness and density. These inherent factors end up in breaking the homogeneity naturally and bringing the variety necessary for urban vitality. Mixed-use is promoted not only horizontally by allocating a mix of functions to adjacent plots, but also vertically by allowing a mix of functional types to emerge into individual buildings. The special attention for urban design is evident in the use of architecturally rich buildings and public spaces along the monumental boulevard that acts as the spine of the district. The contemporary vibrant environment is further strengthened by the historical artistic cultural identity attached to the site. Heritage is weaved with modern fabric and industrial chimneys are displayed as public art (Fig. 18.3).

Fig. 18.3 22@Barcelona (Google Earth)

People: Organisation of symposiums, special service for companies called 22@plus and monthly events like 22@breakfast act in exchange of ideas and innovation between firms and knowledge workers and also acquaint the participants with the activities and environment in other firms in proximity. The 22@inside event is a 3-day event conceived as Barcelona Festival of Visual Arts and Contemporary Music. The economic support like incentives to the small and medium ventures and social factors like vocational training, integration of knowledge-based learning in schools, enhanced digital accessibility for all age groups along with the lifestyle factors like proximity of site to the famous city beaches and entertainment centres has acted as the driving force for attracting firms and people in 22@ (Pancholi et al. 2015).

18.5 Arabianranta, Finland

Context: Helsinki has a strong innovation economy, but the long historical tradition in design, strong international ties strengthened by its strategic position in the Baltic, highly educated and diversely talented workforce have been the reasons that contributed in diversifying its economy. The presence of institutions, such as Finnish Society of Arts and Crafts, University of Art and Design Helsinki (TaiK), and Finnish Innovation Agency (TEKES), gained Helsinki a global reputation as 'showroom of design'. The site under consideration, i.e., Arabianranta, historically had the oldest porcelain factory in Finland. The economic recession in the 1990s led to the closure of the Arabia factory. Later, the traditional design character attached with the site inspired the shift of the University of Art and Design Helsinki into the site (Van Winden et al. 2007; Yigitcanlar and Lönnqvist 2013). The precinct also neighbours the Kumpula campus of the University of Helsinki.

Management: The traditional design-based identity, shift of the university, waterfront of the site and improved connectivity with the rest of the city led to the selection of Arabianranta as the site for knowledge precinct development. The vision was to make it 'a leading centre of art and design in the Baltic area' (Van Winden et al. 2012). The development started as one of the largest public-private ventures of Finland between various stakeholders such as the local and state governments, Iitalla group, i.e., the owner of historical Arabia factory. The management of the site is under Art and Design City (ADC) Helsinki, which is a triple helix partnership system. This partnership was created by City of Helsinki and TaiK, and other partners in ADC are Ministry of Trade and design industry companies and academic institutes situated in Arabianranta.

Firm: Many design-based institutes and industries shifted subsequently to the emergence of TaiK as the anchor in Arabianranta. Presently, a wide array of firm types is seen to be emerging in this KIS. This includes media, digital content, ICT, and design firms. The presence of TaiK as the anchor project act as a major attraction factor for the new firms adding to the factors like the creative identity transposed from its designer past, creative built environment and the possibility of proximity with other creative firms. TaiK also acts as a source of fresh creative talent. Apart from ADC, though no formal institutionalised collaboration is evident, the interaction between firms is mostly informal and depends on the social networks and proximity.

Space: Keeping 'art' at its core, Arabianranta has been designed as a mixed-use development. The design has remained a central statement in the development to maintain its identity of the creative precinct. In order to make design integral, all the companies were selected on the basis of their architectural design and their use of art in public space. This was undertaken to an extent that developers were required to integrate 1–2% of the total construction costs into art projects. The green spaces are an important part and they have public access. With a special concentration on living environment, it has been ensured to provide a mix of different housing typologies accommodating diverse population choices (Fig. 18.4). Attention has also been paid to let distinct architectural style to emerge and add value on to the creative environment (see www.arabianranta.fi).

People: Because of the 'living lab' concept many regular international business visitors get attracted to this location. Even though the location has not been developed for tourist purpose, business visits help in advertising of location at international level. The institutional structure is strong for the design-promotion. Lead by the Finnish Society of Crafts and Design, Design Forum Finland promotes design by organising regular award functions and events with participant universities like TaiK, other design institutes and organisations (Pancholi et al. 2015).

Fig. 18.4 Arabianranta (Google Earth)

18.6 Strijp-S, The Netherlands

Context: Targeted policies and programs aimed at the development of innovation activities have been the driving factor in the success of knowledge economy of Eindhoven. The consecutive implementation of the 1993 Stimulus, 2001 Horizon, and 2013 Brainport Navigator Programs developed by the Foundation Brainport are the major concomitant policy interventions. Along with these, the cultural policy 'total culture' in 2008 acted for encouraging talents and bringing the social groups together. Arising as a strong endogenous asset originating as a tradition from Design Academy of Eindhoven (DAE), i.e., the former renowned design school, Eindhoven has a strong design orientation that spreads along a range of design-oriented interventions.

Management: Strijp-S is an urban redevelopment project developed with the objective of revitalising heritage industrial area into a dynamic experimental city merging culture with technology and conserving its strong historical identity (Van Winden et al. 2012). The vision behind it was the refurbishment of former Philips industrial complex into a knowledge-based high-tech centre. A joint venture between public and private parties, this is managed by a specifically dedicated management company 'Park Strijp Beheer' as a public-private partnership.

Firm: The broad theme of the area for selection includes the firms related to culture, design and technology. The major share of the firms in this knowledge precinct, therefore, relates to creative industries, art and culture, consultancy and business services with a smaller share of technology and R&D as well. In spite of the redundancy of many activities, Philips still has a major share and is the main anchor firm along with Bosch. The BALTAN Lab is a major R&D facility located in the knowledge precinct. It acts in networking the technology into art and provides the facilities for experiments. It acts as a bridge between design institutes and researchers. Under the program of 'Eindhoven Laboratory City', many empty buildings have been acquired and made available for SMEs.

Space: This knowledge precinct has been considered as a demonstrable example of the explicit integration of arts as a tool for urban regeneration. From its transformation from 'forbidden city' to the 'open laboratory', Strijp-S exemplifies the integration of place making in quite an explicit manner with its innovative and people-oriented design of the built environment (Fig. 18.5). This has been developed as a multi-functional urban area (see www.strijp-s.nl). Heritage is integrated by the transformation of the old monumental chimney into an iconic landmark and revitalisation of the Clock Building as a cultural factory to accommodate creative businesses. This reinforces the connection between current Eindhoven and its industrial past (Fernandez-Maldonado 2012).

People: The precinct is a renowned example of how the use of innovative digital techniques (in this case, smart public lighting) can enhance the experience of place. The legacy of its innovation-based past has been carried forward by creating a unique environment as a 'living laboratory' or 'experience community'. This acts as a breeding ground to experiment, develop and research new ideas and concepts bringing government, students, scientists, artists, business and industry together.

Fig. 18.5 Strijp-S (Google Earth)

Glow next, 1001 light rain, crystals and scenario public lighting are amongst the few experimental initiatives. Light-S is another initiative in Strijp-S whereby creating a public lighting experience the place is made more interactive and informative for the users (see www.light-s.nl). Many events, concerts and festivals such as Dutch Design week are also being held in this knowledge precinct (Pancholi et al. 2015).

18.7 Digital Hub, Ireland

Context: Digital Hub of Ireland is a good example of economically successful knowledge precinct with a strong social component in development. The two primary objectives of the project are: (a) Steer the regional economic development, and; (b) Strategically act as the catalyst to develop the areas that surround its premises. As a strategy to recover from the economic recession of the year 2000, the creation of Digital Hub is considered a significant step following the recent policy bend. Though not arising from local tradition, the global nature suiting the openness of Irish economy and the complementarity with other sectors led the shift of focus on the digital media sector. The series of events led to the creation of Digital Hub Development Agency (DHDA), the main stakeholder of the knowledge precinct.

Management: DHDA, the government agency in collaboration with two major private developers, started the development on a former brewery site (i.e., Guinness Brewery) that was lying in a dilapidated area of the city, just adjacent to the city core. DHDA, which is set up by the government acts like a private developer with its autonomous nature of rights in the management of financial assets. It also acts like the main managing body. The Digital Hub development plan is the result of community-public-private partnership (CPPP) with the active participation of 12 stakeholders including the community organisations.

Firm: The digital hub majorly consists of SMEs (79%) with few large-size firms. The firms at the time of their arrival need to fit into the criteria of being a digital media company. 78% of the firms, therefore, are digital media companies. The National Digital Research Centre (NDRC) act in bringing the various actors together and promotes the collaborative research between companies and research institutes. This also acts as the main anchor. Firms are also provided with a variety of business support by the special organisations that are devoted to the support. 80% of the firms indicated some type of collaboration with another firm as evident from the study conducted by Van Winden et al. (2012). The collaboration is majorly in the form of sharing of information on innovations in digital media and general business matters with few even leading to the product development.

Space: Planning for the site targeted to achieve a mix including firms, commercial, residential, community and learning spaces. The knowledge precinct is a well-integrated development interwoven with the existing urban fabric and is not an isolated enclave development. Not only the old trees, but also the historical character of the site is preserved. It is evident in the conservation of St. Patrick that acts as a landmark for the city and the traditional features in many buildings. Informal spaces in the form of outdoor breakout spaces and cafes provide a chance to relax and interact (Fig. 18.6). The provision of spaces at flexible terms and reasonable prices act as a luring factor for start-ups (see www.thedigitalhub.com).

People: At the regional level, education has been used an instrument aimed at making the population digital media-literate and accessible to all. The creation of 'Creative Dublin Alliance' also gives a common platform at the city level to all public and private stakeholders. At the knowledge precinct level, the proximity and shared facilities provide the opportunity for face-to-face interactions between knowledge workers. A lot of interaction in the Digital Hub apart from this happens digitally via the intranet. It exists as a common platform where employees meet and share ideas. Online conferences and information sharing via common business networks are gradually emerging. The CPPP program ensures an active encouragement towards the participation of community majorly through education and as an active consulting member in development (Van Winden et al. 2012; Pancholi et al. 2015).

18.8 Macquarie Park, Australia

Context: Today, Macquarie Park is seen as Australia's Silicon Valley and hosts Australia's top-100 companies and many top global companies' Asia-Pacific headquarters or branches. In the 1960s, the selection of the site by the State Government for the establishment of Sydney's third major university—Macquarie University—was

Fig. 18.6 The Digital Hub (Google Earth)

one of the prime defining moment in its history. Later on, conceived to be developed on the lines of the Stanford model—anticipating knowledge exchange between university and businesses—the area was identified as industry growth area. In order to accomplish that, earlier there was a mandatory requirement for businesses to have an R&D component—a regulation that was changed later on. The companies were keen to choose the site as their headquarters by consolidating their offices spread across different sites into one primarily due to the availability of big blocks of land and also due to its connectivity and accessibility from the CBD. Housing started growing at the periphery of the university and then the Macquarie Centre (shopping mall) was opened in 1981. Construction of some significant infrastructure such as Chatswood to Epping rail line further added to its popularity (Pancholi et al. 2018b).

Management: The knowledge precinct does not have an overarching body corporate (see http://macquariepark.com.au). Instead, each building or building complexes has their own management. This prevents the establishment of common social spheres and public spaces in the precinct. At present, the shopping mall and university campus are the main social gathering spaces. The new proposed planning scheme of the City of Ryde for Macquarie Park Corridor aims to address the traffic and open space challenges of the precinct (City of Ryde 2013).

Firm: Macquarie Park has established itself to a nationally acclaimed, Australia's largest, research and business hub. With Macquarie University—one of Australia's leading research universities—as its key anchor, it is home to many global players as well as head office locations for many of Australia's top 100 companies. Few of its key tenants across pharmaceutical, technology, electronics and telecommunications industries are Johnson & Johnson, Microsoft, Sony, Optus, Cochlear and

Foxtel. Ranked as the area with Australia's 10th highest economic output and with its exceptional growth rate of 6.8%—highest in Sydney—it is soon predicted to outnumber other locations in Australia (PWC 2014). Being recognised as a 'specialist centre' under NSW Government's Metropolitan Strategy for Greater Sydney, the strategy plans that this area will continue to grow as an internationally significant economic hub by continuing to attract top global and Australian companies (Pancholi et al. 2018b).

Space: Spatially at a distance of only 12 km from the city centre, it has a significant location on the Global Economic Corridor of Sydney. This corridor extends from Sydney Airport and Port Botany in the south through the major employment centres of the Sydney Central Business District (CBD), North Sydney, Chatswood, Macquarie Park, and towards Parramatta and Norwest Business Park (Fig. 18.7). The adjacent location of Lane Cove National Park provides an open, green environment and ample options for cycling or walking. One of the biggest attractions for companies to Macquarie Park was that it has a real competitive advantage on the CBD due to lower property prices and availability of lots of car parking spaces. However, car park availability encouraged driving and turned the knowledge precinct into a car-dependent one with congestion in peak-hours. In order to address this issue, a formal group 'Connect Macquarie' is established—as a joint initiative funded by the NSW government, City of Ryde and local businesses. Recent initiatives for employees and society include co-hop carpooling, setting up of bike committee, purpose-built tools and apps, trip planners, centralised transport information, special discounts, and free safety equipment. Moreover, originally a typical business park characterised by low-scale developments, the form of Macquarie Park is constantly evolving. However, it still has the key characteristics of a Technology Park—dispersed precinct with weakly connected buildings, lack of public realm besides the university campus and the shopping centre—and the transformation into a knowledge community precinct will take a lot of affords and government intervention (Pancholi et al. 2018b).

People: Originally designed as a technology park, Macquarie Park is Australia's biggest knowledge precinct and employs more knowledge workers than any other precincts in the country. In recent years, a more people-oriented approach, for the development of the precinct, is envisaged. In order to transform the technology park into a knowledge community precinct and with an aim to develop live-work-learn-play community, NSW government plans to enhance housing options in an economically, socially and environmentally sustainable manner by delivering up to 5800 new homes by 2031 (NSW Government 2014). While there is no formal network in the precinct to provide social and technical activities, the newsletter communicates social activities at the precinct level. There are also relatively strong informal connections at the university, and free movie events by the lake along with conferences taking place at the Macquarie University create an opportunity for the precinct residents to meet (Pancholi et al. 2018b).

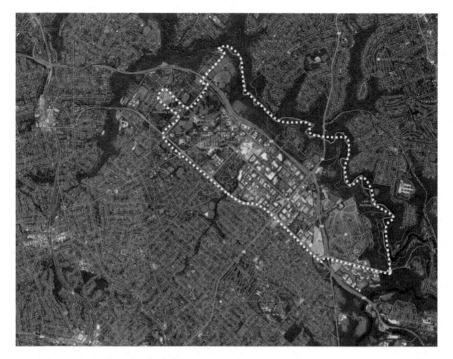

Fig. 18.7 Macquarie Park (Google Earth)

18.9 Monash Employment Cluster, Australia

Context: Monash Employment Cluster is located approximately 20 km from the Melbourne CBD with a strong connectivity through major transport corridors—Monash freeway, and Clayton, Blackburn and Dandenong roads. It is the second biggest knowledge industry agglomeration in Australia after the Macquarie Park in Sydney. History of the area—dating back to late 1980s as an industrial park—provides a strong context, as the existing infrastructure and supportive conditions laid the foundation for its later development as a knowledge precinct. It witnessed its first major shift from traditional heavy manufacturing industries to lighter ones in the 1990s. With the onset of the digital era, the second major shift in the early 2000s was towards the knowledge-based industries. The site is home to the Monash University—member of the Group of Eight Universities (top-eight universities of Australia). It has very quickly attracted many global names and kept on growing spontaneously to become a cluster. In recent years, Monash Employment Cluster—an area that grew spontaneously with no government intervention for decades—has received increased recognition as a National Employment Centre, and its development is now prioritised in the State government policy (Pancholi et al. 2017a).

Management: Industrial activities at the Monash Employment Cluster started to appear more evidently after the establishment of the Clayton Campus of Monash

University in 1981. However, the development was purely organic and the university provided, as an anchor, significant attraction to private companies and public research institutes (Australian Synchrotron being the most well-known one located in the cluster in 2007) for being located in the rapidly expanding cluster. The organic development of knowledge generating businesses and scientific research and R&D institutes resulted in the cluster not having an overall managing body. However, the university and individual private multilevel office buildings and office parks have their own body corporates. This piecemeal development hampered community building efforts as suitable virtual or actual interaction spaces could not be developed due to lack of a management and funding. Nevertheless, in recent years the Victorian planning and policy context started to provide strong support for the cluster. Organisationally, setting up a dedicated administrative body to bring stakeholders exhibiting a variety of mutual interests together under a single platform is still yet to be achieved (Pancholi et al. 2017a).

Firm: The cluster hosts a number of world-class institutions and research organisations including Monash University, Monash Medical Centre, Australian Synchrotron, Australian Stem Cell Centre, Monash Science Technology Research and Innovation Precinct, CSIRO, Bosch, Telstra Research Laboratories, Australian Regenerative Medicine Institute, Monash University Accident Research Centre, Monash Centre for Synchrotron Science, Centre for Human Bioethics, Melbourne Centre for Nanofabrication, John Monash Science School, Monash Health Translation Precinct, and Monash Business Park. Furthermore, designation of the cluster, by the Victorian government, as one of the national employment clusters makes it attractive for other prestigious companies to locate (Pancholi et al. 2017a). Australia's first driverless car was built in the cluster by Bosch, and it will be put into the real-world trial at the CityLink, Monash and Tullamarine freeways in 2017 (Carey 2016). Currently these trials are in place.

Space: The cluster was developed following the 'garden city' model based on a grid layout, providing big blocks of land with a mandatory requirement to have large green setbacks and retain the tree canopy. This low-density development and greenery, due to zoning it as a 'special-use zone', gave the area a competitive edge over other locations (Pancholi et al. 2017a). While the development of the cluster spontaneously in consistency with the market forces helped its growth, it also brought a number of spatial challenges. One of the main challenges is the integration of the numerous precincts of the cluster—i.e., Monash Technology Precinct, Health and Education Precinct, Clayton/Springvale Employment Precinct, Clayton Business Park, Clayton South Industrial Precinct, and Huntingdale Industrial Precinct. Other challenges include: limited social amenities and active public open spaces, inadequate internal and external connections, and lack of onsite residential land provision. Fortunately, the new planning scheme led by the Metropolitan Planning Authority (see https://vpa.vic.gov.au/project/monash-employment-cluster) is targeting to address these issues to turn the business-friendly cluster into a user-friendly one (Fig. 18.8).

People: The limited social activities and disintegration between precincts hampered the creation of the vibe of the cluster. Moreover, due to the absence of public art, heritage or an innovative element, the area lacks uniqueness of character neces-

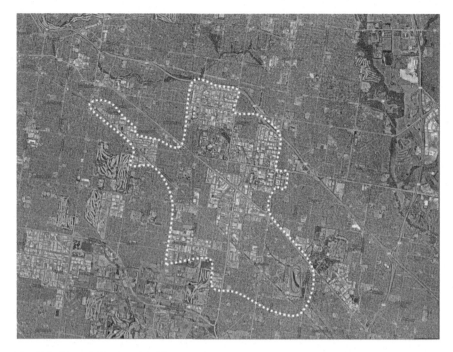

Fig. 18.8 Monash Employment Cluster (Google Earth)

sary for developing a sense of place for the local community (Pancholi et al. 2017a). The university campus is the only place that such social vibe exists in the cluster. However, outside the semester even the campus goes extremely quiet. In recognition of this issue the new planning scheme for the cluster—that is being prepared—are believed to be factoring in the creation of a social atmosphere for the cluster. For example, Plan Melbourne (Victorian Government 2014) has selected some key areas, i.e., around Huntingdale, Clayton and Springvale as focal points to be enhanced as the activity generator cores for the cluster.

18.10 Dutton Park Knowledge Precinct, Australia

Context: Dutton Park Knowledge Precinct is an idea conceived as a part of the larger city-level development of knowledge corridor supported and announced by the Queensland Government under the following strategies: 'Smart State', and 'Smart Cities'. This knowledge corridor—that hosts Dutton Park at its southern-most axis—runs as a spine from north to south of Brisbane, i.e., Bowen Hills to Woolloongabba, interconnecting and integrating a number of health and research precincts, tertiary institutions and cultural facilities. Historically, the area of Dutton

18.10 Dutton Park Knowledge Precinct, Australia

Park was renowned as a leading health precinct owing to the location of regional-level healthcare service providers, i.e., Princess Alexandra Hospital (PAH), Mater Hospital and others, in its vicinity (Pancholi et al. 2018a).

The potential generated by existing collocation of these eminent hospitals with world-class institutions, such as University of Queensland (UQ), Queensland University of Technology (QUT) and Griffith University (GU) paved the pathway for the conception of Ecosciences Precinct—as a seed project for Dutton Park Knowledge Precinct. A number of Queensland Government departments and Commonwealth Scientific and Industrial Research Organisation (CSIRO) jointly invested with a vision to develop it as Queensland's premier research developments—by bringing together over 1000 scientists from different locations in and around Brisbane to work in an interdisciplinary and collaborative environment with world-class research facilities. The project gained further momentum by the establishment of other famed institutions such as Translational Research Institute (TRI) on the same site—funded by state and federal governments in collaboration with institutions such as UQ, QUT and Atlantic Philanthropies (Pancholi et al. 2018a).

A draft master plan was initiated in 2012 by Department of State Development, Infrastructure and Planning for providing a framework to the precinct's current and future development. It was developed in collaboration with the major stakeholders located in Dutton Park. These include PAH, Pharmacy Australia Centre of Excellence (PACE), TRI, Biopharmaceutical Australia, UQ and CSIRO. Brisbane Diamantina Health Partners played an important role by acting as a common organisational platform having a virtual oversight over the project. In terms of policy and planning, the precinct received a supportive context at the state, city and local levels for its development as a knowledge and innovation cluster. Policies—such as Smart State Queensland (Queensland Government 2005), Smart Cities (Queensland Government 2007), and Smart Communities (Queensland Government 2009a)—support innovation, identify the area as a 'super-precinct', and provides guidance to its development (see Hortz 2016).

Documents like South East Queensland Regional Plan (Queensland Government 2009b) resulted in strengthening the public transport connectivity of Dutton Park. Other local government documentations such as Brisbane City Plan (Brisbane City Council 2014), City Shape (Brisbane City Council 2006), West End-Woolloongabba District Local Plans, Woolloongabba Urban Development Schemes, and Eastern Corridor Neighbourhood Plan also recognise the priority of the area as a health precinct, expanding the current rail and bus infrastructure as well as strengthening its accessibility. The precinct will also highly benefit from the planned cross river rail project—Queensland Government's highest priority infrastructure project, with a 10.2-km link from Dutton Park to Bowen Hills, including 5.9 km of tunnel under the Brisbane River and CBD, connecting northern and southern rail networks (Franzmann and Lodge 2016).

Management: In Dutton Park Knowledge Precinct, two of the sub-precinct—Ecosciences and PAH—are managed separately. While each sub-precinct functions efficiently and has their social events and professional activities, the lack of an overarching body corporate makes connectivity and integration of the two sub-precincts much harder. Besides space related issues, the lack of an overarching management body also limits the networking and participation of workers from one sub-precinct in the activities of the other. Most recently, several of design and development companies were engaged by the Queensland Government to prepare a holistic strategy to guide the staged transformation of the precinct into a leading centre of integrated healthcare, research and education. This strategy built upon the precinct's role as a nationally significant health precinct and major economic anchor for Queensland, delivering local, regional and state services, as well as contributing to national and global health solutions. This strategy rebranded the precinct with a new name—Diamantina Knowledge Precinct (see www.lat27.com.au).

Firm: Dutton Park Knowledge Precinct has two anchors Ecosciences precinct and PAH. Both of these anchors are managed to attract several major public and private establishments. These include CSIRO, TRI (world leader in head and neck cancer), Biopharmaceutical Australia, Patheon, and Pharmacy Australia Centre of Excellence. However, for development at this size and high government support, the precinct has not been highly successful in attracting private industries and businesses. While the inner-city location and close proximity to amenities generating a magnet for large businesses, particularly for start-ups and SMEs the amount of rent and limited short-term leasing availability are seen as possible reasons for much less interest in locating the businesses in the precinct than desired (Pancholi et al. 2018a).

Space: Dutton Park Knowledge Precinct—recently rebranded as Diamantina Knowledge Precinct—is located 3.5 km south of the city centre making it a highly strategic urban location. The precinct is characterised by an orthogonally interconnected street grid pattern surrounded by a low-density residential neighbourhood (Fig. 18.9). The site is well serviced by major infrastructure projects such as Pacific Motorway, railway line, and the Eastern busway. This will be further strengthened by the proposed underground bus and train project, i.e., BaT tunnel, planned to connect Dutton Park with inner city bypass. Strong connectivity and high accessibility of the precinct with the rest of the city are the contributing factors to make state's leading centre of integrated health, research and academic activity appealing for workers and businesses. However, this advantage comes at an expense. The boundary by major heavy traffic roads, such as, Pacific Motorway, Ipswich Road, Gladstone Road, and Annerley Road becomes a barrier in the spatial integration of the site with its surroundings, i.e., its permeability. This detachment from its surroundings isolates the precinct as well as declines the scope to expand. Lack of permeability limits interweaving of land uses and forming mixed-use environments. Additionally, the precinct is strictly zoned and do not allow precinct-wide land use integration—two parts of the precinct Ecosciences and PAH sub-precincts are not connected. Equally, significant issue is the order, i.e., how the buildings relate to each other and their surroundings. The architectural design of the buildings is largely introvert in character. Despite being collocated, they do not respond to surrounding streets or buildings. Propitiously, climate-sensitive design and heritage integration strengthen the innovative character of the precinct (Pancholi et al. 2018a).

People: Both physical and specialisation separation of the two sub-precincts of the Dutton Park Knowledge Precinct formed a duality. Ecosciences precinct with

18.10 Dutton Park Knowledge Precinct, Australia

Fig. 18.9 Dutton Park Knowledge Precinct (Google Earth)

an environmental science and PAH with health science focus formed two distinctive communities. Furthermore, spatial separation of the two sub-precincts with the railroad turns a crow fly of 400 m distance into a 1.5 km walk. In other words, the lack of walkability, way finding and availability of active and passive green public spaces negatively impacts on the people-orientedness of the precinct. On the other hand, Ecosciences precinct building, rather than isolated and cabin-like research offices, with its contemporary interior design provides open-plan workplaces and common spaces that encourage interaction and mingling of knowledge workers. Moreover, the lack of on-site accommodation availability along with nonexistence of a formal network for employees also limits the community formation chances within the precinct. However, even in silos—in each precinct—there exist strong informal networks due to organised social and technical activities and attractions including conferences, precinct tours, science poster competitions, science talks, science week, art exhibitions, fun run fund raising event, and the Boggo Road Gaol (Pancholi et al. 2018a).

18.11 Findings from the Comparison Study

In the light of the analysis conducted Table 18.1 highlights the comparative characteristics of seven benchmark best practices from Australia and overseas with a knowledge precinct from Brisbane—i.e., Dutton Park Knowledge Precinct. In comparison to its contemporary best practice knowledge precinct examples, Dutton Park Knowledge Precinct is an emerging knowledge precinct with potential for further development. The findings of the comparative study indicate both various prospects and constraints of Dutton Park Knowledge Precinct. These are briefly elaborated below.

The main prospects of the Dutton Park Knowledge Precinct include government's strong support at the planning and development stages, subtropical architectural design, inner-city location, proximity to socio-cultural amenities, benefiting from high-quality urban infrastructures, high external accessibility, having anchor institutes as a drawcard, and various social and technical activities to form a larger community. Dutton Park Knowledge Precinct can be seen as one of the first comprehensive attempts of the city with reasonable success in developing a knowledge precinct—other projects include Brisbane Technology Park, Herston Health Precinct, Kelvin Grove Urban Village, Health and Food Sciences Precinct, Sippy Downs Knowledge Precinct, and Gold Coast Health and Knowledge Precinct with various success levels where some are still in the developmental stage (Pancholi et al. 2017b).

The main constraints of the Dutton Park Knowledge Precinct include limitations in attracting private companies, physical growth limits to expand, disconnect between two sub-precincts, limited internal connectivity, lack of permeability from neighbouring areas, high property and rental prices, lack of formal networking opportunity, lack of a managing body for the entire precinct, and lack of on-site accommodation, and limited social activity availability. In meeting these challenges and turning them into opportunities, Dutton Park Knowledge Precinct will need support from public, private and academic sectors—ideally their collaboration in the form of triple-helix model partnership with onsite knowledge worker community along with wider local community support (Pancholi et al. 2017b).

The following section of the chapter focuses on the insights generated from the study on Brisbane's knowledge precincts that would be invaluable in bridging Brisbane's knowledge precinct policy with the global best practice cities.

18.12 Insights into Brisbane's Knowledge Precincts

Inspired by Silicon Valley, KBUD has become a widespread goal since the 1970s. Fostering knowledge precincts that concentrate on knowledge generation and innovation has become a priority for many cities all across the globe (Yigitcanlar 2014b). Brisbane is one of these cities. Brisbane, as a city aiming to become a thriving

18.12 Insights into Brisbane's Knowledge Precincts

Table 18.1 Comparative findings (derived from Pancholi et al. 2015)

	Cambridge Science Park (Cambridge)	22@Barcelona (Barcelona)	Arabianranta (Helsinki)	Strijp-S (Eindhoven)	Digital Hub (Dublin)	Macquarie Park (Sydney)	Monash Employment Cluster (Melbourne)	Dutton Park Knowledge Precinct (Brisbane)
Management								
Vision and brand	A research and development engine	An economic and urban regeneration engine	An arts and cultural regeneration engine	An economic and industrial area revitalisation engine	An economic and downtown regeneration engine	Australia's Silicon Valley hosting Australia's top-100 companies	A knowledge economy employment centre with national significance	A vibrant inner-city eco-science and health precinct
Partnership	Private-academia	Public-private	Public-private-academia	Public-private	Public-private	Private-academia	Private-academia	Dominantly public-some private
Managing body	Trinity College (university managed)	22@Barcelona private municipal corporation (public-private)	Art and Design city (public-private-academia-community)	Park Strijp Beheer (public-private)	Digital Hub Development Agency (community-public-private)	None for the entire cluster (individual units have their own body corporates)	None for the entire cluster (individual units have their own body corporates)	None for the entire cluster (individual units have their own body corporates)
Firm								
Anchor	Trinity College	No single firm/institute as the anchor	TaiK (university)	Philips	No single firm as the main anchor; 79% SMEs	Macquarie University	Monash University and Australian Synchrotron	PA hospital and eco-sciences precinct

(continued)

Table 18.1 (continued)

	Cambridge Science Park (Cambridge)	22@Barcelona (Barcelona)	Arabianranta (Helsinki)	Strijp-S (Eindhoven)	Digital Hub (Dublin)	Macquarie Park (Sydney)	Monash Employment Cluster (Melbourne)	Dutton Park Knowledge Precinct (Brisbane)
Diversity	Computer science, telecom, bio-medical companies	Design, technology, bio-medicine, media, energy companies	Media, digital content, ICT, design companies	Arts and culture, consultancy, companies	Digital media companies	ICT, medical research, pharmaceutical companies	ICT, health, manufacturing and materials companies	Health and eco-science institutes
Networking	Social and conference facilities	22@network; 'Creation factories' program	Design Forum Finland	BALTAN laboratories encourages joint R&D	No formal network at the knowledge precinct level but NDRC is present at the national level	No formal network at the knowledge precinct level	Victorian platform technologies network formed by the state	No formal network at the knowledge precinct level
Space								
Functional use	Work-learn-play	Live-work-learn-play	Live-work-learn-play-experiment	Live-work-play-experiment	Live-work-learn-play	Live-work-learn-play	Live-work-learn-play	Work-learn-play
Design and uniqueness	Integration of nature (35–40% green), Low-density park like settings	Proximity to beaches; heritage display as public art	Integration of Arabia factory and public arts in built space	Heritage integration, e.g., Witte Dam, The Clock	St. Patrick as landmark; heritage elements of buildings preserved	Lack of a unique character as each building was designed disjointedly	Lack of a unique character as each precinct and buildings were designed disjointedly	Subtropical design principles were adopted in eco-sciences precinct design, heritage buildings are revitalised

(continued)

18.12 Insights into Brisbane's Knowledge Precincts

Table 18.1 (continued)

	Cambridge Science Park (Cambridge)	22@Barcelona (Barcelona)	Arabianranta (Helsinki)	Strijp-S (Eindhoven)	Digital Hub (Dublin)	Macquarie Park (Sydney)	Monash Employment Cluster (Melbourne)	Dutton Park Knowledge Precinct (Brisbane)
Public realm	Introvert character but an active realm for internal people	Central boulevard as public spine; congregating spaces; open	Accessible public realm; green spaces; Pedestrian-oriented	Innovative and experimental public realm, e.g., Light-S; formal guided tourist tours	Informal spaces, cafes, outdoor break-out spaces	Dispersed precinct with weakly connected buildings, public realm at the university campus and the shopping centre	Dispersed cluster with weakly connected precincts containing public realm at the university campus and the shopping centre	Buildings of the precincts are disconnected and only individual building provide social space
People								
Image and perception	Science and research park	Knowledge-city model	Centre of art and design; living lab	Creative city and open laboratory	High-tech space	Organically developed knowledge cluster	Organically developed knowledge cluster	Planned inner-city knowledge cluster
Networks and participation	Strong informal connections between people	Artistic informal networks, 22@network (formal)	Strong community networks, e.g., e-moderators, ICT platforms	Foundational Alice; no formal website; experimental living lab	CPPP; intranet for informal interactions and; newsletters	Newsletter communicates social activities at the precinct level, relatively strong informal connections at the university	Not much connections at the cluster level, relatively strong informal connections at the university	Relatively strong informal connections between people in the silos of eco-sciences precinct and PA hospital and TRI

(continued)

Table 18.1 (continued)

	Cambridge Science Park (Cambridge)	22@Barcelona (Barcelona)	Arabianranta (Helsinki)	Strijp-S (Eindhoven)	Digital Hub (Dublin)	Macquarie Park (Sydney)	Monash Employment Cluster (Melbourne)	Dutton Park Knowledge Precinct (Brisbane)
Attractions	Industry events; conferences	22@inside event and others	The Masters of Arts festival	Dutch Design week (international event); art festivals	Majorly industry events; award functions	Free movie events by the lake, conferences taking place at the Macquarie University	Conferences taking place at the Monash University and Australian Synchrotron	Conferences, precinct tours, science poster competitions, science talks, science week, art exhibitions, fun run fund raising event, Boggo Road Gaol

knowledge city, over the years invested in the development of a number of knowledge precincts in and around the city. The research project reported in this book has concentrated on one of them, Dutton Park Knowledge Precinct (hosting Ecosciences Precinct, PAH, and TRI), and investigated thoroughly. The other notable and large-scale knowledge precincts of Brisbane (and South East Queensland)—some are still at the developmental or early uptake stage—include:

- Brisbane Technology Park;
- Herston Health Precinct;
- Kelvin Grove Urban Village;
- Health and Food Sciences Precinct;
- Sippy Downs Knowledge Precinct, and;
- Gold Coast Health and Knowledge Precinct.

Brisbane has a continuously strengthening its base of knowledge, talent, and innovation as the city provides world-leading facilities, universities and researchers work in collaboration with industry. City's abovementioned knowledge precincts provide space and place for knowledge-based activities. Brisbane's journey to develop knowledge precincts dates back to 1980s. The foundations of knowledge precinct developments in Brisbane were seeded in the mid-1980s. The very first knowledge precinct development was the Brisbane Technology Park—a Queensland Government initiative established to foster the growth and collaboration of companies in knowledge-based industries in 1986. This is followed by the initiation of a number of aforementioned knowledge precinct projects in South East Queensland.

In the light of the literature review, and global best practice and knowledge precinct comparative analyses the overall insights of Brisbane's knowledge precincts—exemplifying Dutton Park Knowledge Precinct, Kelvin Grove Urban Village, and Sippy Downs Knowledge Precinct—are discussed below.

Size: Even being generally accepted that size only matters in financing and maintaining the development, smaller areas seem to be of advantage in establishing cohesiveness in the community, as exemplified by Kelvin Grove Urban Village, and Dutton Park Knowledge Precinct. Sippy Downs Knowledge Precinct is a large and ambitious project—the Australian answer to Silicon Valley—and it aims to take the knowledge precinct to a 'knowledge town' level. With its large size, Sippy Downs Knowledge Precinct is likely to have voluminous challenges ahead, especially when considering the limited financial public-sector support behind this development (Yigitcanlar et al. 2016).

Firm: The type and nature of knowledge-intensive businesses/industries and R&D activities undertaken in knowledge precincts are quite important for long-term success. For example, if the knowledge-intensive activities were based on an endogenous asset, the chance of success is higher (Romer 1994). Kelvin Grove Urban Village invests in biosciences and creative industries based on the strengths of the university and its stakeholders. Likewise, Dutton Park Knowledge Precinct invests in ecosciences and health sciences specialising in fields that benefit and are supported by the State of Queensland. The business and industrial specialisation of Sippy Downs

Knowledge Precinct is not as clear as the other two. Even though investing in subtropical design and lifestyle options for knowledge workers sounds good to the ear, it is not clear how Sippy Downs Knowledge Precinct intends to attract knowledge communities when employment opportunities are generally limited in the region (Yigitcanlar et al. 2016).

Anchor: Having an anchor institution is critical for the performance and attractiveness of a knowledge precinct. Fortunately, there exist one or more anchor institutions in all of Brisbane's knowledge precincts. These anchors are as follows: for Dutton Park Knowledge Precinct—various Queensland Government departments, CSIRO, PAH, TRI, and UQ; for Kelvin Grove Urban Village—QUT; and for Sippy Downs Knowledge Precinct—University of the Sunshine Coast, and Innovation Centre Sunshine Coast (Yigitcanlar et al. 2016).

Support: Political support is an essential driver as well as an external precondition for knowledge precinct formation. All three knowledge precincts were initiated during the Queensland Government's 'Smart State Strategy' initiative, which has expired in 2015 (Queensland Government 2005). With no strong political drive behind knowledge precincts, KBUD in the Brisbane region might be at risk. Considering the completion and take-up rates, Sippy Downs Knowledge Precinct is the one that seems to be the most disadvantaged by this. On the other hand, recent global financial crises and the end of the mining boom have brought attention to the need for diversification of the Australian economy; this might mean bringing back the policy of moving towards a knowledge economy (Yigitcanlar et al. 2016).

Partnership: The triple-helix model partnership plays an important role in the formation of knowledge communities. In this regard, all three knowledge precincts have received support from public-private-academic partnerships by being clustered around major universities in the greater Brisbane area. However, it must be considered that University of the Sunshine Coast's budget as a regional university is much lower in comparison to the sandstone Group of Eight University member UQ, Australian Technology Network of Universities member QUT, or even Griffith University that is included in Australia's top 15 universities (Yigitcanlar et al. 2016).

Community: It is true that developing a knowledge precinct requires a large investment; however, the real cost associated with it is to maintain its attractiveness and competitive edge. This generally requires onsite management that is supported by onsite stakeholders. Healthy social interactions and strong identity are essential to achieve the goals of forming a knowledge community worth living in. Kelvin Grove Urban Village is a step ahead in forming such an active community due to the efforts of its onsite management in comparison to the other two knowledge precincts—although Kelvin Grove Urban Village is still quite far from reaching an ideal level of cohesive community formation. Dutton Park Knowledge Precinct and Sippy Downs Knowledge Precinct are still in their infancy in terms of community formation. The changing of focus on the branding promotions observed both in Kelvin Grove Urban Village and Dutton Park Knowledge Precinct can be seen as a response to these dynamics of maintaining attractiveness and competitive edge. This might be what Sippy Downs Knowledge Precinct needs to do in order to get off the ground (Yigitcanlar et al. 2016).

18.12 Insights into Brisbane's Knowledge Precincts

Planning: Unlike brands, planning is static and lacks much dynamism. This is a serious contradiction as it signals that planning efforts might not follow the branding or vice versa. Planning and branding need to work as complementary activities where the branding offers the dynamism that the planning cannot, and therefore, planning policies could be prepared with this in mind, hence providing possible evolutionary adjustments. In Kelvin Grove Urban Village, the vision and spirit of two plans remain the same apart from some additional new activities and urban design features. In Dutton Park Knowledge Precinct, from the preliminary to final plan only the architectural design was changed. In Sippy Downs Knowledge Precinct, the approved master plan has not been fully implemented yet. In all, efforts have been made so that the design matches the features and the cultural aspects of these places; furthermore, the design principles suggested in the planning documents seem to align with the branding strategy. Kelvin Grove Urban Village and Dutton Park Knowledge Precinct have an inner-city design character, where Sippy Downs Knowledge Precinct has beach-style relaxed tourism site design characteristics. All cases follow the same layout principles, featuring a main street, a square, and central urban design features. However, they all present a major disadvantage as being an enclosed development. Although they are not gated communities and there is access, flow, and 'communication' with surrounding areas, the connection is artificial and not authentically integrated. The edges are blocked with busy roads or train tracks, which make it hard to walk in and out, and marks them less permeable, despite their promises (Yigitcanlar et al. 2016).

Design: Brisbane's knowledge precincts seem to be accommodated with good design principles—including transit-oriented development, New Urbanism, crime prevention through environmental design, and water sensitive urban design. Kelvin Grove Urban Village's compact urban village development concept has been recognised with a national design excellence award. The University of the Sunshine Coast library building has won numerous design accolades including the Australian building of the year. Dutton Park Knowledge Precinct employs award-winning architectural style and sustainable design principles to create a highly sustainable and collaborative environment for the community. Ecosciences Precinct in Dutton Park Knowledge Precinct has been recognised by national, state, and regional level awards including the awards for commercial architecture at the national and regional levels and Brisbane regional building of the year in 2011. These achievements demonstrate the design success and the degree of attractiveness that it entails for these knowledge precincts in terms of their architectural features (Yigitcanlar et al. 2016).

Brand: In all investigated knowledge precincts brand is not static; it evolves over time and is adapted to the needs of the market, audience, and development phases. For instance, in Kelvin Grove Urban Village, it was clear that at the beginning of the project, 'knowledge' was the main message; then it is shifted to 'urban lifestyle', and once the knowledge precinct was established, the branding message was returned back to the original vision and focused on communicating 'knowledge'. In Dutton Park Knowledge Precinct, it started as 'knowledge' and then due to the expansion of the project it became 'urban lifestyle'. However, in Sippy Downs Knowledge Precinct, due to the infancy of the project, 'beach-style (Bohemian) lifestyle' so far is the unchanged brand. This constant evolution of branding and promotional strategies can be understood as a response to the dynamics of maintaining attractiveness and competitive edge (Yigitcanlar et al. 2016).

Image: Establishment of the 'image' varies in all cases, due to the fact that image is a social construction that depends on the acknowledgement of the places and respective brands by each of their audiences. As a consequence of the age of each analysed precinct, it is needless to say that Kelvin Grove Urban Village and Dutton Park Knowledge Precinct have better-established images, while Sippy Downs Knowledge Precinct is still a vision embedded in the brand and has not yet achieved an actual consistent image. Promoting a similar desired future and environment including authenticity and uniqueness is a question for the cases. The policy documents of all three cases say they are 'unique'; however, the features they promote are pretty much the same—vibrant community, workable environment, safe, sustainability, promoting a lifestyle. The differentiation lays mainly on each specific flagship development and unique location—Kelvin Grove Urban Village's Creative Industries Precinct and inner-city location, Dutton Park Knowledge Precinct's Ecosciences Precinct and inner-city location, and Sippy Downs Knowledge Precinct's Innovation Centre Sunshine Coast and close to the beach location. Kelvin Grove Urban Village's rebranding indicates to a certain degree failure in the previous brand's appeal to the market. Dutton Park Knowledge Precinct was launched as a knowledge precinct without living facilities and as it is now evolving into a true knowledge precinct; its promotional efforts are following and so might its future perceived image. The branding and promotions of Sippy Downs Knowledge Precinct are still in initial stages; it might prove an interesting case as the way this brand is managed might be one of the internal factors that define the success of this development despite the scale of the envisioned project (Yigitcanlar et al. 2016).

Challenges: In the journey from space to place—through brand, legibility, image, and identity—all three knowledge precincts stand at different points. Kelvin Grove Urban Village, being the most mature precinct, is a good example of using place branding at the planning stage to form a knowledge community with strong support from university and government. However, at the initial stage stronger support from the private sector was missing. Such involvement is essential in becoming a globally recognised knowledge precinct. Dutton Park Knowledge Precinct has strong government backing that helps its establishment as a knowledge precinct. Although initially it was planned to include surrounding areas offering wider residential and lifestyle facilities, in its current formation as a knowledge precinct, it fails to provide a home for knowledge communities. Thus, the main challenge for Dutton Park Knowledge Precinct is to evolve into a healthy knowledge precinct by providing an onsite living, shopping, cybering, and recreation activities. Sippy Downs Knowledge Precinct, on the other hand, is an ideal knowledge precinct conception located on a prime coastal position. Due to the State and Local Governments' economic development programs, limited economic activities in the region might forbid or significantly delay greater knowledge community gain (Yigitcanlar et al. 2016).

Drawbacks: There are some interesting inconsistencies between the promoted brand (intended image) and actual place identity and possible perceived image of the place in Brisbane's knowledge precincts. For example, Kelvin Grove Urban Village branding promises a vibrant local community, but nowadays Kelvin Grove hosts a large population of undergraduate students who do not reside in the precinct for more than a few years, or who do not live in the area at all and simply travel to and from Kelvin Grove Urban Village. Due to the limited industry take up and initiatives, future employment opportunities for these young knowledge workers in the same environment where they studied are very slim. This means Kelvin Grove Urban Village does not benefit adequately from the talent it has fostered. Dutton Park Knowledge Precinct's brand aims at the formation of a knowledge community but provides limited non-work (residential) land uses. This limits the community formation, despite the branding efforts promoting a lively environment that makes no necessary reference to the existing scientific work/precinct. Sippy Downs Knowledge Precinct has not yet been developed enough to be able to host a knowledge community. Additionally, even though all knowledge precincts aim ambitiously to become 'a place to visit' by people outside their communities, unfortunately, none has managed to achieve that yet (Yigitcanlar et al. 2016).

Lastly, this investigation, to generate useful insights, reveals that Brisbane's signature knowledge precinct initiatives demonstrate various levels of potential for success in terms of place making for knowledge and innovation economy and communities. However, despite over three decades of experience and investment, Brisbane has not been entirely successful in placing any of its knowledge precincts in the national or global maps. The biggest challenge for the city seems to be establishing a vibrant venture capital environment to nurture local start-ups growth and attract major global and national companies into Brisbane's knowledge precincts.

18.13 Summary

This chapter of the book concentrated on providing an understanding on the international and national standing of one of Brisbane's knowledge precinct in the league of prosperous knowledge precincts with a central question in mind. The question, how Brisbane's knowledge precincts compare to the other reputable ones in Australia and overseas, is addressed through assessing one of Brisbane's knowledge precinct in comparison to five international and two Australian knowledge precincts with the help of the place making facilitators of knowledge and innovation spaces framework.

The literature indicates that knowledge precincts often have two general objectives. The first is to play an incubator role nurturing the development and growth of new, small, high-technology firms, facilitating the transfer of university know-how to tenant companies, encouraging faculty-based spin-offs, and stimulating innovative products and processes. The second objective is to act as a catalyst for regional economic development that promotes economic growth and contributes to the development of the city as a knowledge city (Yigitcanlar 2014b). The background research highlights that the intention behind Brisbane's knowledge precinct development concerns both of these objectives.

The findings of the comparison analysis reveal, at least in the case of Dutton Park Knowledge Precincts, the city started to showcase promising knowledge precinct practice—despite its downfalls in some areas. Due to earlier mentioned challenges (or weaknesses) in Brisbane's over 30 years of KBUD journey, the city has just managed to pave its path with some useful learnings and experiences. However, there is still so many needs to be done to bring Brisbane's knowledge precincts at par with the global best practice cases. These recommendations are discussed in the next chapter of the book in detail.

Lastly, in interpreting the findings of the comparison study the following limitations needs to be noted: (a) Lack of in-depth prior research studies on the investigated case studies may have an impact on the accuracy of the findings; (b) Data collection limitations may have an impact on the accuracy of the findings; (c) Differing socio-political contexts across the investigated cases may have an impact on the interpretation of the findings, and; (d) Researcher's unintentional cultural, professional and other type of bias might have an impact on the findings. Despite the abovementioned limitations and constraints, this analytical study serves as a basis of curiosity, some level of scepticism, learning, future development, and triggers and encourages further studies to place Brisbane's knowledge precincts under the KBUD microscope.

Despite abovementioned limitations, the study generates many invaluable insights. These insights are not only useful for the case city of Brisbane to develop prosperous knowledge precincts, but also useful for many other cities of the world that are attempting to formulate policies to establish prosperous knowledge precincts.

References

Adams, N., Cotella, G., & Nunes, R. (2014). *Territorial development, cohesion and spatial planning: Knowledge and policy development in an enlarged EU*. London: Routledge.
Berkes, F. (2009). Evolution of co-management: Role of knowledge generation, bridging organizations and social learning. *Journal of Environmental Management, 90*(5), 1692–1702.
Bradfield, J. (1981). The Cambridge Science Park. *Trends in Analytical Chemistry, 1*(2), 5–9.
Brisbane City Council. (2006). *Brisbane city shape 2026*. Brisbane, Australia.
Brisbane City Council. (2014). *Brisbane city plan 2014*. Brisbane, Australia.
Carey, A. (2016). *Trial of driverless cars to begin on Melbourne's freeways early next year*. Accessed March 21, 2017 from http://www.theage.com.au/victoria/trial-of-driverless-cars-to-begin-on-melbournes-freeways-early-next-year-20161214-gtbk1j.html.
Chesbrough, H. (2003). The era of open innovation. *MIT Sloan Management Review, 44*(3), 35–41.
Chesbrough, H., Vanhaverbeke, W., & West, J. (2006). *Open innovation: Researching a new paradigm*. Oxford: Oxford University Press.
City of Ryde. (2013). *Planning proposal for Macquarie Park Corridor*. City of Ryde, NSW: Urban Planning Unit.
Ergazakis, K., Metaxiotis, K., & Psarras, J. (2004). Towards knowledge cities: Conceptual analysis and success stories. *Journal of Knowledge Management, 8*(5), 5–15.
Fernandez-Maldonado, A. (2012). Designing: Combining design and high-tech industries in the knowledge city of Eindhoven. In T. Yigitcanlar, K. Metaxiotis, & J. Carrillo (Eds.), *Building

prosperous knowledge cities: Policies, plans and metrics (pp. 175–194). Northampton: Edward Elgar.

Fernandez-Maldonado, A., & Romein, A. (2010). The role of organisational capacity and knowledge-based development. *International Journal of Knowledge-Based Development, 1*(1–2), 79–96.

Franzmann, L., & Lodge, T. (2016). *Ideas for future urban mobility: Inner city Brisbane strategy*. Brisbane: Corview & Hassell.

Gonzalez, O., & Carrillo, F. (2012). Cities-benchmarking algorithm: A meta-ranking exercise. *International Journal of Knowledge-Based Development, 3*(1), 367–387.

Gottdiener, M. (2000). Lefebvre and the bias of academic urbanism: What can we learn from the 'new' urban analysis? *City, 4*(1), 93–100.

Hong, P., Hong, S. W., Jungbae Roh, J., & Park, K. (2012). Evolving benchmarking practices: A review for research perspectives. *Benchmarking: An International Journal, 19*(4/5), 444–462.

Hortz, T. (2016). The smart state test: A critical review of the smart state strategy 2005–2015's knowledge-based urban development. *International Journal of Knowledge-Based Development, 7*(1), 75–101.

Kumar, A., Antony, J., & Dhakar, T. S. (2006). Integrating quality function deployment and benchmarking to achieve greater profitability. *Benchmarking: An International Journal, 13*(3), 290–310.

Martí-Costa, M., & Miquel, M. P. (2014). The knowledge city against urban creativity? Artists' workshops and urban regeneration in Barcelona. *European Urban and Regional Studies, 19*(1), 92–108.

Maynard, N. (2008). Municipal ICT policy goals and technology choices: A decision framework. In T. Yigitcanlar, et al. (Eds.), *Creative urban regions: Harnessing urban technologies to support knowledge city initiatives* (pp. 95–113). Hersey: IGI Global.

McCoy, J., & Evans, G. (2002). The potential role of the physical environment in fostering creativity. *Creativity Research Journal, 14*(3–4), 409–426.

Musterd, S., & Zoltan, K. (2013). *Place-making and policies for competitive cities*. Somerset: Wiley.

NSW Government. (2014). *Herring Road Macquarie Park urban activation precinct planning report*. Department of Planning and Environment. Accessed April 7, 2016 from www.planning.nsw.gov.au.

Oksanen, K., & Ståhle, P. (2013). Physical environment as a source for innovation: Investigating the attributes of innovative space. *Journal of Knowledge Management, 17*(6), 815–827.

Pancholi, S., Yigitcanlar, T., & Guaralda, M. (2014). Urban knowledge and innovation spaces: Concepts, conditions, and contexts. *Asia Pacific Journal of Innovation and Entrepreneurship, 8*(1), 15–38.

Pancholi, S., Yigitcanlar, T., & Guaralda, M. (2015). Place making facilitators of knowledge and innovation spaces: Insights from European best practices. *International Journal of Knowledge-Based Development, 6*(3), 215–240.

Pancholi, S., Yigitcanlar, T., & Guaralda, M. (2017a). Governance that matters: Identifying place-making challenges of Melbourne's Monash Employment Cluster. *Journal of Place Management and Development, 10*(1), 73–87.

Pancholi, S., Yigitcanlar, T., & Guaralda, M. (2017b). Place making for innovation and knowledge-intensive activities: The Australian experience. Technological Forecasting and Social Change, https://doi.org/10.1016/j.techfore.2017.09.014.

Pancholi, S., Yigitcanlar, T., & Guaralda, M. (2018a). Attributes of successful place making in knowledge and innovation spaces: Evidence from Brisbane's Diamantina Knowledge Precinct. *Journal of Urban Design, 23*(5), 693–711.

Pancholi, S., Yigitcanlar, T., & Guaralda, M. (2018b). Societal integration that matters: Place making experience of Macquarie Park Innovation District, Sydney. *City, Culture and Society, 13*, 13–21.

Peschl, M., & Fundneider, T. (2012). Spaces enabling game-changing and sustaining innovations: Why space matters for knowledge creation and innovation. *Journal of Organisational Transformation and Social Change, 9*(1), 41–61.

PWC. (2014). *Big city analytics*. Australia: Price Waterhouse Coopers (PWC).
Queensland Government. (2005). *Smart Queensland: Smart state strategy*. Brisbane, Australia.
Queensland Government. (2007). *Smart cities: Rethinking the city centre*. Brisbane, Australia.
Queensland Government. (2009a). *Smart communities*. Brisbane, Australia.
Queensland Government. (2009b). *South East Queensland regional plan*. Brisbane, Australia.
Romer, P. (1994). The origins of endogenous growth. *Journal of Economic Perspectives, 8*(1), 3–22.
Saxenian, A. (1994). *Regional advantage: Culture and competition in Silicon Valley and Route 128*. Cambridge: Harvard University Press.
Smith, K. (1995). Interaction in knowledge systems: Foundations, policy implications and empirical methods. *STI Review, 16*(1), 69–102.
Van Winden, W., de Carvalho, L., Van Tuijl, E., Van Haaren, J., & Van den Berg, L. (2012). *Creating knowledge locations in cities: Innovation and integration challenges*. London: Routledge.
Van Winden, W., Van Den Berg, L., & Pol, P. (2007). European cities in the knowledge economy: Towards a typology. *Urban Studies, 44*(3), 525–549.
Victorian Government. (2014). *Plan Melbourne*. Melbourne: Victorian Government.
Yigitcanlar, T. (2009). Planning for knowledge-based urban development: Global perspectives. *Journal of Knowledge Management, 13*(5), 228–242.
Yigitcanlar, T. (2011). Position paper: Redefining knowledge-based urban development. *International Journal of Knowledge-Based Development, 2*(4), 340–356.
Yigitcanlar, T. (2014a). Position paper: Benchmarking the performance of global and emerging knowledge cities. *Expert Systems with Applications, 41*(12), 5549–5559.
Yigitcanlar, T. (2014b). Valleys, alleys and roundabouts: Innovating beyond a precinct. *The Conversation*. Accessed March 17, 2017 from http://theconversation.com/valleys-alleys-and-roundabouts-innovating-beyond-a-precinct-24290.
Yigitcanlar, T., & Dur, F. (2013). Making space and place for knowledge communities: Lessons for Australian practice. *Australasian Journal of Regional Studies, 19*(1), 36–63.
Yigitcanlar, T., Guaralda, M., Taboada, M., & Pancholi, S. (2016). Place making for knowledge generation and innovation: Planning and branding Brisbane's knowledge community precincts. *Journal of Urban Technology, 23*(1), 115–146.
Yigitcanlar, T., & Lönnqvist, A. (2013). Benchmarking knowledge-based urban development performance: Results from the international comparison of Helsinki. *Cities, 31*(1), 357–369.
Yigitcanlar, T., O'Connor, K., & Westerman, C. (2008). The making of knowledge cities: Melbourne's knowledge-based urban development experience. *Cities, 25*(2), 63–72.

Chapter 19
Recommendations and Strategic Directions for Knowledge Precinct Place Making

Abstract There are numerous key challenges to construct successful knowledge precincts, which make knowledge precinct development a risky business. At the same time, in a highly globalised and competitive world it is critical to invest on knowledge and innovation activities that are mainly housed by knowledge precincts. The chapter proposes invaluable recommendations and strategic directions in the areas of context, space and place, companies, and users particularly for a specific knowledge precinct of Brisbane city from Australia—i.e., Dutton Park Knowledge Precinct. However, the mentioned directions and recommendations are also useful for other city knowledge precincts that are aiming to establish prosperous and sustainable knowledge and innovation spaces.

Keywords Knowledge-based urban development · Knowledge city · Smart city · Sustainable urban development · Knowledge and innovation spaces · Knowledge precinct · Innovation district · Place making

Reviewed academic and grey literature, investigated global best practice cases, and the benchmarking exercise of one of Brisbane's knowledge precincts against the global knowledge precincts, have generated invaluable findings. This chapter of the book aims to address the following critical question by providing insights, suggestions and potential strategic directions in the light of these conceptual and practical findings targeting both hard and soft factors/measures.

- What are the key issues concerning planning and design of Brisbane's knowledge precincts in a city-wide holistic and integrated perspective?
- How can those recommendations and strategic directions be considered for other cities aiming to establish successful knowledge precincts?

In order to address this question, the chapter proposes some recommendations and strategic directions concerning the key elements related to the knowledge-based development of and place making in knowledge precincts—i.e., context, space and place, knowledge-intensive companies, users, and prospective investigations.

19.1 Context

In a knowledge precinct setting, context refers to the processes that are embedded in a broader set of social, political, economic, historical, and spatial conditions of that knowledge precinct and its surrounding areas and regions. Recommendations of the book in regards to contextual issues include the followings:

- *Keeping path-dependence in mind*: Policies and decisions should not be made solely for short-term but considering actions for the long run—to create a path—as history matters for current policy and decision-making situations. Furthermore, past achievements have a strong influence on strategic planning—i.e., competences that have been built in the past define the options range for the present time. Successful global knowledge precincts—such as Silicon Valley, One-North—have been building on their past achievements.
- *Establishing a tradition of building upon local*: A tradition focusing on unique local/endogenous strength identification—and building upon them to provide a strong base and giving knowledge precincts (also to the city) a competitive edge over other knowledge precincts (and their cities)—should be established. Successful global knowledge precincts—such as Barcelona and Arabianranta—have been building on their local identities.
- *Practising positive intervention*: Competencies should be built in inducing the emergence and development of knowledge precincts through dedicated authorities that oversee the coordination of collaboration, fundraising and development responsibilities. Successful global knowledge cities—such as Manchester, Melbourne and Singapore—have orchestrated their knowledge precinct emergence, support and development processes via a dedicated authority.
- *Practising perpetual prioritisation*: In plans and policies, knowledge precinct project development should be given continued prioritisation irrespective of change in governments, and local and global economic climate. Successful global knowledge cities—such as Barcelona, Melbourne and Singapore—have prioritised their knowledge precinct vision via effective plans and policies.
- *Excelling in collaborative and motivated governance*: Vertically (intergovernmental) and horizontally (interdepartmental) public institutions should be able to collaborate effectively in prioritising and supporting KBUD and place making attempts in knowledge precinct projects. Successful global knowledge cities—such as Helsinki, Melbourne and Barcelona—have accomplished successful collaborative public-sector practices to support their knowledge precincts.
- *Providing a supportive environment*: Governments should invest into emergence and growth of knowledge precincts along with other supporting institutions that strengthen formal networks in diverse sectors of the knowledge precincts as well as informal social networking. Moreover, branding and marketing knowledge precincts effectively and recognising them as key projects in policies are also critical. Successful global knowledge cities—such as San Francisco, and Singapore—have been promoting their knowledge precincts effectively in the global arena.

19.2 Space and Place

In recent years knowledge-intensive companies have started to pay an increasing attention to the physical environments in which their creative and innovative activities take place. These environments—such as knowledge precincts—reflect the companies' strategic intentions towards innovation and provide a physical embodiment of their desired space, place and modes of working (see Moultrie et al. 2007). Recommendations of the book in regards to space and place include the followings:

- *Building on unique local characteristics*: Preservation of the unique characteristics of the site the knowledge precinct planned to be built (also the surroundings and the city) and reflecting them on the knowledge precinct would give the development a genuine and distinctive identity. Successful global knowledge precincts—such as Arabianranta and Strijp-S—have been preserved their former industrial heritage identity.
- *Integrating knowledge precincts with the city*: Designating dedicated land-use zones to promote mixed and diverse uses in and around knowledge precincts that are spatially integrated with the rest of the city is important. Successful global knowledge precincts—such as One-North and Arabianranta—are well integrated with their surroundings and rest of the city.
- *Providing developmental incentives*: In knowledge precincts, developmental incentives, such as relaxation in regulations, can be used as a tool for achieving desired physical character such as ensuring flow and connectivity within precinct, vertical permeability or provision of affordable housing. Furthermore, providing world-class hard and soft amenities that enhance the vibrancy of the knowledge precincts and generate a pull factor is of importance. Successful global knowledge precincts—such as One-North, 22@Barcelona and Digital Hub—have greatly benefited from premium hard and soft amenity offerings and developmental incentives to attract investment and talent.
- *Performing people-oriented design practice*: A fine-grained mix of land use diversity with timely designated casual spaces for knowledge workers and community contributes in enhancing vibrancy necessary for the development of a sense of place. Safety and walkability are also among these vital factors. As every knowledge precinct needs to have an active public realm catering to various needs and requirements, a step ahead is to creatively engage them by interactive physical displays, exhibitions, and so on. Moreover, setting up local bodies for designing and making knowledge precincts more people and community oriented is a useful practice. Successful global knowledge precincts—such as One-North and 22@Barcelona—offer public-realm for interaction and recreation.
- *Improving connectivity*: The flow and connectivity within the site and between the buildings are critical to avoid isolation of campuses and a collaborative environment. Additionally, establishing networking organisations and a common management can contribute to ensuring interaction and strengthening the formal and informal networks between tenants as well as outside knowledge precincts. Successful global knowledge precincts—such as One-North,

22@Barcelona and Digital Hub—have high-level physical and digital connectivity.
- *Generating housing intermix*: A genuine and fine intermix of diversity in housing options (including affordability) will not only cater to the needs of various segments of society, but also will effectively integrate them into a knowledge community. Successful global knowledge precincts—such as One-North and 22@Barcelona—generate housing intermix in and around their premises. However, pretty much in every prosperous knowledge cities, and their precincts, affordability remains a major challenge.

19.3 Companies

High growth companies, such as fast-growing start-ups, are recognised for their job creation and value in driving change and introducing innovations (Hornery Institute 2017). One of the biggest challenges, if not the most important one, of Brisbane's knowledge precincts, is attracting global or Australian knowledge-intensive high-growth companies, along with nurturing and fostering innovative start-up and retaining them in the city. In order to help with this issue, the chapter puts forward some recommendations that include the followings:

- *Boosting venture capital and incentive availability*: A business climate, with strong funding availability and risk-taking opportunities, is highly essential for nurturing of new start-ups and SMEs and attracting and retaining them in the city—as for many innovators, generation of an innovative idea and turning it into an innovative product/process/service require financial support (Sabatini-Marques et al. 2015a, b). Venture capital and incentive availability are paramount for knowledge precincts (Yigitcanlar et al. 2017, 2018). Successful knowledge cities—such as San Francisco and Singapore—provide an entrepreneurial environment with various and rich funding and support mechanisms.
- *Promoting triple-helix partnership*: Stated by Etzkowitz (2003, p. 293), "innovation is increasingly based upon a Triple Helix of university-industry-government interactions". Public, private and academic partnership in and out knowledge precincts are essential to improve innovation and knowledge generation process and funding of it. Successful global knowledge precincts—such as Silicon Valley and One-North—have a long history of fruitful triple-helix partnerships.
- *Encouraging incubator and accelerator formation*: Availability of affordable physical and digital infrastructure with collaborative spaces along with and training and marketing support opportunities for start-ups and SMEs in knowledge precincts are highly important. Successful global knowledge precincts—such as Silicon Valley and Australian Technology Park (Sydney)—host global incubator and accelerator best practices.

19.3 Companies

- *Developing a profile*: Established institutions, firms and leading researchers develop a profile and brand identity for knowledge precincts. Therefore, well-planned marketing and branding strategies should be used as promotional tools for defining an integrated identity to knowledge precincts. Successful global knowledge precincts—such as Silicon Valley, Massachusetts Route 128, and One-North—have been promoting (and exporting by being widely replicated) their brands effectively all across the globe.
- *Establishing incentivised infrastructure*: For attracting knowledge-intensive companies—and fostering start-ups and SMEs—planning and provision of cost-effective high-quality physical and virtual resources within knowledge precincts are necessary. At their emergence stage successful global knowledge precincts—such as Singapore Science Park and Hsinchu Science Park (Taiwan)—were providing subsidised world-class facilities to the resident companies. Today, however, the subsidies are only available for start-ups.
- *Utilising university as an anchor*: Universities have a major role in the emergence and development of knowledge precincts by adding intellectual know-how to private sector commercialisation capabilities. They also play as the moderator between society and private sector by organising events that bring them together, making knowledge precincts an extension of campus or research centre life. Successful global knowledge precincts—such as Silicon Valley, One-North, Macquarie Park, and Monash Employment Cluster—have one or more globally top-ranking universities in or adjunct to their blurring borders.

19.4 Users

A knowledge precinct is not solely a workspace and successful contemporary exemplar precincts provide living, playing, learning, and cybering spaces for knowledge workers, their families and visitors and local communities. Recommendations of the book in regards to users—that are resident knowledge workers and their families (i.e., knowledge communities), and local communities using facilities or participation in activities—include the followings:

- *Developing positive image and perceptions*: Perceptions of on-site knowledge and surrounding local communities of knowledge precincts should be taken into consideration in the design and development stages as their views will contribute to the formation of the image for these knowledge precincts. Successful global knowledge precincts—such as Arabianranta and 22@Barcelona—have considered knowledge and local community views in their developmental stage.
- *Encouraging and facilitating coordination and collaboration*: Ensuring coordination between various tiers of government and stakeholders to deliver the best knowledge precinct outcomes for the knowledge and local communities is impor-

tant. Additionally, public and knowledge worker community engagement is important to form a diverse but cohesive community. Successful global knowledge precincts—such as Arabianranta and 22@Barcelona—have performed well in facilitating coordination and collaboration efforts concerning internal and external communities of their knowledge precincts.

- *Establishing democratic participation opportunities*: Knowledge precincts should pioneer democratic participation of all stakeholders in decision-making processes to develop a sense of ownership in them. Strengthening equal democratic participation of stakeholders and community in the decision-making process is paramount. For fruitful knowledge exchange from knowledge precinct, the participation of the community in the activities of knowledge precinct should be encouraged. Functionally, it refers to bringing out the research happening behind the walls and making people aware of and participative in it. Organising useful intranet, newsletters, talks, seminars, visual demonstrations, markets and festivals are few examples of successful global knowledge precincts—such as Digital Hub, Arabianranta and Cambridge Science Park.
- *Practising transparent community consultation*: During various stages of planning, it is necessary to keep the community well informed which, on the one hand, develops a sense of trust and assurance in them regarding the plans and, on the other, lubricates the process of development. Transparency, in regards to the plans and proposals related to knowledge precinct and explicitness in interpreting KBUD, is the main requisite. Successful global knowledge precincts—such as Monash Employment Cluster and Macquarie Park—have transparent planning and expansion practices.
- *Creating effective communication options*: Initiatives such as answering the queries in clear messages translated into multiple languages (when appropriate) and a passionate communicating team are few examples of effective communication. Furthermore, using networking organisations as a catalyst is required in order to build a cohesive and resilient community—through the establishment of formal and informal groups that help in lubricating the process. Successful global knowledge precincts—such as 22@Barcelona and Cambridge Science Park—practices highly effective communication methods.
- *Practising participatory planning and execution*: In theory, integrating inputs in future plans and proposal from the businesses, knowledge workers, and the local community is crucial. Involving them in the decision-making process and assigning significant networking roles to these groups in the post-development stage of knowledge precincts can develop a sense of belonging/ownership. In practice, the resident businesses of successful global knowledge precincts—such as Singapore Science Park 1 and 2—are involved in the future development decisions. However, this study has not encountered any global knowledge precinct practice involving workers and users in the development decisions affecting them directly or indirectly.

19.5 Prospective Investigations

Even though this chapter generates invaluable insights and informed understanding of a number of key issues relating emergence and development of prosperous knowledge precincts in Brisbane; it barely scratches the surface of the complex and complicated knowledge precinct phenomenon. There are still a lot of unanswered questions critical for Brisbane's KBUD. Some of these questions are listed below:

- What are the key attributes, contextual features, conditions and factors of success in creating and stimulating knowledge precinct developments in Brisbane?
- How does the local asset base of Brisbane influence the development and management of knowledge precincts?
- Which aspects of the local context frame the direction of policy supporting the formation of knowledge precincts in Brisbane?
- How can knowledge-based economic development and urban spatial policymaking be interconnected for the formation of knowledge precincts in Brisbane?
- What are the possible unintended consequences of policies promoting the formation of knowledge precincts in Brisbane?
- How can knowledge precincts be planned and successfully integrated into the urban fabric in Brisbane?

We believe investigating these issues further in the case of Brisbane in prospective research projects will shed light on better utilisation of knowledge precinct policies and practices in the city and its metropolitan region. Thus, we propose the following future investigations in the case of Brisbane to be considered by the city—these are also fundamental studies that any city in the context of knowledge economy should consider:

- As part of the *contextual analysis*, mapping Brisbane's critical local knowledge assets and their interconnections and interdependencies;
- As part of the *path-dependency analysis*, investigating space-time-policy dynamics of Brisbane's KBUD;
- As part of the *system analysis*, developing a framework of understanding to interconnect Brisbane's innovation and production system with policy and planning system;
- As part of the *performance analysis*, conducting a thorough place-making and innovation performance investigation on knowledge precincts of Brisbane.

19.6 Summary

This chapter of the book focused on producing some generic recommendations and strategic directions for supporting Brisbane's knowledge precinct agenda. While doing so the following question was kept in mind: In a citywide holistic and integrated perspective, what the key issues concerning planning and design of Brisbane's knowledge precincts are.

The recommendations put forward in this chapter are generic and aim to create an interest in the State and local governments to consider developing a strong knowledge precinct perspective and strategy. However, incorporation of these recommendations through effective governance will help in the facilitation of a:

- Robust context that provides strong planning, policy, and institutional support for emergence and growth of knowledge precincts;
- Uniquely featured knowledge precincts to enhance their competitiveness and resourcefulness;
- Strengthened connectivity and urban character of knowledge precincts and their surrounding environment;
- Open, diverse, connected and collaborative socioeconomic environment to encourage knowledge sharing and exchange, and;
- The sense of identity, ownership and positive image perceived by resident workers of knowledge precincts and surrounding local communities.

Implementation of these recommendations would likely to contribute to the success of knowledge precincts in perpetual knowledge generation and innovation and lead to a sustainable KBUD. Nevertheless, in order to achieve desired outcomes, in addition to the study reported in this book, more thorough investigations need to be conducted. These studies should include scrutinising Brisbane's knowledge precincts in a comprehensive way from multiple angles, while benchmarking against the other prosperous knowledge precincts of the world. These recommendations are also highly relevant to the many cities of the world aiming a knowledge-based development through formation of knowledge and innovation spaces in their cities.

References

Etzkowitz, H. (2003). Innovation in innovation: The triple helix of university-industry-government relations. *Social Science Information, 42*(3), 293–337.

Hornery Institute. (2017). *Brisbane benchmarked: Exploring its relative performance on global indicies*. Brisbane: The Hornery Institute.

Moultrie, J., Nilsson, M., Dissel, M., Haner, U. E., Janssen, S., & Van der Lugt, R. (2007). Innovation spaces: Towards a framework for understanding the role of the physical environment in innovation. *Creativity and Innovation Management, 16*(1), 53–65.

Sabatini-Marques, J., Yigitcanlar, T., & Costa, E. (2015a). Incentivizing innovation: A review of the Brazilian federal innovation support programs. *Asia Pacific Journal of Innovation and Entrepreneurship, 9*(1), 31–56.

Sabatini-Marques, J., Yigitcanlar, T., & Costa, E. (2015b). Australian innovation ecosystem: A critical review of the national innovation support mechanisms. *Asia Pacific Journal of Innovation and Entrepreneurship, 9*(2), 3–28.

Yigitcanlar, T., Sabatini-Marques, J., Costa, E., Kamruzzaman, M., & Ioppolo, G. (2017). Stimulating technological innovation through incentives: Perceptions of Australian and Brazilian firms. *Technological Forecasting and Social Change*, https://doi.org/10.1016/j.techfore.2017.05.039.

Yigitcanlar, T., Sabatini-Marques, J., Kamruzzaman, M., Camargo, F., Costa, E., Ioppolo, G., & Palandi, F. (2018). Impact of funding sources on innovation: Evidence from Brazilian software companies. *R&D Management, 48*(4), 460–484.

Chapter 20
Conclusion to Part IV

Abstract The chapter forms the concluding section of this part that focused on the planning, design, development and management of knowledge precincts as the nexus of knowledge-based urban development in prosperous knowledge cities. Generated conclusions in this chapter, in the light of the analysis reported in the earlier chapters, are not only useful for the Brisbane city, but also invaluable for other emerging knowledge cities of the world that aspires a prosperous knowledge precinct formation. In addition to generalised findings of the part, the chapter also suggests prospective conceptual and empirical studies to better formalised the knowledge and innovation space theory and practice. This in turn also will help further advancement of the knowledge-based urban development as a new discipline.

Keywords Knowledge-based urban development · Knowledge city · Smart city · Sustainable urban development · Knowledge and innovation spaces · Knowledge precinct · Innovation district · Place making

The part of the book placed Brisbane under the microscope to scrutinise its knowledge precincts achievements, progress, and potentials from the lens of knowledge-based urban development and place making.

This timely research aimed to provide fundamental underpinnings and background information to policy makers—as stated by Esmaeilpoorarabi et al. (2018) across the globe policy makers are becoming seduced by the knowledge precinct concept without sufficient understanding of its fundamental underpinnings. The research results also indicate a number of key strategies and actions that are critical for successfully transforming Brisbane's knowledge precincts into thriving and globally recognised precincts. This study, therefore, suggests the following recommendations—in addition to the recommendations made earlier chapters of the book—that are useful to be taken into consideration by the state and local governments for implementation in Brisbane and South East Queensland:

- Engaging public sector—i.e., three tier government—as the committed principal initiator and enabler in the emergence and development processes of knowledge precincts is crucial;

- Founding a dedicated agency for overseeing the knowledge precincts' lifecycle—and establishing coordination and collaboration with other public departments and private and academic sectors—is beneficial;
- Developing a long-term pathway and a roadmap—through policy-making, planning and investment—for sustainable emergence and development of knowledge precincts is critical;
- Establishing a tradition of building upon local assets and strengths, which will also determine areas to specialise in (so-called smart specialisation), is essential;
- Creating a vibrant and diverse entrepreneurial environment—supported by venture capital funding, and public-sector incentives—is necessary;
- Constructing a vibrant urban environment—supported by world-class hard and soft infrastructures and amenities—is paramount;
- Designing public places and participation mechanisms that encourage and increase knowledge communities' interaction with local communities and having say about the growth and future directions of their knowledge precincts is essential;
- Investing further in networking and pull factor creation to attract (and retain) investment and talent in knowledge precincts—that also includes branding and marketing—is needed;
- Keeping a close eye on the progress of competitor cities and their knowledge precincts—while networking and collaborating with them—through constant benchmarking exercises is useful, and;
- Undertaking further comprehensive investigations of Brisbane and its existing, emerging and prospective knowledge precincts is needed.

Lastly, as this chapter aimed to tackle a complex and complicated issue of knowledge precincts in a rather limited time frame with limited resources, it only provided generic and concise outputs to the city. The chapter, therefore, points out for more detailed investigations on Brisbane to precisely understand its context, path-dependency, innovation-production and policy-planning systems, knowledge precinct performances, and the most appropriate directions for the city to follow in its knowledge precinct development journey—along with finding suitable ways to operationalise the recommendations put forward in this study. The recommendations put forward in this chapter are also invaluable suggestions for other cities, aiming a knowledge-based development through establishing prosperous knowledge precincts, to consider and adopt after a careful evaluation and customisation.

Reference

Esmaeilpoorarabi, N., Yigitcanlar, T., Guaralda, M., & Kamruzzaman, M. (2018). Does place quality matter for innovation districts? Determining the essential place characteristics from Brisbane's knowledge precincts. *Land Use Policy, 79*, 734–747.

Afterword

In today's new millennium socioeconomic landscape cities play a central role as drivers of value creation dynamics and more generally of human prosperity for local, regional and national systems. Indeed, cities are increasingly becoming the centre of the governance of business development as well as of social and cultural prosperity. For this reason, the understanding of the factors affecting the wealth creation dynamics of cities is essential for the definition of urban plans and initiatives aimed at enhancing the tangible and intangible infrastructures as well as the working mechanisms of cities that in turn impact on the quality of citizens' life at large.

This authoritative book "Geographies of disruptions: place making for innovation in the age of knowledge economy", by Tan Yigitcanlar and Tommi Inkinen, deepens our understanding of the relevance of knowledge and innovation as both resources and sources of the competitiveness of cities and provides fundamental insights on how to design, develop and assess urban strategy development initiatives to shape knowledge cities and knowledge precincts.

Adopting a knowledge-based perspective the notion of geographies of disruptions is analysed with the scope of providing a theoretical and practical framework to support urban planners, managers and policy makers in the understanding of how to promote and facilitate urban transformations by shaping the conditions for prosperity and sustainability. The creation and facilitation of the conditions to support the formation of spatial nexus in which knowledge is generated, socially diffused, applied and combined represent an antecedent for the activation of innovation processes driving local development. The understanding of how to create a vibrant urban life, attract talented people and continuously support knowledge workers, promote creativity and diversity by developing social cohesion and tolerance, and define the tangible and intangible knowledge-based infrastructure for economic and socio-cultural development represent a priority for cities, regions and nations.

Geographies of disruption navigates the readers through four main parts that define the conceptual and practical framework to elucidate the characteristics of innovation geography and how to make knowledge-based places, i.e. cities and precincts, so that they can foster innovation in today's knowledge age. It outlines how to shape the conditions for the planning, designing and management of knowledge-based places providing evidences, managerial and policy implications, and strategic and leadership insights to facilitate the transformation of cities into knowledge cities and, particularly, by promoting the formation and of development of knowledge precincts.

The definition, support and development of knowledge precincts can be seen as a key process and/or action to shape knowledge cities. Indeed, knowledge precincts define milieu of innovation that create and enable the circulation, transfer and creation of information and knowledge that are essential for the development of innovations. They generate multiple impacts by creating a systemic effect through which the stakeholders of a knowledge precinct gain benefits by being immersed into a knowledge atmosphere which is conducive of creativity, of the formation of a dense network of supply-chain partnerships as well as of clusters of actors whose relationships are characterised by co-petition, i.e., the coexistence of competition and collaboration, and, more generally, of the creation of a collective intelligence which ultimately drives situated and collective learning processes and systemic decision making. Subsequently, knowledge precincts enable their actors to define disruptive innovations, to foresight future development trends, to recognise great challenges and to identify cutting edge solutions. They can be seen both as an engine for innovation by boosting the innovative capacity of their actors, and as a flywheel that enhances its stakeholders 'catch up ability' to keep the pace of innovation into a socioeconomic age characterised by a stringent acceleration of change.

The authors develop their arguments in the field of economic geography with the aim to define the conceptual pillars of the innovation geography explaining the dimensions and actions to promote the innovative development of cities. Such understanding is supported by a thoroughly investigation of the geography and regional literature as well as by an important analysis of case examples drowned from various locations and cities globally. A key starting conceptual dimension is the notion of city. In the last decades, a number of alternative and generally overlapping definitions have been coined to point out the gradual transformation of cities towards new competitive systemic models that recognise the central role of knowledge and technology as drivers of success. Therefore, labels such as digital, intelligent, innovative, smart, creative and knowledge-based have been adopted to denote the growing importance of the knowledge assets for the competitiveness and sustainability of cities.

This book acknowledges the different perspectives distinguishing the specific focus of the alternative conceptualisations and proposes the notion of knowledge city to point out that the significant shift of the post-modern cities characterising today's socioeconomic landscape is, indeed, the relevance of the intangible and cognitive assets. The most competitive cities have moved from an industrial nature to a service based nature, and their capacity of developing and safeguarding the quality of citizens' life is tied to their ability to continually generate and acquire knowledge that is translated into technologies, infrastructures, routines, administrative and governance mechanisms, relational and social capital, identity and image and reputation, and agglomeration of specialised vertical and horizontal knowledge firm networks. According to this view, the approach of knowledge-based urban development is suggested as a key interpretive lens to understand the economy, the environment, the wellbeing, and urban governance and administration of cities. It presupposes that the development dynamics of a city and specifically the innovation processes are the results of stakeholders' activities and relations taking place in a specific place and space. The model of the triple/quadruple-helix suggests that business organisations (including the not-for-profit entities), universities/research agencies, public institutions, and society at large are responsible for the generation, transfer, combination and application of knowledge which fuel innovation processes.

Empirical evidences of the relevance of managing knowledge in accordance with the quadruple helix model and to the knowledge-based view of the collaborative and networking relations are collected by the analysis of four case studies, respectively: Helsinki, Stockholm, Tallinn, and Copenhagen. The analysis of these case examples shows the practical elements adopted and deployed by the cities in order to facilitate innovative development. Besides, it is worth to mention that one of the most relevant knowledge-based dimension emerging from the comparative analysis of the case studies is the way how local knowledge and citizen information are integrated into the urban planning, and how the use of technologies aims to enable citizens to participate and co-create. This points out the fundamental importance of place making for knowledge cities. Indeed, the planning, designing, and development of knowledge precincts presuppose a focus on the creation, diffusion and transfer, combination, application of knowledge. This equals to apply the knowledge-based urban development point of view to understand the economic, socio-cultural, spatial and institutional development dynamics.

The book concludes with an outline of the strategies and actions that cities can put in place either to design their transformation into geography of disruptions or to assess their journey and revise their plans and initiatives. This is further developed with the analysis of the strategies and actions for transforming urban spaces for innovation into knowledge precincts that enable geography of disruptions.

Although each city has to find its identity and formulate its specific value propositions that define and make the uniqueness of the urban space, place, life and atmosphere, there are some fundamental tacit rules that can help any city to discover and reflect as well as to design and plan its living and dynamic systemic nature. This book 'crack the codes' for understanding how to strategically assess and manage the transformation of a city towards a platform of sustainable wealth creation by leveraging on knowledge assets and innovation dynamics. It is a must reading for both scientists and practitioners that want to have a comprehensive understanding of the links between innovation geography, knowledge cities and knowledge precincts as well as are searching guidelines and instruments that can be deployed to transform cities and innovation hubs into dynamics sources of innovative transformation.

Professor Giovanni Schiuma

Department of Mathematics, Informatics and Economics
University of Basilicata
Viale dell'Ateneo Lucano, Potenza, Italy
giovanni.schiuma@unibas.it

Definition of Terms

Agglomeration economy	A localised economy in which a large number of companies, services, and industries exist in close proximity to one another and benefit from the cost reductions and gains in efficiency that result from this proximity (Merriam-Webster 2017).
Agglomeration	An economic term that is used to refer to the phenomenon of firms being located close to one another. It is also a critical mass of people and economic activity, which carries the benefits, such as higher productivity and wages (Lambooy 1997).
Cluster theory	Cluster theory argues that co-location with other firms does not imply clustering when associated benefits like prolific innovation, increase in productivity, and/or the presence of externalities cannot be demonstrated (Kuah 2002).
Cluster	A cluster consists of groups of associated and interconnected proximate firms that are linked vertically and/or horizontally through their commonalities and complementariness in products, services, inputs, technologies, or outputs activities (Kuah 2002).
Economic development	A process for improving the financial well-being of a community, through efforts focused on investment attraction, job creation, knowledge generation, and

	forming a good 'business climate' leading to improvement in quality of life for the community (Carrillo et al. 2014).
Digitalisation	A generic concept describing the changes taking place in all forms of human interaction enabled by the expansion and societal embedding of digital technologies including the development of hardware (devices) and software (applications) (Inkinen et al. 2018).
E-governance	A dynamic administrative process enabling and enhancing interactions between citizens, public administration, private sector, and third sector. Digitalisation is centre driving force to foster these interactions (Jauhiainen and Inkinen 2009).
E-government	It includes the processes and structures of public sector organisations to provide (deliver) digitalised electronic services to the target groups, and to conduct electronic transactions within and between organisational entities. The external objective of e-government is to simplify the citizen's interaction with various online political and administrative services with public administration (Jauhiainen and Inkinen 2009).
Information society	A societal development stage where main characteristics of society are defined and transformed through information, its dissemination, distribution and integration. Information is the enabler of new products and developments and they are relying on digital technologies (Inkinen and Jauhiainen 2006).
Innovation systems	The network of institutions in the public and private sectors whose activities and interactions initiate, import, modify, and diffuse new innovations—in the form of product, process or service (Freeman 1987).

Definition of Terms

Institutional development	An integral part of an urban planning, development and management process that produces enablers for knowledge-based development in an urban region through forming a good 'governance climate' (Carrillo et al. 2014).
Knowledge city	A city that searches for the creation of value in all its areas and develops high standards of life, cultural support and economic development, among other aspects including higher level of income, education, training and research, at the same time it is a regional knowledge economy driven locality with high value-added exports created through research, technology, and brainpower and purposefully designed to encourage the nurturing of knowledge (Yigitcanlar 2015).
Knowledge community precinct	A mixed-use post-modern urban setting—e.g., flexible, decontextualized, enclaved, fragmented—including a critical mass of knowledge enterprises and advanced networked infrastructures, developed with the aim of collecting the benefits of blurring the boundaries of living, shopping, recreation and working facilities of knowledge workers and their families—i.e., knowledge community (Yigitcanlar 2015).
Knowledge economy	An economy that encourages its organisations and people to acquire, create, disseminate and use knowledge—i.e., codified and tacit—more effectively for greater economic and social development and it is also regarded as a separate section of the economy, the one in which new and in most cases technological knowledge is generated (Yigitcanlar 2015).
Knowledge precinct	A, in most cases an urban, cluster that knowledge is generated, exchanged and marketed by talented knowledge workers. The contemporary practice moves from

	work focused knowledge precincts—e.g., science and technology parks, innovation parks—multi-activity focused knowledge community precincts—i.e., work, live, play, and cyber (Yigitcanlar 2015).
Knowledge society	A new society formed as a result of the contemporary societal change pushed by technological innovation and institutional transformation, which is not only about technological innovations, but also about human beings, their personal growth and their individual creativity, experience and participation in the generation of knowledge. The primary role of cities in a knowledge society is to ensure that their knowledge sources are passed on and advanced by each generation (Yigitcanlar 2015).
Knowledge worker	Someone with an occupation in a post-industrial economy that is the most portable work of all, in which knowledge is of the higher value than any mere product or commodity and is the most highly paid of all workers because s/he knows stuff. Knowledge worker can also be seen as someone who is computer adept, scientist, symbolic analyst, financial wizard, writer, artist, and later-day Bohemian and someone who gets to decide what s/he does each morning (Yigitcanlar 2015).
Knowledge-based development	A humanistic perspective and development process uses a variety of knowledge management systems and approaches based on a shared vision and value sets to capture new opportunities, advance the economy and society, compete successfully in sustainable and global knowledge economy and achieve progress in the evolution of human civilisation (Yigitcanlar 2015).
Knowledge-based urban development	A process with a set of policies targeting of building a place to form perfect climates for business, people,

	space/place and governance, and emphasise on the balance and integration of these climates (Yigitcanlar 2014).
National innovation system	A national innovation system is the flow of technology and information among people, enterprises and institutions which is key to the innovative process on the national level. According to innovation system theory, innovation and technology development are results of a complex set of relationships among actors in the system, which includes enterprises, universities and government research institutes (Freeman 1995).
Organisational learning	Organisation-wide continuous process that enhances its collective ability to accept, make sense of, and respond to internal and external change. It requires systematic integration and collective interpretation of new knowledge that leads to collective action and involves risk taking as experimentation (Business Dictionary 2017).
Path dependence	Path dependence is the idea that decisions we are faced with depending on past knowledge trajectory and decisions made and are thus limited by the current competence base. In other words, history matters for current decision-making situations and has a strong influence on strategic planning (Financial Times 2017).
Regional innovation system	A regional innovation system is a set of networks between public and private agents that interact and give mutual feedback in a specific territory by taking advantage of their own infrastructure to adapt, generate and extend knowledge and innovation (Cooke et al. 1997).
Smart city (in a broader sense)	A smart city could be an ideal model to build the cities of the 21st century, in the case, its practice involves a system of systems approach and a sustainable and balanced view on the economic, societal, environmental and institutional development domains (Yigitcanlar et al. 2018).

Smart city (in a narrow sense)	A technology-led urban utopia permeated with top-down and centrally controlled technological infrastructures, with the aim to improve the urban environment in terms of efficiency, security and sustainability (Niaros 2016).
Smart specialisation	An approach that combines industrial, educational and innovation policies to suggest that cities or regions identify and select a limited number of priority areas for knowledge-based investments, focusing on their strengths and comparative advantages (OECD 2013).
Spatial development	A process that encompasses interventions in both natural and built environments for achieving an ecologically sensitive urban development along with maintaining a high quality of life and place and forming a good 'spatial climate' (Carrillo et al. 2014).
Spill-over effect	A positive externality that happening; for example, the transmission of an advanced technology from a foreign-owned firm (thus foreign direct investment) to domestic firms (Investorwords 2017).
Spin-off company	A company that develops through the sale or distribution of new shares of an existing business or division of a parent company (Investopedia 2017).
Sustainable urban development	Improving the quality of life in a city, including ecological, cultural, political, institutional, social and economic components without leaving a burden, e.g., the result of a reduced natural capital and an excessive local debt, on the future generations—and thus forming the sustainable city (Yigitcanlar and Teriman 2015).

References

Business Dictionary. (2017). *Organizational learning*. Accessed on March 10, 2017 from http://www.businessdictionary.com/definition/organizational-learning.html.

Carrillo, F. J., Yigitcanlar, T., García, B., & Lönnqvist, A. (2014). *Knowledge and the city: Concepts, applications and trends of knowledge-based urban development*. New York: Routledge.

Cooke, P., Uranga, M. G., & Etxebarria, G. (1997). Regional innovation systems: Institutional and organisational dimensions. *Research Policy, 26*(4–5), 475–491.

Financial Times. (2017). *Definition of path dependence*. Accessed on March 8, 2017 from http://lexicon.ft.com/Term?term=path-dependence.

Freeman, C. (1987). *Technology and economic performance: Lessons from Japan*. London: Pinter.

Freeman, C. (1995). The national system of innovation in historical perspective. *Cambridge Journal of Economics, 19*(1), 5–24.

Inkinen, T. & Jauhiainen, J. (Eds.). (2006). *Geography of information society* [In Finnish: Tietoyhteiskunnan maantiede]. Helsinki: Gaudeamus.

Inkinen, T., Merisalo, M., & Makkonen, T. (2018). Variations in the adoption and willingness to use e-services in three differentiated urban areas. *European Planning Studies, 26*(5), 950–968.

Investorwords. (2017). *Spillover*. Accessed on March 8, 2017 from http://www.investorwords.com/17778/spillover.html.

Investopedia. (2017). *What is a 'spinoff'*. Accessed on March 8, 2017 from http://www.investopedia.com/terms/s/spinoff.asp.

Jauhiainen, J. S. & Inkinen, T. (2009). E-governance and the information society in periphery: Case study from Northern Finland. In C. Reddick (Ed.), *Handbook of research on strategies for local e-government adoption and implementation: comparative studies* (pp. 496–513). Hershey: IGI Global.

Kuah, A. T. H. (2002). Cluster theory and practice: Advantages for the small business locating in a vibrant cluster. *Journal of Research in Marketing and Entrepreneurship, 4*(3), 206–228.

Lambooy, J. G. (1997). Knowledge production, organisation and agglomeration economies. *GeoJournal, 41*(4), 293–300.

Merriam-Webster. (2017). *Agglomeration economy*. Accessed on March 10, 2017 from https://www.merriam-webster.com/dictionary/agglomeration%20economy.

Niaros, V. (2016). Introducing a taxonomy of the "smart city": Towards a commons-oriented approach? *TripleC: Communication, Capitalism & Critique, 14*(1), 51–61.

Yigitcanlar, T. (2015). Knowledge based urban development. In M. Khosrow-Pour (Ed.), *Encyclopedia of information science and technology* (3rd ed.) (pp. 7475–7485). Hersey, PA: IGI Global.

Yigitcanlar, T., Kamruzzaman, M., Buys, L., Ioppolo, G., Sabatini-Marques, J., Costa, E., & Yun, J., (2018). Understanding 'smart cities': Intertwining development drivers with desired outcomes in a multidimensional framework. *Cities, 81*, 145–160.

Yigitcanlar, T., & Teriman, S. (2015). Rethinking sustainable urban development: Towards an integrated planning and development process. *International Journal of Environmental Science and Technology, 12*(1), 341–352.

Index

A

Accessibility, 49, 63, 116, 127, 150, 187, 193, 222, 226, 254, 265, 269, 274, 279, 280, 282

Administration, 5, 7, 43, 45, 46, 52, 56, 66, 68–70, 76, 136, 137, 149, 228, 229, 242, 246

Agglomeration, 8, 11, 14, 17, 18, 52, 90, 106, 211, 217, 228, 229, 276

Agglomeration theory, 11, 16

Arabianranta, 106, 212, 218, 233, 269, 270, 283–286, 296, 297, 299, 300

Asset, 188, 194, 232, 271, 287, 301

Austin, 106, 108, 113, 114, 135, 139–144, 153–155, 200, 201, 212, 218

Australia, 106–108, 113, 135, 148, 149, 151, 152, 160–163, 186–188, 193–195, 203, 207, 212, 213, 218, 239, 251, 252, 262, 273–277, 279, 280, 282, 283, 288, 291, 295

Automation, 20, 47

Autonomous vehicle, 42

B

Barcelona, 106, 108, 112, 114, 135–137, 139, 153–155, 204, 212, 213, 240–243, 256, 257, 261, 262, 267, 269, 283–286, 297–300

Benchmarking, 12, 66, 70, 152, 159–162, 164, 199, 205, 207, 261, 295, 302, 304

Best practice, 8, 48, 106–108, 116, 135, 139, 154–156, 172, 174, 199, 200, 202, 213, 217, 233, 239, 240, 247, 256, 257, 264, 282, 287, 292, 295

Birmingham, 108, 112, 172–187, 191, 192, 195

Boston, 106, 108, 113, 114, 139, 161, 163, 172–186, 188, 191, 195, 212, 218, 243, 262

Branding, 60, 63, 155, 188, 191, 194, 197, 201, 266, 288–291, 296, 299, 304

Brisbane, 106–108, 113, 121, 135, 136, 154–156, 160, 162, 164, 171 204, 207, 208, 212, 213, 218, 239, 240, 256, 257, 261, 262, 278, 279, 282–292, 295, 298, 301–304

Built environment, 116, 219, 250, 269, 271

Business climate, 189, 196, 200, 257, 298

Business-University-Government (B-U-G), 64, 65

C

Cambridge Science Park, 118, 218, 233, 265, 267, 283–286, 300

Canada, 113

Capital systems, 76

Carrillo, 105, 112, 113, 116, 120–122, 162–165, 216, 217, 220–223, 233, 263

City, 5, 7, 8, 18, 42–44, 46, 52, 55–57, 60–79, 85, 97, 101, 105–107, 109–112, 115–128, 135–137, 139–141, 143–150, 152–156, 159–164, 166–172, 174, 175, 177–183, 185–204, 207, 208, 211–213, 216, 217, 221–223, 226–228, 231, 232, 240–243, 245–247, 251, 253, 256, 257, 261, 265, 267, 269, 271–275, 277–280, 282, 283, 285, 287, 289–292, 295–298, 301, 303, 304

City branding, 125, 127, 139, 169, 171, 176, 178, 179, 183, 185, 191, 192, 197, 208
Climate change, 5, 126, 128, 189, 196
Cluster, 4, 8, 12, 14, 18–20, 52, 64, 102, 115, 142, 143, 146, 172, 174, 180, 181, 185–187, 215, 217, 218, 222, 228, 243, 244, 261, 267, 276–279, 283, 285
Cluster analysis, 171–174, 182, 188
Cluster theory, 217
Community, 115, 120–122, 125, 127, 137, 140, 141, 143, 144, 149, 150, 155, 169, 171, 189, 190, 196, 201, 215, 219, 220, 223, 244, 246, 248, 253–255, 257, 268, 271, 273, 275, 277, 278, 281–283, 285, 287–291, 297–300
Community engagement, 125, 169, 171, 176, 178, 179, 183, 186, 192, 207, 299
Competition, 15, 16, 18, 60, 61, 93, 160, 162, 186, 190–192, 196, 197, 243, 262
Competitiveness, 15, 77, 120, 142, 149, 154, 161, 163, 215, 217, 222, 242, 255, 302
Composite index, 70
Connectivity, 8, 17, 62, 65, 69, 73, 78, 223, 254, 265, 269, 274, 276, 279, 280, 282, 297, 298, 302
Copenhagen, 41, 56, 63, 69–74, 79–81, 83, 86, 89–92, 96, 101, 114, 218
Cost of living, 5, 124, 140, 144, 161, 163, 168, 170, 175, 177, 178, 182, 189, 200
Creative destruction, 8, 14, 15, 224
Creative industry, 241
Creativity, 12, 14, 72, 77, 96, 115, 119, 121, 143, 144, 201, 211, 215, 219, 226, 254, 262, 263, 265
Cultural diversity, 124, 147, 166, 168, 174, 176, 177, 180, 181, 188, 189, 196

D

Data, 4–8, 15, 19, 40–43, 45–47, 55, 57, 60, 64–66, 68, 70–75, 79–83, 85, 86, 93–98, 102, 143, 162–165, 172–175, 192, 197, 198, 247, 249, 292
Decentralisation, 21
Decision making, 44, 48, 232
Denmark, 8, 56, 57, 63, 69, 70, 72, 80–95, 218
Desired policy outcome, 117
Development, 3–7, 11–13, 15, 17–21, 39–43, 45, 47–49, 51, 52, 55–57, 59–78, 82, 84, 85, 88, 89, 91–96, 98, 102, 106, 107, 110–112, 114–123, 125, 127, 128, 136, 137, 139, 141–155, 159–165, 167, 168, 170, 171, 173, 179, 187–198, 200, 201, 204, 207, 208, 211, 212, 215–233, 239, 240, 242, 244, 247–251, 254, 256, 257, 261–263, 265–273, 275–280, 282, 283, 287, 289–292, 295–297, 299–301, 303, 304
Diamantina Knowledge Precinct, 280
Digital divide, 13
Digital Hub, 233, 272–274, 283–286, 297, 298, 300
Digitalisation, 3, 4, 6, 13, 20, 42, 43, 45, 60–63, 67, 68, 70, 73, 77, 92, 93, 96, 97
Disruption, 3, 97, 305–307
Distance, 19–22, 52, 56, 78, 188, 194, 202, 275, 281
Diversity, 62, 70, 71, 80, 84, 115, 116, 122, 124, 127, 143, 151, 153, 163, 165, 166, 168, 178, 179, 181, 187, 193, 208, 263, 265, 266, 268, 284, 297, 298
Dutton Park Knowledge Precinct, 107, 121, 212, 261, 262, 278–291, 295

E

Economic development, 8, 13, 15, 44, 51, 56, 69, 95, 105, 109, 111, 112, 123, 137, 141, 142, 145–148, 164, 166, 172, 174, 177–180, 184, 185, 187, 188, 190, 194, 200, 201, 204, 216, 219, 220, 225, 243, 251, 268, 272, 290, 291, 301
Economic geography, 3, 8, 11, 13, 14, 17, 19, 22, 40, 51, 90, 97, 216
Economic growth, 48, 64, 76, 83, 106, 109, 112, 118, 120, 145, 146, 211, 216, 217, 220, 251, 291
Economy, 3–8, 11, 17, 39, 41, 52, 55, 60–62, 69, 71, 75, 77, 80, 84, 87–89, 105, 106, 109, 118–120, 122, 128, 136, 137, 141, 142, 144, 147, 149–151, 153, 189–191, 195, 196, 200, 204, 216, 220, 222, 228–230, 232, 242, 245, 248, 251, 253, 267, 269, 272, 288
Education, 3, 6, 12, 17, 43, 44, 62–64, 71, 75, 76, 78–81, 88, 90, 96, 105, 112, 115–117, 123, 127, 128, 144, 146–148, 150, 152, 154, 165, 167, 174, 176, 177, 180, 181, 187, 189, 190, 193, 196, 219, 223, 248, 249, 273, 277, 280
Effective leadership, 125, 169, 171, 176, 178, 179, 183, 192, 198
E-health, 49, 60
Electronic governance (E-governance), 42, 43, 46, 75, 125, 168, 170, 176, 178, 179, 183, 192

Index 321

Electronic government (E-government), 5, 7, 45–47, 52, 61, 63, 67, 70, 74, 75, 91, 92, 125, 168, 170
Environment, 5, 22, 41, 43, 46, 56, 62, 64, 67–69, 71, 75, 76, 78, 84, 102, 112, 117, 136, 137, 139, 141, 143, 148, 189, 196, 212, 220–222, 225, 226, 229–232, 248–250, 253–256, 263, 266–271, 275, 279, 289–291, 296–298, 302, 304
Environmental impact, 124, 167, 169, 175, 177, 178, 181, 182, 189
Environmental innovation, 85, 97, 102
Estonia, 8, 56, 57, 74–77, 80–87, 89–96
European Union (EU), 57, 62, 69, 70, 76, 79–96, 98, 266
Eurostat, 57, 79, 81–94

F

Finland, 8, 40, 43, 44, 56, 57, 59, 62, 63, 69, 78, 80–95, 218, 269, 270, 284
Flexible specialisation, 21
Foreign direct investment (FDI), 123, 164, 166, 173, 175, 177, 179
Functional city, 62

G

Geographical information systems (GIS), 19, 41
Global economy, 5, 39, 40, 189, 196, 232, 242
Global financial crisis (GFC), 107, 126, 192, 198, 243, 288
Global knowledge economy, 106, 107, 110, 119, 120, 128, 149, 159, 161, 171, 173, 211, 215, 221, 222, 226, 250, 267
Global village, 3
Google, 42, 66, 97, 148, 153, 245, 246, 265, 267, 268, 270, 272, 274, 276, 278, 281
Governance, 7, 8, 43, 45, 46, 55, 74, 98, 106, 108, 115–117, 120, 125–127, 138, 144, 146, 147, 153–155, 160–163, 165, 168, 170, 179, 182, 183, 189, 191, 192, 195–197, 202–204, 207, 208, 216, 219, 223, 226, 228, 229, 243, 257, 262, 264, 296, 302
Government, 4, 8, 43, 45, 46, 55, 59, 61, 64, 65, 68, 69, 75, 88, 91, 93, 97, 106, 107, 127, 136, 139–141, 143–147, 149–152, 154, 155, 186, 188–191, 194–197, 201, 203, 204, 212, 218, 220, 226, 231, 242–244, 246, 248, 251, 253, 254, 256, 262, 266, 267, 271–273, 275–280, 282, 287, 288, 290, 298, 299, 303
Government effectiveness, 125, 168, 170, 175, 177, 178, 182, 183, 192
Gross domestic product (GDP), 60, 77, 79, 123, 145, 164–167, 173, 175, 177, 179, 187, 232
Growth, 3, 12, 14–17, 45, 52, 56, 59, 61, 62, 64, 70–72, 75, 77, 79–96, 102, 106, 111, 112, 116, 117, 139, 141–145, 148–151, 153, 171, 173, 187–190, 193, 194, 196, 201, 202, 204, 208, 211, 212, 215–217, 219, 220, 222–224, 229–231, 243, 244, 248, 251–253, 262, 274, 275, 277, 282, 287, 291, 296, 298, 302, 304

H

Helsinki, 41, 42, 44, 56, 60–64, 67, 69–73, 77–81, 83, 84, 86–92, 95, 96, 101, 106, 108, 136, 172–186, 191, 195, 200, 202, 212, 213, 218, 261, 262, 269, 283–286, 296
High technology sector, 145, 243
Housing affordability, 5, 125, 168, 170, 175, 177, 178, 182, 187, 189, 190, 193, 196, 200, 255
Human capital, 56, 62, 81, 115, 121, 201, 219, 224

I

Index, 69, 123, 143, 161–166, 170, 172, 174, 177, 179, 232
Indicator, 5, 52, 57, 69–71, 79, 87, 88, 93, 94, 96, 162–173, 178, 180, 181, 185–187, 192, 198
Industrial districts, 111
Information, 3–8, 13, 14, 19, 20, 39, 42, 45–47, 49, 51, 57, 67, 68, 70, 72–74, 76, 77, 93, 94, 96, 102, 105, 109, 113, 116, 119, 121, 128, 136, 139, 144, 148, 151, 153, 160, 162, 163, 165, 215, 216, 241, 253, 266, 273, 275, 303
Information and communication technology (ICT), 4, 6, 14, 21, 59, 60, 64, 66, 70, 77, 87, 96, 102, 137, 149, 223, 242, 243, 247, 250, 269, 284, 285
Information society, 3, 4, 6, 13, 19, 47, 59, 63, 67, 75, 90, 92
Infrastructure, 5, 16, 20, 56, 64, 72, 73, 76, 98, 113, 115, 120, 122, 127, 128, 136, 142,

146, 150, 151, 153, 155, 160, 162, 187, 189, 190, 193, 196, 201, 203, 204, 208, 222, 223, 230, 231, 240, 242, 246, 249, 251, 254, 256, 257, 274, 276, 279, 280, 298, 299
Inkinen, 6, 12, 13, 16, 17, 42, 44, 46, 60, 78, 83, 88, 90, 217, 218
Innovation, 4, 6, 8, 11–21, 39, 40, 43–45, 48, 51, 52, 55, 56, 59–62, 64–67, 69–72, 74–86, 88, 90, 95–97, 101, 102, 106, 107, 110, 112, 115–122, 127, 128, 136, 137, 140, 141, 143–146, 148–151, 153, 155, 161, 163, 186, 191, 195, 197, 200–203, 208, 211, 215–220, 222, 225–230, 232, 233, 242–244, 249–251, 255–257, 261–263, 266, 269, 271, 277, 279, 282, 287, 288, 290, 291, 295, 297, 298, 301, 302, 304
Innovation district, 106, 137, 144, 153, 212, 240, 254
Innovation economy, 123, 150, 164, 166, 173, 175, 177, 179, 187, 269, 291
Innovation park, 251
Innovation platforms, 61, 128
Innovation strategies, 57, 225
Innovation system, 12, 55, 78, 153
Innovative cities, 64, 143, 161, 163
Institutional development, 107, 122, 125, 168, 170–174, 178, 179, 182–185, 187, 191, 192
Intangible, 121, 122, 129, 139, 192, 198
Intellectual capital, 111
Intellectual property rights (IPR), 12, 115
Intelligent city, 110, 111
Interface, 18, 46, 141
Internet, 4, 6, 7, 13, 69, 70, 91, 144, 188, 194, 245
Internet-of-Things (IoT), 7, 43, 74, 98
Invention, 148
Ireland, 112, 218, 272
Istanbul, 108, 113, 172–187, 191, 192, 195

K

Know-how, 61, 146, 215, 223, 291, 299
Knowledge, 4, 6, 8, 12, 13, 16, 19, 20, 22, 41–43, 46, 48, 49, 51, 52, 61, 63, 75–78, 80, 87–90, 96, 97, 105–123, 126–129, 135–137, 139–141, 145, 146, 149–156, 160, 162, 172, 174, 186–197, 199–205, 207, 208, 211–213, 215–233, 239–244, 247–250, 252–257, 261–267, 269, 271, 273–279, 281, 282, 285, 287–292, 295–304
Knowledge and innovation space, 146, 208, 218, 239, 243, 303
Knowledge asset, 13, 114, 115, 118, 122, 204, 301
Knowledge base, 43, 116, 127
Knowledge-based development, 5, 8, 13, 48, 56, 88, 107, 113, 117, 119, 120, 144, 149, 155, 183, 189, 202, 208, 221, 295, 302, 304
Knowledge-based urban development, 62, 106, 107, 112, 113, 116, 119, 121, 123, 126, 128, 138, 164, 166, 179, 184, 189–191, 195–197, 207, 208, 211, 212, 216, 303
Knowledge business, 136, 219, 243, 277
Knowledge city, 57, 60, 105–108, 110–116, 118–120, 126, 128, 129, 135–137, 139, 144, 147–156, 160–164, 172, 174, 180, 181, 185–187, 189, 191, 192, 195–197, 199–201, 203, 204, 207, 208, 212, 213, 216, 233, 240, 247, 282, 291
Knowledge community precinct, 250, 254, 275
Knowledge economy, 105, 110–112, 120, 123, 128, 136, 149, 153–155, 160–166, 177, 178, 180, 187–190, 194, 196, 200, 202, 211, 215, 216, 220, 256, 271, 283, 288, 301
Knowledge economy era, 107, 110, 120, 128, 159, 161, 211
Knowledge generation, 106, 149, 155, 200, 203, 211, 215–217, 222, 226, 230, 257, 282, 298, 302
Knowledge industry, 106, 122, 140, 146, 211, 217, 276
Knowledge precinct, 106, 139, 151, 154, 202, 212, 213, 215, 216, 220–225, 227–233, 239, 240, 244, 249, 250, 255–257, 261–264, 266, 268, 269, 271–276, 282, 284, 287–291, 295–297, 299–304
Knowledge production, 71, 220, 221, 250
Knowledge resources, 74, 78, 102, 118, 119, 228
Knowledge society, 116, 119, 136, 208
Knowledge transfer, 8, 12, 20, 61, 150, 217, 250

Knowledge worker, 120, 121, 123, 127, 165, 167, 174, 176, 177, 179, 180, 187, 188, 194, 255, 282, 299

L

Labour or labour markets, 6, 16, 63, 90, 115, 242
Land use, 215, 254, 266, 280, 297
Legislation, 22, 40, 44, 47, 66, 75, 95
Lifecycle, 223, 224, 304
Living lab, 137, 270, 285
Local innovation, 225, 229
Location, 5, 6, 11, 12, 15, 19–22, 39, 40, 43, 44, 48, 49, 51, 52, 56, 57, 64, 67, 69, 74, 76, 79, 82, 95–97, 101, 102, 111, 115, 122, 147, 148, 150, 155, 161, 163, 188, 194, 227, 229, 231, 248, 251–255, 265, 266, 270, 275, 279, 280, 282, 290
Location-based services, 7, 42
Logistics, 7, 18, 21, 40, 41, 77, 92, 96, 146, 150, 223, 251
London, 118, 218
Lundvall, 18

M

Macquarie Park, 213, 233, 239, 251–255, 273–276, 283–286, 299, 300
Macquarie Park Innovation District, 251
Management, 5, 12, 13, 20, 40–42, 46–48, 51, 52, 55, 61, 65, 66, 73, 75, 93–95, 114–117, 119, 120, 137, 150, 153, 160, 162, 211, 225, 229–231, 242, 255, 257, 261–263, 265–267, 269, 271–274, 276, 277, 279, 280, 283, 288, 297, 301, 303
Manchester, 108, 112, 114, 152, 154, 172–188, 191, 195, 203, 296
Market economy, 4, 15, 52
Marshall, 11
McLuhan, 3
Melbourne, 106, 108, 112–114, 135, 148–155, 172–188, 191, 193–195, 203, 204, 213, 261, 262, 276–278, 283–286, 296
Mobile, 17, 18, 21, 42, 43, 83, 88, 91, 92, 96, 143, 219, 242, 244, 245
Model, 21, 40, 56, 64, 65, 95, 111, 136, 137, 139, 144, 147, 152, 154, 155, 161–164, 186, 189, 195, 196, 201, 203, 208, 218, 233, 241, 242, 245, 252, 262, 267, 274, 277, 282, 285, 288
Monash Employment Cluster, 151, 233, 276, 278, 283–286, 299, 300

Mountain View, 245

N

National innovation system (NIS), 16, 27–29, 51, 84, 313
Natural environment, 128, 146
Network, 6, 13, 14, 16, 19, 40, 45, 73, 83, 87, 97, 102, 127, 128, 141, 187, 193, 217, 222–225, 231, 245, 247, 253, 255, 268, 275, 281, 284, 285, 288
Networking, 13, 19, 45, 64, 68, 78, 125, 139, 169, 171, 176, 178, 179, 183, 186, 192, 223, 230, 231, 249, 255, 266, 271, 280, 282, 284, 296, 297, 300, 304
Nordic countries, 42–44, 70, 81, 82, 84–86, 88, 91–94
Normalisation, 163, 165
Northern Europe, 64, 80, 86, 91

O

One-North, 106, 122, 146, 212, 213, 218, 220, 221, 233, 239, 247–251, 255, 296–299
Open data, 68, 70, 72, 73
Open innovation, 233, 262
Organisation, 45, 71, 111, 112, 136, 153, 154, 203, 217, 220, 225, 227, 228, 240, 255, 266, 268, 279
Organisation for Economic Cooperation and Development (OECD), 63, 119, 147, 154, 161, 163
Organisational learning, 228, 229

P

Parkville knowledge precinct, 151
Participation, 43, 52, 65, 73, 95, 116, 125, 155, 169, 171, 189, 196, 229, 255, 263, 265, 273, 280, 285, 299, 300, 304
Patent application, 82, 83, 85
Patents, 79, 81–85, 97, 115, 145, 232
Path dependency, 8, 14, 15, 52
People climate, 263
Performance analysis, 160, 162, 164, 301
Place dynamics, 34
Place making, 106, 122, 146, 211–213, 239, 261–264, 266, 271, 291, 295, 296, 303
Planning, 7, 12, 15, 20, 43, 48, 51, 52, 56, 59, 63, 66, 72, 73, 79, 93, 94, 101, 106, 107, 112, 116, 118, 119, 125, 128, 129, 139, 146, 147, 149, 150, 153, 154, 160, 162, 168, 170, 171, 173, 179, 183, 186, 189, 190, 192, 195, 196, 202–204, 207, 208,

211–213, 218–223, 226–230, 232, 233, 239–241, 246–251, 254, 256, 261, 262, 265, 268, 273, 274, 277–279, 282, 288–290, 295, 299–304
Policy, 22, 44, 45, 56, 57, 59–63, 65, 68, 70, 81, 88, 93, 98, 106, 107, 110, 112, 116, 117, 119, 123–125, 129, 137, 140, 141, 144, 146, 150, 154, 155, 159–162, 190, 191, 196, 197, 200–203, 208, 211, 213, 216–219, 221–223, 226–229, 232, 233, 240, 243, 262, 263, 271, 272, 276, 277, 279, 282, 288, 290, 296, 301–304
Porter, 19, 111, 217
Privacy, 40, 45, 66, 94, 98, 102
Production chain, 21, 40
Productivity, 3, 77, 106, 211, 217, 223, 253
Proximate relatedness, 20, 21
Proximity, 17, 20–22, 40, 41, 48, 51, 52, 56, 118, 119, 148, 190, 196, 223, 231, 242, 248, 250, 252, 256, 263, 265, 268, 269, 273, 280, 282, 284
Public policy, 56, 219
Public-private-partnership (PPP), 72, 73
Public sector, 45, 46, 48, 49, 52, 55, 62, 68, 70, 77, 81, 86, 87, 97, 200, 204, 303

Q
Quadruple-helix, 52, 64, 65, 68, 74, 95
Quality of life, 5, 70, 116, 119, 122, 124, 128, 142, 143, 160–163, 165, 167, 169, 171, 173, 175, 177, 178, 182, 188, 189, 222, 223, 230, 256
Quality of place, 163, 165, 178, 182, 188, 222, 228, 256
Quantitative analysis, 172, 174

R
Real-world, 277
Region, 12, 14, 21, 42, 43, 48, 55, 57, 60–62, 77, 78, 80, 81, 83, 84, 88–92, 96, 106, 107, 114, 121, 127, 128, 139, 142, 144, 149, 150, 152, 154, 188, 190, 192, 194, 196, 198, 202, 204, 208, 211, 212, 217, 222, 232, 246, 248, 250, 262, 288, 290, 301
Regional actor, 128
Regional innovation system (RIS), 16
Regulation, 95, 274
Relational capital, 113, 115
Research and development (R&D), 12, 15, 17, 42, 62, 79, 81, 86–88, 97, 116, 123, 127, 128, 145, 155, 165, 167, 173, 175, 177, 179, 208, 218, 223, 230, 231, 242, 244, 247, 248, 266, 271, 274, 277, 284, 287

Rogers, 6

S
Safety, 5, 70, 125, 127, 161, 163, 168, 170, 175, 177, 178, 182, 190, 223, 253, 275, 297
San Francisco, 18, 106, 108, 113, 139, 154, 172–186, 188, 191, 195, 204, 212, 218, 243, 245–247, 256, 257, 296, 298
Schumpeter, 14–17
Science and technology, 89, 90, 120, 128, 145, 146, 148, 215, 218, 232, 243, 265
Science and technology park, 120
Silicon Valley, 13, 18, 106, 118, 139, 148, 212, 213, 218, 220, 221, 233, 239, 243–247, 251, 255, 257, 262, 273, 282, 283, 287, 296, 298, 299
Singapore, 106, 108, 113, 114, 118, 122, 135, 144–148, 153–155, 200–202, 212, 213, 218, 221, 239, 247, 248, 250, 251, 256, 257, 296, 298–300
Skilled workforce, 144, 150, 246
Small and medium-sized enterprises (SMEs), 15, 127, 128, 151, 155, 188, 194, 200, 230, 248, 266, 271, 273, 280, 283, 298, 299
Smart city, 6, 13, 48, 49, 52, 55–57, 60–75, 78, 84, 85, 91, 98, 102, 110, 139, 161, 163, 204, 240, 246, 247
Smart community, 279
Smart precinct, 97, 106, 107, 151
Smart specialisation, 304
Social capital, 115, 117, 187, 193, 201, 257
Social cohesion, 62, 63, 125, 170, 172, 176, 178, 179, 184, 186, 192
Social condition, 5, 44, 56
Social equity, 116, 263
Social media, 40, 65, 68, 96, 98, 245, 246, 249
Social tolerance, 124, 143, 166, 168, 174, 176, 177, 180, 181, 188
Societal development, 102, 165, 167, 172, 174, 177–179, 181, 184, 185, 187, 190, 200, 215
Society, 4, 6, 13, 14, 45, 52, 56, 69, 70, 74, 82, 88, 105, 106, 110, 112, 119, 120, 137, 141, 149, 153, 156, 161, 163, 189–191, 196, 197, 200, 204, 220, 241, 242, 253, 262, 263, 269, 270, 275, 298, 299
Socioeconomic dependency, 181, 188
Spain, 112, 136, 213, 218, 239, 240
Spatial development, 5, 8, 14, 17, 48, 61, 63, 167, 169, 172, 174, 177, 178, 181, 182, 185, 188, 191, 201, 204, 219, 221
Spatiality, 3, 6, 13, 40

Spill-over, 227, 228
Spill-over effect, 314
Spin-off company, 314
Stakeholder, 48, 71, 76, 272
Start-up, 7, 12, 64, 76, 78, 142, 143, 149, 244, 245, 247, 256, 298
Statistical analysis, 171, 173
Stockholm, 5, 41, 44, 56, 63–73, 79–81, 83, 86–92, 96, 101
Strategic partnership, 125, 169, 171, 176, 178, 179, 183, 186, 192
Strategic planning, 118, 125, 154, 160, 162, 168, 170, 176, 178, 179, 183, 184, 192, 296
Strategy, 52, 57, 60–78, 80, 107, 112, 119, 120, 129, 136, 142, 145, 148–151, 153, 154, 189, 190, 195, 196, 200–204, 225, 241, 243, 251, 268, 272, 275, 280, 288, 289, 302
Strengths, weaknesses, opportunities and threats (SWOT), 76, 186, 187, 192, 193, 199
Strijp-S, 106, 212, 218, 233, 271, 272, 283–286, 297
Supply-chain, 13, 41
Sustainability, 41, 62, 65, 70–74, 79, 84, 98, 112, 119, 120, 216, 229, 231, 232, 262, 290
Sustainable development, 114, 119, 121, 219, 242, 257
Sustainable transport, 5, 124, 167, 169, 175, 177, 178, 181, 182, 188
Sustainable urban development, 5, 124, 155, 163, 165, 167, 169, 178, 182, 188, 243
Sweden, 8, 43, 56, 57, 63, 64, 69, 80–95
Sydney, 108, 113, 151, 172–188, 191–195, 202, 203, 213, 251–253, 256, 257, 261, 262, 273, 275, 276, 283–286, 298
System, 16–19, 44–46, 49, 63, 66, 74, 75, 94, 136, 141, 142, 149, 187, 189, 190, 193, 195, 196, 202, 223, 226–229, 231, 240, 244, 247, 266, 269, 301

T
Tallinn, 56, 69, 75–82, 86, 91, 101, 102
Technology, 4, 5, 7, 8, 12–15, 18, 41, 42, 44, 45, 47, 48, 56, 59–61, 64–72, 75, 77, 83, 84, 88, 89, 91, 92, 95–98, 105, 107, 109, 112, 118, 119, 121, 122, 127, 137, 139, 141–145, 148, 152, 204, 212, 215, 217, 218, 220, 223, 225, 231, 241–248, 250, 252, 262, 266–268, 271, 274, 275, 277, 279, 282, 284, 287, 288, 291, 298
Telecommunication, 42

Tertiary education, 43, 80, 81, 89, 90, 97, 149, 187, 188, 193, 194, 201, 225
The Knowledge-Based Urban Development Assessment Model (KBUD/AM), 162, 164, 186, 195
The Most Admired Knowledge Cities Award Model (MAKCi), 113, 136, 139, 144, 148, 152
The Netherlands, 72, 111, 112, 218
Toronto, 108, 113, 139, 161, 163, 172–187, 191, 195
Transparency, 7, 41, 42, 46, 52, 65, 75, 116, 137, 242, 254, 255, 300
Transport, 7, 42, 43, 56, 66, 67, 75, 77, 93, 150, 187–190, 193, 194, 196, 202, 222, 240, 248, 253, 267, 275, 276, 279
Triple-helix, 48, 61, 64, 65, 95, 125, 169, 171, 189, 196, 282, 288, 298
Turkey, 113

U
Unemployment, 44, 49, 83, 106, 124, 127, 147, 167, 169, 174, 176, 178, 181, 187, 188, 193, 211, 217, 232, 265
United Kingdom, 265
United States of America, 85, 87, 139, 213, 218, 239, 243
University, 60, 63–65, 78, 86, 90, 119, 124, 140–142, 144, 152, 154, 155, 161, 163, 166, 168, 174, 176, 177, 180, 181, 188, 201, 215, 220, 231, 244, 248, 250, 252, 254–256, 262, 265, 266, 269, 273–279, 283, 285–291, 298, 299
Urban competitiveness, 105, 123, 128, 161, 163, 164, 166, 173, 175, 177, 179, 187, 217
Urban density, 5, 124, 167, 169, 175, 177, 178, 181, 182, 189, 254
Urban design, 7, 51, 187, 189, 193, 196, 257, 268, 289
Urban development, 8, 43, 44, 48, 62, 67, 72, 76, 79, 98, 101, 106, 110, 113, 118, 119, 123, 124, 128, 136, 139, 141, 149, 155, 211, 215, 217, 219, 220, 232, 250, 279
Urban economy, vii
Urban form, 5, 7, 124, 167, 169, 175, 177, 178, 181, 182, 189, 248
Urban innovation system (UIS), 68
Urban marketing, 60, 78, 139
Urban planning, 8, 43, 46, 52, 62, 63, 67, 68, 72–74, 96, 118, 137, 148, 202, 203, 215, 221, 226
Urban technology, 41, 42, 66, 190, 196, 243
Urbanisation, 62, 102, 216

V

Value creation, 116
Vancouver, 108, 113, 172–188, 191, 195
Vibrancy, 105, 150, 187, 193, 232, 254, 297
Virtual, 13, 144, 223, 277, 279, 299
Virtual innovation system, 32
Vision, 65, 71, 77, 107, 108, 113, 116, 136, 137, 139, 148–150, 152–154, 191, 192, 195, 197, 199, 202–204, 207, 208, 225, 226, 229, 256, 262, 265, 266, 269, 271, 279, 283, 289, 290, 296

W

Waste management, 41, 56, 60, 62, 68, 74, 85, 86
Weighting, 162–165, 172, 174
weighting, 180, 185

Wi-Fi, 42, 73, 116, 148, 188, 194
WLAN, 42
Work force, 14, 41, 57, 71, 89, 90, 96, 142, 143, 145, 150, 153, 200, 246, 247, 253, 269
World class, 116

Y

Yigitcanlar, 5, 8, 56, 105, 106, 110–120, 122, 126, 136, 137, 139–141, 143, 146, 149, 152, 159–166, 171, 173, 177, 179, 202, 211, 215–221, 226, 233, 242, 243, 246, 247, 261–263, 267, 269, 282, 287–291

Z

Z-score, 163, 165

Printed by Printforce, the Netherlands